ATOMIC UNITS

Quantity	Symbol or Expression	CGS Equivalent	Important Related Properties
Mass	m_e	9.10939×10^{-28} g	m_e = electron mass
Charge	e	4.803207×10^{-10} stat C	$-e$ = electron charge
Angular momentum	$\hbar$	1.05457×10^{-27} erg s	
Length (bohr)	$a_0 = \hbar^2/m_e e^2$	0.5291772×10^{-8} cm	a_0 = radius of lowest energy Bohr orbit
Energy (hartree)	$E_h = e^2/a_0$	4.35975×10^{-11} erg	$\frac{1}{2}E_h$ = ionization energy of hydrogen
Time	$\tau_0 = \hbar^3/m_e e^4$	2.41887×10^{-17} s	$2\pi\tau_0$ = orbital period of lowest Bohr orbit
Frequency	$m_e e^4/\hbar^3$	4.13416×10^{16} s^{-1}	
Velocity	$e^2/\hbar$	2.18770×10^8 cm/s	$e^2/\hbar$ = velocity of electron in lowest Bohr orbit
Force	e^2/a_0^2	8.23873×10^{-3} dyne	
Electric field	e/a_0^2	1.71526×10^7 stat C/cm^2 (5.14221×10^9 V/cm)	e/a_0^2 = field experienced by electron in lowest Bohr orbit
Electric potential	e/a_0	9.07675×10^{-2} stat C/cm (27.2114 V)	
Fine structure constant	$\alpha = e^2/\hbar c$	1/137.036	
Power	$m_e^2 e^8/\hbar^5$	1.80239×10^6 erg/s (0.180239 W)	
Magnetic moment	$\beta_e = e\hbar/2m_e c$	0.92740×10^{-20} erg/gauss	β_e = magnetic moment of orbital motion in lowest Bohr orbit
Magnetic field	$e\hbar/m_e c a_0^3$	1.25168×10^5 gauss	Magnetic field at nucleus due to orbital motion in first Bohr orbit

Quantum Mechanics in Chemistry

QUANTUM MECHANICS
in CHEMISTRY

GEORGE C. SCHATZ and MARK A. RATNER

NORTHWESTERN UNIVERSITY

PRENTICE HALL
Englewood Cliffs, New Jersey 07632

Library of Congress Cataloging-in-Publication Data

Schatz, George C., 1949–
    Quantum mechanics in chemistry/George C. Schatz and Mark A.
Ratner.
       p.   cm.
    Includes bibliographical references and index.
    ISBN 0-13-747585-3
    1. Quantum chemistry. 2. Quantum mechanics. I. Ratner, Mark A.,
1942–   . II. Title.
QD462.S33 1993
541.2'8--dc20

92-38805
CIP

Editor in Chief, Tim Bozik
Senior editor, Diana Farrell
Senior managing editor, Jeanne Hoeting
Production, Nicholas Romanelli
Prepress buyer, Paula Massenaro
Manufacturing buyer, Lori Bulwin
Editorial assistant: Lynne Breitfeller

 © 1993 by Prentice-Hall, Inc.
A Simon & Schuster Company
Englewood Cliffs, New Jersey 07632

Printed in the United States of America
10 9 8 7 6 5 4 3 2 1

ISBN 0-13-747585-3

9 780137 475858

Prentice-Hall International (UK) Limited, *London*
Prentice-Hall of Australia Pty. Limited, *Sydney*
Prentice-Hall, Inc., *Toronto*
Prentice-Hall Hispanoamericana, S.A., *Mexico*
Prentice-Hall of India Private Limited, *New Delhi*
Prentice-Hall of Japan, *Tokyo*
Simon & Schuster Asia Pte. Ltd., *Singapore*
Editora Prentice-Hall do Brasil, Ltda, *Rio de Janeiro*

To Margaret and Nancy

Contents

5   *Interaction of Radiation with Matter*   78

6 Occupation Number Representations 111

Preface

This book is intended as a follow-up to the many introductory quantum chemistry texts used in teaching graduate and advanced undergraduate students. It is the result of our many years of teaching at Northwestern University, in courses taken by hundreds of students. Like many universities, we have a one-term introductory quantum chemistry course taken by graduate and advanced undergraduate students, for which several excellent introductory texts exist. This is followed by an additional term (or sometimes two) in which more advanced topics are considered.

We have always been frustrated by the lack of texts that treat many of the advanced topics, so over the years we developed an extensive set of lecture notes to remedy the problem, and the present book evolved from those notes. There are a large number of physics-oriented texts that cover many of the topics we consider, but the absence of chemical, solid-state, and materials applications has always been a problem for our students. One advanced topic, advanced methods for describing the electronic structure of molecules (i.e., beyond Hartree–Fock), is covered in depth in a few textbooks, so this topic is not dealt with directly in this text. Instead, we emphasize areas of dynamics, of symmetry, and of formalism in quantum mechanics that contain essential tools for both experimental and theoretical students working in a wide variety of subdisciplines of chemistry and materials science. In addition, many of the topics in this book are relevant to the interests of students in certain areas of physics, biology, and engineering. One venerable but unfortunately outdated text that provided much inspiration for our text is *Quantum Chemistry* by Eyring, Walter, and Kimball (1944).

Our choice of topics has several themes, which can roughly be grouped as follows: symmetry and rotations in quantum mechanics, time-dependent quantum mechanics and its applications to spectroscopy, collisions and rate processes, occupation number representations of quantum mechanics, and the use of correlation functions and density matrices in quantum mechanics. After an introductory chapter that reviews basic concepts from introductory quantum mechanics, our second and third chapters focus on symmetry and rotations. Chapter 2 is partially a review of elementary concepts associated with point groups and is partially a consideration of the symmetry properties of many-electron wavefunctions. In Chapter 3 we consider the two- and three-dimensional rotation groups as they apply to electronic structure, and examine the related topic of angular momentum addition. In Chapter 4 we introduce the basic formalism of time-dependent quantum mechanics with an emphasis on time-dependent perturbation theory and Fermi's golden rule, and in Chapter 5 provide applications of this formalism to the interaction of radiation and matter (light absorption, emission, scattering). In Chapter 6 we introduce occupation number representations, including applications to both quantized radiation fields and electronic structure; it is lengthy and detailed, because the material is relatively new to most scientists. In Chapter 7 we present an introduction to scattering theory, especially as it applies to chemical problems. These concepts are then used in Chapter 8 to develop basic theories of chemical reaction rates. Along the way we discover that rates can be obtained from correlation functions, and in Chapters 9 and 10 the subject of correlation functions is extended in many other directions with an emphasis on spectroscopy and on theories of electron transfer. Finally, in Chapter 11 we combine many of topics of the previous chapters to describe electronic structure, optical and magnetic resonance spectroscopy, and condensed phase dynamics using density matrices.

The ordering of these chapters follows largely from our own need to divide the material covered between two courses that are taken by students with different interests. Chapters 1 to 6 cover one course that is taken by a fairly broad spectrum of students from all areas of chemistry. These chapters emphasize topics in symmetry, spectroscopy, and electronic structure that find widespread application, and in addition introduce elements in the formalism of quantum mechanics that have become part of the "language" of chemistry. Chapters 7 to 11 cover topics that are of more specific interest to physical chemists and materials scientists, with an emphasis on dynamical processes. There are several subthemes that make the pairs of chapters 4 and 5, 7 and 8, 9 and 10, and 6 and 11 closely connected, so these four chapter pairs could easily be presented in any order, and one could also omit pairs according to the needs of the students taking the material. Chapters 8, 9, 10, and 11· go beyond a straightforward consideration of quantum-mechanical methods: Statistical considerations are important in all four, and all point strongly toward forefront areas of current research.

With each chapter we include problems that we have used in our courses, designed to illustrate further applications of the theory as it is developed. Many of these problems extend our development in directions that represent important areas of modern research, while others provide classic examples that illustrate important physical effects. Some of the problems are quite lengthy and challenging.

An attempt is made throughout the book to be concise and to make the formal development useful to chemists and materials scientists. The relatively informal wording in parts of the book reflects the pedagogic nature of much of the material. Extensive discussions of interpretation, as well as digressions to deal with special cases and pedagogic asides, have been held to a minimum in an attempt to produce a useful and rational guide to quantum mechanics for chemists and materials scientists.

The development of this book would have been very difficult without the assistance of an unusually talented typist, Jan Goranson. We are extremely indebted to her for the countless hours of careful, painstaking effort needed to get all the formulas right. A change in word-processing software halfway through development of the manuscript required that most equations be typed twice. Despite this, Jan persevered, and the result is a testimony to her efforts.

We should also acknowledge Dan Joraanstad, Diana Farrell, and Lynne Breitfeller of Prentice Hall for their encouragement during this long project. In addition, a large number of graduate students at Northwestern have used portions of this book in their courses, and they have made countless suggestions for the improvement of both text and problems. We have further benefitted from the suggestions of B. Whalley and N. Snider, both of whom carefully read the text, from Stacy Ratner's careful proofreading, and from Daniel Ratner's and Matt Todd's exemplary work with the figures. The presentation of the occupation number formalism for electrons is based on ideas presented by Jan Linderberg. Finally, the patience, inspiration, and encouragement of our wives, Margaret and Nancy, have been crucial to the book since its inception.

George Schatz / Mark Ratner

Quantum Mechanics in Chemistry

1

Review of Basic Concepts in Quantum Mechanics

1.1

Fundamental Definitions

Our starting point is the set of basic rules or postulates by which quantum mechanics is developed. The reader who is interested in the historical or philosophical underpinnings of these rules should refer to a large number of excellent texts that cover this material. In the quantum description of the motions of particles, one describes the system using a wavefunction. Ψ. Ψ depends on the coordinates $x_1, x_2, \ldots, x_N$ of all the particles and on time t. (Here we consider motion of N indistinguishable particles in one dimension.) The physical significance of Ψ is determined by the following fundamental definition:

$$|\Psi(x_1, x_2, \ldots, x_N, t)|^2 \, dx_1 \, dx_2 \cdots dx_N$$
$$\equiv \text{probability of finding a}$$
$$\text{particle between } x_1 \text{ and} \qquad (1.1)$$
$$x_1 + dx_1, \text{ and a particle}$$
$$\text{between } x_2 \text{ and } x_2 + dx_2, \ldots \text{ at time } t$$

The time evolution of Ψ for a system governed by a Hamiltonian H is determined by the Schrödinger equation:

$$i\hbar \frac{\partial \Psi}{\partial t} = H\Psi \qquad (1.2)$$

where $\hbar$ is Planck's constant divided by 2π.

The operator H in Eq. (1.2) as well as any other quantum operator with a classical analog can be constructed by writing down the corresponding classical mechanical dynamical variable in terms of cartesian coordinates (x) and momenta (p) and making the replacements:

Classical variable	Quantum operator
x	x
p_x	$-i\hbar \dfrac{\partial}{\partial x}$

Generalizations of this procedure to noncartesian coordinates and to complicated dynamical variables follow straightforwardly. The rewriting of the classical hamiltonian in quantal form occasionally requires symmetrization to ensure linearity and hermiticity. Sometimes we will put a "hat" on quantum operators, as in $\hat{x}$, to distinguish them from other symbols.

The quantum operator corresponding to any dynamical variable must be a linear hermitian operator. A linear operator G satisfies

$$G(C_1\Psi_1 + C_2\Psi_2) = C_1 G\Psi_1 + C_2 G\Psi_2$$

for any wavefunctions Ψ_1 and Ψ_2, and any constants C_1 and C_2. A hermitian operator G satisfies

$$\int \Psi^* G\Phi \, d\tau = \int \Phi(G\Psi)^* \, d\tau \qquad (1.3)$$

where Φ and Ψ are any "well-behaved" functions (i.e., functions that are twice differentiable and for which $\int \Psi^*\Psi \, d\tau$ is finite). In Eq. (1.3), $d\tau$ stands for the appropriate integration volume element. Often we will use the so-called bracket notation, whereby Eq. (1.3) becomes

$$\langle \Psi | G\Phi \rangle = \langle G\Psi | \Phi \rangle \qquad (1.4)$$

For a system governed by a wavefunction Ψ, one can determine the average or expectation value of any dynamical variable by evaluating

$$\langle G \rangle = \frac{\langle \Psi | G | \Psi \rangle}{\langle \Psi | \Psi \rangle} \qquad (1.5)$$

The denominator in Eq. (1.5) is called the normalization integral and is often required to be unity. From Eqs. (1.4) and (1.5) one can easily show that hermitian operators always have real expectation values.

1.2

Eigenvalues and Eigenfunctions

If the hamiltonian H is time independent, the Schrödinger equation can be separated into coordinate- and time-dependent parts. (In Chapter 4 we discuss the case where H depends on time.) Writing

$$\Psi(x_1, x_2, \ldots, x_N, t) = \psi(x_1, x_2, \ldots, x_N)\chi(t) \tag{1.6}$$

one can separate variables in Eq. (1.2) to get

$$i\hbar \frac{d\chi}{dt} = E\chi \tag{1.7}$$

and

$$H\psi = E\psi \tag{1.8}$$

where E is a separation constant. Equation (1.7) is easily solved to give

$$\chi = e^{-iEt/\hbar} \tag{1.9}$$

Equation (1.8) is known as the *time-independent Schrödinger equation*, and its solution gives a time-independent wavefunction. Note that the probability density $|\Psi|^2$ is the same as $|\psi|^2$ and is stationary in this case.

Equation (1.8) is an example of an eigenfunction/eigenvalue equation which generally has the form

$$G\psi = g\psi \tag{1.10}$$

where G is any operator and g is a number. A wavefunction that satisfies Eq. (1.10) is known as an eigenfunction of the operator G, and g is its eigenvalue.

If we evaluate Eq. (1.5) for $G = H$ and use Eqs. (1.6), (1.8), and (1.9), we find that

$$\langle H \rangle = E \tag{1.11}$$

This implies that the separation constant E is just the expectation value corresponding to the energy of the system. Equation (1.8) tells us that E is the eigenvalue of the operator H, and Eq. (1.11) tells us that expectation values calculated using eigenfunctions are the corresponding eigenvalues.

It is often useful to ask when a wavefunction can be an eigenfunction simultaneously of two operators, such as G and H. To answer this we form the product $(GH - HG)\psi$. If ψ is an eigenfunction of G with eigenvalue g, and of H with eigenvalue E, then

$$(GH - HG)\psi = (gE - Eg)\psi = 0$$

If this is required to be true for all eigenfunctions of G and H, we must have

$$GH - HG \equiv [G, H] = 0 \qquad (1.12)$$

The symbol $[G, H]$ is an abbreviation for the left-hand side of Eq. (1.12) and is called a commutator. Equation (1.12) implies that operators must commute if they are to have simultaneous eigenfunctions. Note that the converse of this statement does not always work: that is, just because two operators commute does not guarantee that an eigenfunction of one will be an eigenfunction of the other. However, if two operators commute it is always possible to find simultaneous eigenfunctions. An important example of two operators that do not commute is x and p (for which $[x, p] = i\hbar$). Clearly, the uncertainty principle would be violated if these operators had simultaneous eigenvalues.

Now consider two eigenfunctions of the operator H. Call them ψ_1 and ψ_2, and let the corresponding eigenvalues be E_1 and E_2. Then since H is hermitian,

$$\langle \psi_1 | H \psi_2 \rangle - \langle H \psi_1 | \psi_2 \rangle = 0 \qquad (1.13)$$

But this quantity also equals $(E_2 - E_1) \langle \psi_1 | \psi_2 \rangle$, which implies that either

$$E_2 - E_1 = 0 \qquad (1.14a)$$

or

$$\langle \psi_1 | \psi_2 \rangle = 0 \qquad (1.14b)$$

These equations say that for any two eigenfunctions of H (or, indeed, of any other hermitian operator), either E_1 and E_2 are *degenerate* [Eq. (1.14a)] or the eigenfunctions must be *orthogonal* [Eq. (1.14b)]. Degenerate eigenvalues can also be constructed to be orthogonal, but this is not required. Eigenfunctions that satisfy Eq. (1.14b) and are normalized are said to be *orthonormal*.

Another property of eigenfunctions of the hamiltonian is *completeness*. This means that any well-behaved function $|f\rangle$ can be represented as an expansion in a set of eigenfunctions that is defined in the same coordinate region and which satisfies the same boundary conditions. If the eigenfunctions are labeled $|n\rangle$ and they are orthonormal, the expansion is

$$|f\rangle = \sum_n c_n |n\rangle \qquad (1.15)$$

with $c_n = \langle n | f \rangle$.

1.3

Approximate Methods

1.3.1 Time-Independent Perturbation Theory

Often one needs to solve the time-independent Schrödinger equation for a problem that is too difficult or inconvenient to do exactly. One situation where accurate approximate solutions can be developed is when the hamiltonian H can be written as the sum of a zeroth-order hamiltonian H_0, for which exact solutions exist, plus a time-independent perturbation V which is in some sense small. Let us write this as

$$H = H_0 + \lambda V \tag{1.16}$$

where λ is a parameter (that will be set to one at the end) that keeps track of the order of the perturbation. Let us denote the eigenfunctions of H_0 by ϕ_n^0 and the corresponding eigenvalues as E_n^0 (implying that $H_0 \phi_n^0 = E_n^0 \phi_n^0$).

To develop approximate solutions to the full Schrödinger equation, we use the expansion

$$\psi_n = \phi_n^0 + \lambda \phi_n^{(1)} + \lambda^2 \phi_n^{(2)} + \cdots \tag{1.17a}$$

for the wavefunction, and the corresponding expansion

$$E_n = E_n^0 + \lambda E_n^{(1)} + \lambda^2 E_n^{(2)} + \cdots \tag{1.17b}$$

for the energy. Substituting Eqs. (1.16) and (1.17) into $H\psi_n = E_n\psi_n$ and equating like powers of λ leads to

$$H_0 \phi_n^0 = E_n^0 \phi_n^0 \qquad \text{(zeroth order)} \tag{1.18}$$

$$V \phi_n^0 + H_0 \phi_n^{(1)} = E_n^{(1)} \phi_n^0 + E_n^0 \phi_n^{(1)} \qquad \text{(first order)} \tag{1.19}$$

$$V \phi_n^{(j-1)} + H_0 \phi_n^{(j)} = \sum_{k=0}^{j} E_n^{(k)} \phi_n^{(j-k)} \qquad [\,j\text{th order } (j \geq 1)] \tag{1.20}$$

Equation (1.18) is satisfied by assumption. To solve Eq. (1.19), we expand $\phi_n^{(1)}$ in terms of the complete set of zeroth-order states

$$\phi_n^{(1)} = \sum_{k \neq n} C_{nk}^{(1)} \phi_k^0 \tag{1.21}$$

where C_{nk} is a coefficient, and without loss of generality we assume that $\langle \phi_n^0 | \phi_n^{(1)} \rangle = 0$. Substitution of Eq. (1.21) into (1.19), followed by multiplication by ϕ_n^{*0} and integration, then leads to

$$E_n^{(1)} = \langle \phi_n^0 | V | \phi_n^0 \rangle \tag{1.22}$$

$$C_{nk}^{(1)} = \frac{\langle \phi_k^0 | V | \phi_n^0 \rangle}{E_k^0 - E_n^0} \qquad (k \neq n) \tag{1.23}$$

One can similarly develop expressions for higher-order terms in both the energy and wavefunction. For example,

$$E_n^{(2)} = \sum_{k \neq n} \frac{\langle \phi_n^0 | V | \phi_k^0 \rangle \langle \phi_k^0 | V | \phi_n^0 \rangle}{E_n^0 - E_k^0} \tag{1.24}$$

and

$$C_{nk}^{(2)} = \sum_{l \neq n} \frac{V_{ln}(V_{kl} - V_{nn}\delta_{kl})}{(E_n^0 - E_l^0)(E_n^0 - E_k^0)} \qquad (k \neq n) \tag{1.25}$$

Here we have introduced the Kronecker delta function (i.e., $\delta_{nk} = 1$ for $k = n$, $= 0$ for $k \neq n$).

Static perturbation theory is very widely used in chemistry, since so many physical situations (particularly, spectroscopic ones) consist of an unperturbed hamiltonian and a small perturbation; in the case of spectroscopy, the smallness of the perturbation generally arises from the weakness of laboratory fields compared to intramolecular Coulomb potential fields.

1.3.2 Variational Theory

Another approach to determining approximate solutions to the Schrödinger equation is to use the variational theorem. This states that for any well-behaved function ϕ, the following inequality holds:

$$\frac{\langle \phi | H \phi \rangle}{\langle \phi | \phi \rangle} \geq E_0 \tag{1.26}$$

where E_0 is the exact ground-state energy.

To prove this, we expand ϕ in terms of the exact eigenfunctions of H, which we denote ψ_n (with energies E_n):

$$\phi = \sum_n C_n \psi_n \tag{1.27}$$

If we rewrite Eq. (1.26) as

$$\langle \phi | H \phi \rangle - E_0 \langle \phi | \phi \rangle \geq 0 \tag{1.28}$$

and substitute in Eq. (1.27), we can reduce this to

$$\sum_n (E_n - E_0)|C_n|^2 \geq 0 \tag{1.29}$$

Since $E_n - E_0$ is necessarily nonnegative, the inequality in Eq. (1.29) must hold and the theorem is proved.

The most common applications of variational theory in chemistry involve ϕ written as an expansion in a set of functions with coefficients that are to be optimized so as to minimize the left-hand side of Eq. (1.26). If we use the symbol B_n to denote these basis functions, the expansion would be

$$\phi = \sum_n C_n B_n \qquad (1.30)$$

Substitution of Eq. (1.30) into (1.26), and minimization with respect to each coefficient C_k, leads to the equations

$$\sum_n (H - E_0) B_n C_n = 0 \qquad (1.31)$$

Multiplying Eq. (1.31) by B_k^* and integrating, we obtain the linear algebraic equations

$$\sum_n (H_{kn} - E_0 S_{kn}) C_n = 0 \qquad (1.32)$$

where

$$H_{kn} = \langle B_k | H | B_n \rangle \qquad (1.33)$$

is a *hamiltonian matrix element*, and

$$S_{kn} = \langle B_k | B_n \rangle \qquad (1.34)$$

is an *overlap matrix element*.

The nontrivial solutions to Eq. (1.32) are generated by solving the secular equation

$$|\mathbf{H} - E\mathbf{S}| = 0 \qquad (1.35)$$

where $\mathbf{H}$ and $\mathbf{S}$ stand for matrices whose elements are given by Eqs. (1.33) and (1.34), respectively.

1.4

Raising and Lowering Operators

1.4.1 Harmonic Oscillator

The hamiltonian for a mass m moving in one dimension subject to a harmonic potential is

$$H = \frac{1}{2m}p^2 + \frac{1}{2}kx^2 \qquad (1.36)$$

where k is the force constant. To determine the eigenvalues and eigenfunctions associated with H, it is convenient to rewrite H in terms of operators b and b^+, which we define by

$$b = \left(\frac{m\omega}{2\hbar}\right)^{1/2}\left(x + \frac{ip}{m\omega}\right) \qquad (1.37a)$$

$$b^+ = \left(\frac{m\omega}{2\hbar}\right)^{1/2}\left(x - \frac{ip}{m\omega}\right) \qquad (1.37b)$$

where $\omega = (k/m)^{1/2}$ is the classical oscillator angular frequency. If Eqs. (1.37) are inverted to determine x and p and the results substituted into Eq. (1.36), one obtains

$$H = \tfrac{1}{2}\hbar\omega(bb^+ + b^+b) \qquad (1.38a)$$

$$= \hbar\omega(b^+b + \tfrac{1}{2}) \qquad (1.38b)$$

The second form here [Eq. (1.38b)] can be derived from the first by invoking the important commutation relation (which may readily be verified):

$$[b, b^+] = 1 \qquad (1.39)$$

The form of Eq. (1.38b) implies that eigenfunctions of H must also be eigenfunctions of b^+b. Let us use the notation

$$b^+b|\lambda\rangle = \lambda|\lambda\rangle \qquad (1.40)$$

for the eigenfunctions of b^+b, where λ is both a *number* (the eigenvalue) and a *label* (of the eigenfunction). From Eq. (1.39) one can now easily show that

$$b^+b(b^+|\lambda\rangle) = b^+(1 + b^+b)|\lambda\rangle = (\lambda + 1)b^+|\lambda\rangle \qquad (1.41)$$

which implies that $b^+|\lambda\rangle$ is an eigenfunction of b^+b with an eigenvalue of $\lambda + 1$. This means that b^+ acts on $|\lambda\rangle$ to raise the value of the eigenvalue by one. Similarly, from

$$b^+bb|\lambda\rangle = (bb^+ - 1)b|\lambda\rangle = (\lambda - 1)b|\lambda\rangle \qquad (1.42)$$

we infer that b lowers λ by one. There is, however, a lower limit to the value λ. This can be inferred from the fact that $\langle\lambda|b^+b|\lambda\rangle$, which from Eq. (1.40) is just λ, is inherently nonnegative, as it is the absolute square of $b|\lambda\rangle$. To avoid making it possible to apply b so many times to $|\lambda\rangle$ that it produces a state with a negative eigenvalue, we terminate the series by choosing one of the eigenvalues λ to have the value zero. The eigenfunction corresponding to $\lambda = 0$ would then satisfy $\langle 0|b^+ b|0\rangle = 0$, which implies that

$$b|0\rangle = 0 \tag{1.43}$$

Since the eigenvalues above $\lambda = 0$ can differ from $\lambda = 0$ by at most an integer, it is convenient to define an integer quantum number n that is identical to the eigenvalue λ. This implies that $b^+ b|n\rangle = n|n\rangle$, or that

$$H|n\rangle = \hbar\omega(b^+ b + \tfrac{1}{2})|n\rangle = \hbar\omega(n + \tfrac{1}{2})|n\rangle \tag{1.44}$$

Consideration of the expectation value $\langle n|b^+ b|n\rangle$ can also be used to show that

$$b|n\rangle = \sqrt{n}\,|n - 1\rangle \tag{1.45a}$$

and by similar reasoning,

$$b^+|n\rangle = \sqrt{n + 1}\,|n + 1\rangle \tag{1.45b}$$

The specific functional form of the ground-state eigenfunction is determined by solving Eq. (1.43) as a first-order differential equation with the operator form of Eq. (1.37a). The result is $|0\rangle = N \exp(-m\omega x^2/2\hbar)$. Higher eigenfunctions then follow from Eq. (1.45), which can be iterated to give $|n\rangle = (b^+)^n|0\rangle/(n!)^{1/2}$.

1.4.2 Angular Momentum Operators

Raising and lowering operators are also of great use in determining eigenfunctions of the angular momentum operators $\hat{\mathbf{l}}$. These operators are defined in terms of the coordinate $\hat{\mathbf{r}}$ and momentum $\hat{\mathbf{p}}$ via $\hat{\mathbf{l}} = \hat{\mathbf{r}} \times \hat{\mathbf{p}}$. An important property of the components $\hat{l}_x$, $\hat{l}_y$, and $\hat{l}_z$ thus defined is that they do not commute. Instead, they satisfy

$$[\hat{l}_x, \hat{l}_y] = i\hbar\hat{l}_z \quad \text{(and cyclically)} \tag{1.46}$$

The operator $\hat{l}^2 = \hat{l}_x^2 + \hat{l}_y^2 + \hat{l}_z^2$ does commute with $\hat{l}_x$, $\hat{l}_y$, and $\hat{l}_z$, so it is possible to find simultaneous eigenfunctions of $\hat{l}^2$ and one of the other operators. The latter is conventionally chosen to be $\hat{l}_z$. The eigenfunctions of $\hat{l}^2$ and $\hat{l}_z$ are labeled $|lm\rangle$, where

$$\hat{l}^2|lm\rangle = \hbar^2 l(l + 1)|lm\rangle \tag{1.47a}$$

$$\hat{l}_z|lm\rangle = \hbar m|lm\rangle \tag{1.47b}$$

and l and m are the quantum numbers associated with the magnitude and projection of the angular momentum, respectively. (Note that we use the symbol l for both the operator and the eigenvalue; therefore, we use the "hat" notation to clarify.)

Raising and lowering operators $\hat{l}_\pm$ are defined by

$$\hat{l}_\pm = \hat{l}_x \pm i\hat{l}_y \qquad (1.48)$$

From Eq. (1.46) one can show that $[\hat{l}_z, \hat{l}_\pm] = \pm\hbar\hat{l}_\pm$, so it follows that

$$\hat{l}_z\hat{l}_\pm|lm\rangle = \hbar(m \pm 1)\hat{l}_\pm|lm\rangle \qquad (1.49)$$

This means that $\hat{l}_+$ raises m by unity and $\hat{l}_-$ lowers it by unity.

The proportionality constant in the relation $\hat{l}_+|lm\rangle = \text{constant }|lm + 1\rangle$ can be determined by evaluating $\langle lm|\hat{l}_-\hat{l}_+|lm\rangle$, as this is just the square of this constant. Using Eq. (1.46), one can show that $\hat{l}_-\hat{l}_+ = \hat{l}^2 - \hat{l}_z^2 - \hbar\hat{l}_z$, so from Eqs. (1.47) it follows that "constant" $= \hbar[l(l + 1) - m(m + 1)]^{1/2}$ (where we have chosen the constant to be real and positive). This implies that

$$\hat{l}_+|lm\rangle = \hbar[l(l + 1) - m(m + 1)]^{1/2}|lm + 1\rangle \qquad (1.50a)$$

and one can similarly use the operator $\hat{l}_+\hat{l}_-$ to show that

$$\hat{l}_-|lm\rangle = \hbar[l(l + 1) - m(m - 1)]^{1/2}|lm - 1\rangle \qquad (1.50b)$$

Since the square roots in Eqs. (1.50) must have real values ($\langle lm|\hat{l}_-\hat{l}_+|lm\rangle$ must be positive or zero, as it is an absolute square), the allowed range of m is from $-l$ to $+l$. For m to vary in integer steps between these limits, $2l$ must be an integer, which means that l can be either an integer ($l = 0, 1, 2, \ldots$) or a half-integer ($l = \frac{1}{2}, \frac{3}{2}, \ldots$).

An important special case of half-integer angular momentum is the spin angular momentum s of the electron, for which $s = \frac{1}{2}$. Since there are only two possible values ($\pm\frac{1}{2}$) of the projection quantum number m_s, the eigenfunctions $|s, m_s\rangle$ are given the special symbols $|\alpha\rangle$ and $|\beta\rangle$, where $|\alpha\rangle = |\frac{1}{2}\,\frac{1}{2}\rangle$ and $|\beta\rangle = |\frac{1}{2}\,-\frac{1}{2}\rangle$. Note that the analogs of Eqs. (1.50) in this case are

$$\hat{s}_+|\alpha\rangle = 0 \qquad\qquad \hat{s}_+|\beta\rangle = \hbar|\alpha\rangle \qquad (1.50c)$$

$$\hat{s}_-|\alpha\rangle = \hbar|\beta\rangle \qquad\qquad \hat{s}_-|\beta\rangle = 0 \qquad (1.50d)$$

In the case where l is an integer, the wavefunctions $|lm\rangle$ are known as the *spherical harmonics*, and they are often labeled by the symbol Y_{lm}. Explicit expressions for the spherical harmonics in spherical coordinates are found in many textbooks (see the Bibliography). For the special case $l = 1$, the spherical harmonics are related to the angular parts of the cartesian coordinates x, y, z via

$$\frac{x}{r} \propto (-Y_{11} + Y_{1-1}) \qquad (1.50e)$$

$$\frac{y}{r} \propto (Y_{11} + Y_{1-1}) \qquad (1.50f)$$

$$\frac{z}{r} \propto Y_{10} \qquad (1.50g)$$

These formulas determine the angular parts of the well-known p orbitals (p_x, p_y, p_z) of the hydrogen atom.

1.5

Two-Body Problems

1.5.1 Relative-Motion Schrödinger Equation

Two problems in chemistry where the motions of two interacting "bodies" are important are the hydrogen atom (where the "bodies" are the electron and proton) and the diatomic molecule (where the "bodies" are the two nuclei). In either case the hamiltonian is

$$H = \frac{\mathbf{P}_A^2}{2m_A} + \frac{\mathbf{P}_B^2}{2m_B} + V \qquad (1.51)$$

where m_A and m_B are the masses of bodies A and B, $\mathbf{P}_A$ and $\mathbf{P}_B$ are the momenta, and V is the potential. If V depends only on the distance r between A and B, it is convenient to transform to new coordinates $\mathbf{r}$ and $\mathbf{R}$, where $\mathbf{r}$ is the relative coordinate

$$\mathbf{r} = \mathbf{R}_A - \mathbf{R}_B \qquad (1.52)$$

and $\mathbf{R}$ is the center-of-mass coordinate

$$\mathbf{R} = \frac{m_A \mathbf{R}_A + m_B \mathbf{R}_B}{m_A + m_B} \qquad (1.53)$$

Transforming Eq. (1.51) to these coordinates gives

$$H = \frac{\mathbf{P}_R^2}{2M} + \frac{\mathbf{P}_r^2}{2\mu} + V(r) \qquad (1.54)$$

where $M = m_A + m_B$, $\mu = m_A m_B/(m_A + m_B)$ and $\mathbf{P}_R$ and $\mathbf{P}_r$ are the momenta corresponding to $\mathbf{R}$ and $\mathbf{r}$, respectively. The first term in Eq. (1.54) represents the motion of the center of mass, and since V does not depend on $\mathbf{R}$, this motion is simply that of a free particle. We will ignore this in what follows.

The second and third terms in Eq. (1.54) describe relative motion. As long as V depends only on the scalar distance r between A and B, the relative-motion Schrödinger equation may be separated in polar coordinates. Defining these coordinates to be r, θ, and ϕ, the relative-motion Schrödinger equation is

$$\left\{-\frac{\hbar^2}{2\mu}\left(\frac{1}{r}\frac{\partial^2}{\partial r^2}r + \frac{1}{r^2}\left[\frac{1}{\sin\theta}\frac{\partial}{\partial\theta}\sin\theta\frac{\partial}{\partial\theta} + \frac{1}{\sin^2\theta}\frac{\partial^2}{\partial\phi^2}\right]\right) + V\right\}\psi = E\psi$$

(1.55)

The operator in square brackets in this expression is proportional to the l^2 operator in Section 1.4.2, so $\psi(r, \theta, \phi)$ may be written

$$\psi(r, \theta, \phi) = \chi(r)|lm\rangle \qquad (1.56)$$

where $|lm\rangle$ is the angular momentum eigenfunction (spherical harmonic). Substituting Eq. (1.56) into Eq. (1.55), we obtain the following *radial* Schrödinger equation:

$$\left\{-\frac{\hbar^2}{2\mu}\frac{1}{r}\frac{d^2}{dr^2}r + \frac{\hbar^2 l(l+1)}{2\mu r^2} + V(r)\right\}\chi = E\chi \qquad (1.57)$$

This equation can be further simplified by introducing a wavefunction ϕ via

$$\phi = r\chi \qquad (1.58)$$

Substituting this into Eq. (1.57) gives a radial equation that looks like the one-dimensional Schrödinger equation

$$\left(-\frac{\hbar^2}{2\mu}\frac{d^2}{dr^2} + V_{\text{eff}}(r)\right)\phi = E\phi \qquad (1.59)$$

where the effective potential V_{eff} is given as the sum of the real potential V plus the centrifugal potential

$$V_{\text{eff}} = V + \frac{\hbar^2 l(l+1)}{2\mu r^2} \qquad (1.60)$$

The only difference between Eq. (1.59) and a one-dimensional Schrödinger equation is in the range of r (0 to ∞ in three dimensions, $-\infty$ to ∞ in one dimension).

For small-amplitude diatomic vibrational motions, V_{eff} can be Taylor expanded about the equilibrium geometry ($r = r_e$). If this expansion is truncated at the quadratic term and the displacement coordinate $x = r - r_e$ is introduced, one recovers the harmonic oscillator hamiltonian [Eq. (1.36)].

1.5.2 Hydrogen Atom

For the special case of a hydrogen-like atom, the potential V is just the Coulomb potential $-Ze^2/r$, where Z is the atomic number. (Here we use cgs units, as this is the most commonly used system in the chemical community. To convert this to SI units, replace e^2 by $e^2/4\pi\varepsilon_0$, where ε_0 is the permittivity of free space. See the inside cover of this book for values of e, ε_0, and other fundamental constants.) The reduced mass μ is nearly equal to the electron mass m_e, so the approximation $\mu \approx m_e$ can be made. Equation (1.59) can then be solved analytically to give the hydrogenic radial wavefunctions

$$\phi(r) = R_{nl}(r) = N_{nl} \left(\frac{2Zr}{na_0}\right)^l e^{-Zr/na_0} L_{n+l}^{2l+1} \left(\frac{2Zr}{na_0}\right) \tag{1.61}$$

where

$$a_0 = \frac{\hbar^2}{m_e e^2} \tag{1.62}$$

$$N_{nl} = \left[\frac{4Z^3(n-l-1)!}{n^4 a_0^3 [(n+l)!]^3}\right]^{1/2} \tag{1.63}$$

and L_{n+l}^{2l+1} is an associated Laguerre polynomial. The quantum numbers in this case have the values $l = 0, 1, \ldots, n - 1$, and $n = 1, 2, \ldots, \infty$.

Energy eigenvalues for the hydrogen atom are given by

$$E_n = -\frac{Z^2}{2n^2} \frac{e^2}{a_0} \tag{1.64}$$

where a_0 was defined in Eq. (1.62).

1.6

Electronic Structure of Atoms and Molecules

1.6.1 Many-Electron Hamiltonian; Born–Oppenheimer Approximation

It is convenient at this point to switch to the use of atomic units ($e = \hbar = m_e = 1$). Derived constants in this system of units include the unit length a_0 [Eq. (1.62)], known as the bohr and the unit energy E_h [the constant e^2/a_0 in Eq. (1.64)] known as the hartree; 1 bohr $= 0.529177$ Å, 1 hartree $= 27.2114$ eV. Conversion factors for different energy units are listed on the inside cover.

The complete nonrelativistic hamiltonian of an arbitrary free molecule is

$$H = T_N + T_e + V_{eN} + V_{ee} + V_{NN} \tag{1.65}$$

where T_N = kinetic energy of the nuclei

T_e = kinetic energy of the electrons

V_{eN} = electron–nuclear attractive Coulomb potential

V_{ee} = electron–electron repulsive Coulomb potential

V_{NN} = nuclear–nuclear repulsive Coulomb potential

If we used the index α to label the N nuclei and i to label the n electrons, then the explicit form of each term in Eq. (1.65) is (in atomic units)

$$T_N = \sum_{\alpha}^{N} \frac{P_{\alpha}^2}{2M_{\alpha}} \tag{1.66a}$$

$$T_e = \sum_{i}^{n} \frac{P_i^2}{2} \tag{1.66b}$$

$$V_{eN} = -\sum_{\alpha}^{N} \sum_{i}^{n} \frac{Z_{\alpha}}{R_{i\alpha}} \tag{1.66c}$$

$$V_{ee} = \sum_{i}^{n} \sum_{j<i}^{n} \frac{1}{r_{ij}} \tag{1.66d}$$

$$V_{NN} = \sum_{\alpha}^{N} \sum_{\beta<\alpha}^{N} \frac{Z_{\alpha}Z_{\beta}}{R_{\alpha\beta}} \tag{1.66e}$$

The electron–nuclear wavefunction will be denoted by $\Psi(\mathbf{r}, \mathbf{R})$, where $\mathbf{r}$ stands for the collection of all electron coordinates and $\mathbf{R}$ stands for the collection of all nuclear coordinates. The Schrödinger equation is

$$H\Psi = W\Psi \tag{1.67}$$

where W is the total electron–nuclear energy.

For most chemical applications it is a good approximation to assume that the Schrödinger equation can be parametrically separated into a product of electronic and nuclear parts. This approximation, called the Born–Oppenheimer approximation, leads to factorization of the wavefunction Ψ as follows:

$$\Psi(\mathbf{r}, \mathbf{R}) = \psi(\mathbf{r}; \mathbf{R})\chi(\mathbf{R}) \tag{1.68}$$

where ψ is a wavefunction associated with solving the electronic part of the Schrödinger equation for fixed nuclear coordinates, and χ is a wavefunction associated with nuclear motion. The electronic Schrödinger equation includes all the terms in Eq. (1.65) that depend on electronic coordinates and is given by

$$(T_e + V_{eN} + V_{ee})\psi = E_{el}(\mathbf{R})\psi \tag{1.69}$$

where the energy E_{el} (as well as ψ) is a parametric function of the nuclear coordinates **R**. If Eq. (1.68) is substituted into Eq. (1.67) and Eq. (1.69) is applied, we obtain

$$(T_N + V_{NN} + E_{el})\psi\chi = W\psi\chi \qquad (1.70)$$

The Born–Oppenheimer approximation now consists of neglecting the **R** dependence of ψ so that $T_N\psi\chi = \psi T_N\chi$. This allows us to cancel ψ from both sides of Eq. (1.70), giving us

$$(T_N + V)\chi = W\chi \qquad (1.71)$$

where

$$V = V_{NN} + E_{el} \qquad (1.72)$$

is the *electronic potential energy surface* that governs nuclear motion.

Corrections to the Born–Oppenheimer approximation arise from the nuclear dependence of the electronic wavefunction ψ. Terms of the type $\langle \psi | T_N | \psi \rangle$ are often added to the electronic energy E_{el}; these so-called diagonal adiabatic corrections improve the energy of the electronic wavefunction. These can be understood from Eq. (1.22) as the lowest-order perturbation corrections, treating the kinetic energy of the nuclei as a perturbation on the ground-state electronic energy. More important corrections for many types of applications arise when kinetic energy coupling between two or more different electronic states leads to transitions from one state to another while nuclei move. This situation is addressed in Problem 4 of Chapter 4 and throughout Chapter 10.

1.6.2 Pauli Principle; Hartree–Fock Theory

The Pauli principle requires that the electronic wavefunction ψ be antisymmetric with respect to the interchange of any two electrons. One general way to ensure that this happens is to write the wavefunction as a Slater determinant of spin orbitals. Each spin orbital is the product of a spatial orbital S and a spin function α or β, and the general form of the Slater determinant for a closed-shell atom or molecule containing n electrons is

$$\psi = (n!)^{-1/2} \begin{vmatrix} S_1(1)\alpha_1 & S_1(1)\beta_1 & S_2(1)\alpha_1 & S_2(1)\beta_1 & \cdots \\ S_1(2)\alpha_2 & S_1(2)\beta_2 & S_2(2)\alpha_2 & S_2(2)\beta_2 & \cdots \\ S_1(3)\alpha_3 & S_1(3)\beta_3 & S_2(3)\alpha_3 & S_2(3)\beta_3 & \cdots \\ \cdots & \cdots & \cdots & \cdots & \cdots \end{vmatrix} \qquad (1.73)$$

In this equation we assume that each spatial orbital S_i is doubly occupied, so that its product with both α and β spin functions appears. Also, if the S_i's are all orthonormal, the normalization factor is given by the prefactor. Generalizations of Eq. (1.73) to molecules with open shells are straightforward but will not be considered.

If we apply variational theory to the determination of the optimum spatial orbitals S_i in Eq. (1.73), the following one-electron Schrödinger equation [known as Hartree–Fock (HF) or self-consistent field (SCF) equation] may be derived:

$$f_i S_i = \varepsilon_i S_i \tag{1.74}$$

where f_i (the Fock operator) is

$$f_i = -\frac{1}{2} \nabla_i^2 - \sum_{\alpha=1}^{N} \frac{Z_\alpha}{R_{i\alpha}} + \sum_{j=1}^{n/2} [2J_j(i) - K_j(i)] \tag{1.75}$$

ε_i in Eq. (1.74) is the one-electron energy eigenvalue. The Fock operator in Eq. (1.75) includes exact kinetic energy and electron–nuclear attraction terms plus two approximate electron–electron repulsion terms, the Coulomb and exchange operators. These are defined in terms of their action on the orbital S_i as follows:

$$J_j(i)S_i = \langle S_j | \frac{1}{r_{ij}} | S_j \rangle S_i \tag{1.76}$$

$$K_j(i)S_i = \langle S_j | \frac{1}{r_{ij}} | S_i \rangle S_j \tag{1.77}$$

where the integration variable is the coordinate r_j associated with electron j. Note that the solution to Eq. (1.74) must be accomplished self-consistently, as the Coulomb and exchange operators in f_i depend on the orbitals that one is seeking. Note also from (1.77) that the exchange operator, unlike the Coulomb operator, is nonlocal in space.

The total electronic energy E_{el} associated with the Slater determinant (1.73) using orbitals that satisfy Eq. (1.74) is given by

$$E_{el} = 2 \sum_{i=1}^{n/2} \varepsilon_i - \sum_{i}^{n/2} \sum_{j}^{n/2} (2J_{ij} - K_{ij}) \tag{1.78}$$

where J_{ij} is the Coulomb integral:

$$J_{ij} = \langle S_i(1)S_j(2) | \frac{1}{r_{12}} | S_i(1)S_j(2) \rangle \tag{1.79}$$

and K_{ij} is the exchange integral:

$$K_{ij} = \langle S_i(1)S_j(2) | \frac{1}{r_{12}} | S_i(2)S_j(1) \rangle \tag{1.80}$$

1.6.3 LCAO–MO–SCF

A standard method for solving the Hartree–Fock (SCF) equation for molecules involves expanding each molecular orbital (MO) S_i as a linear combination of atomic orbitals (LCAO), with the expansion coefficients to be variationally optimized. Thus we write

$$S_i = \sum_\mu C_{i\mu} B_\mu \qquad (1.81)$$

where B_μ is an atomic orbital. In most molecular electronic structure calculations, the B_μ's are chosen to be gaussian functions (or sums of gaussians), as this facilitates multicenter two-electron integral evaluation, but for the present discussion the functional form and number of B_μ's is irrelevant.

Since Eq. (1.81) is the same kind of expansion as Eq. (1.30), the application of variational theory leads to the same result, namely a secular equation to determine the optimized energies. In the present case this secular equation is

$$|\mathbf{f} - \varepsilon \mathbf{s}| = 0 \qquad (1.82)$$

Here $\mathbf{s}$ is an overlap matrix involving the AO basis functions (i.e., $s_{\mu\nu} = \langle B_\mu | B_\nu \rangle = \langle \mu | \nu \rangle$) and $\mathbf{f}$ is the matrix representation of the Fock operator:

$$\mathbf{f} = \mathbf{h} + 2\mathbf{J} - \mathbf{K} \qquad (1.83)$$

In (1.83), $\mathbf{h}$ includes the one-electron terms in $\mathbf{f}$:

$$h_{\mu\nu} = \langle \mu | -\frac{1}{2}\nabla_i^2 - \sum_\alpha \frac{Z_\alpha}{R_{i\alpha}} | \nu \rangle \qquad (1.84)$$

and $\mathbf{J}$ and $\mathbf{K}$ are Coulomb and exchange integrals given by

$$J_{\mu\nu} = \sum_j \sum_{\lambda\sigma} C_{j\lambda} C_{j\sigma} \langle \mu(1)\lambda(2) | \frac{1}{r_{12}} | \sigma(2)\nu(1) \rangle \qquad (1.85)$$

$$K_{\mu\nu} = \sum_j \sum_{\lambda\sigma} C_{j\lambda} C_{j\sigma} \langle \mu(1)\lambda(2) | \frac{1}{r_{12}} | \nu(2)\sigma(1) \rangle \qquad (1.86)$$

Note that the Coulomb and exchange integrals depend on the molecular orbital coefficients $\mathbf{C}$, which are known only after the secular equation (1.82) is solved. Evidently, an iterative solution is required, starting with an initial set of coefficients that is usually obtained from an approximate molecular orbital method.

1.6.4 Electronic Structure Methods

Molecular orbital methods that solve Eq. (1.82) using integrals that are calculated accurately are called *ab initio* methods, while methods that parametrize the integrals are known as *semiempirical* methods. In *ab initio* methods the only approximations made are in the LCAO expansion (1.81) and in using a single Slater determinant to represent the wavefunction. If the LCAO expansion is converged then the Hartree–Fock equations are being solved exactly (the so-called Hartree–Fock limit). This still can give electronic wavefunctions that are substantially in error due to the neglect of electron correlation effects in the single-determinant wavefunction. There are a number of methods that go beyond Hartree–Fock, including perturbation theory methods and variational methods. For example, one variational method, configuration interaction, uses an expansion in Slater determinants D_k:

$$\psi = \sum_k A_k D_k \tag{1.87}$$

where the expansion coefficients A_k are to be optimized. Since this expansion is like (1.30), variational theory applied to (1.69) leads to a secular equation

$$|\mathbf{H}_{el} - E\mathbf{S}| = 0 \tag{1.88}$$

where the electronic hamiltonian and overlap matrices now involve the full many-electron determinantal wavefunctions.

Further details associated with both *ab initio* and semiempirical molecular orbital methods are given in Chapter 6.

BIBLIOGRAPHY FOR CHAPTER 1

The material in this chapter is covered in more detail in a large number of textbooks on quantum chemistry. Among them are:

Atkins, P. W., *Molecular Quantum Mechanics*, 2nd Ed. (Oxford University Press, New York, 1983).

Dykstra, C. E., *Quantum Chemistry and Molecular Spectroscopy* (Prentice Hall, Englewood Cliffs, N.J., 1992).

Flurry, R. L., Jr., *Quantum Chemistry* (Prentice Hall, Englewood Cliffs, N.J., 1983).

Flygare, W. H., *Molecular Structure and Dynamics* (Prentice Hall, Englewood Cliffs, N.J., 1978).

Hameka, H. F., *Quantum Mechanics* (Wiley, New York, 1981).

Hanna, M. W., *Quantum Mechanics in Chemistry* (Benjamin-Cummings, Menlo Park, Calif., 1981).

Hehre, W., L. Radom, P. v. R. Schleyer, and J. A. Pople, *Ab Initio Molecular Orbital Theory* (Wiley, New York, 1986).

Karplus, M., and R. N. Porter, *Atoms and Molecules* (W. A. Benjamin, New York, 1970).

Kauzmann, W., *Quantum Chemistry* (Academic Press, New York, 1957).

Levine, I. N., *Quantum Chemistry*, 4th Ed. (Prentice Hall, Englewood Cliffs, N.J., 1991).

Lowe, J. P., *Quantum Chemistry* (Academic Press, New York, 1978).

McQuarrie, D. A., *Quantum Chemistry* (University Science Books, Mill Valley, Calif., 1983).

Pilar, F. L., *Elementary Quantum Chemistry*, 2nd Ed. (McGraw-Hill, New York, 1990).

PROBLEMS FOR CHAPTER 1

1. Which of the following operators are hermitian?

   (a) * (complex conjugate operator)

   (b) $ix \, \partial/\partial y$

   (c) $x + i \, \partial/\partial x$

   (d) $e^{ix} + e^{-ix}$

   (e) $x \, d/dx \, x$

2. Consider a particle in the one-dimensional box pictured in Fig. 1.1, having a step of height V_0 inside it, starting at a point $x = L/2$.

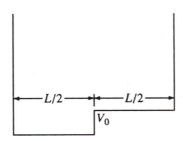

Figure 1.1. One-dimensional box with step in right half.

(a) Consider energies of less than V_0. What are the general solutions to the Schrödinger equation for $0 \leq x \leq L/2$ and $L/2 < x < L$, which have $\psi(x) = 0$ at $x = 0$ and $x = L$?

(b) What boundary conditions should be applied to the wavefunctions at $x = \frac{1}{2}L$? Derive a formula that determines the energy eigenvalues in terms of known constants such as m, L, and V_0.

(c) Suppose that V_0 is small compared to the zero-point energy. Show that the exact expression for the ground-state energy reduces to that derived from first-order perturbation theory.

(d) Suppose that V_0 is very large compared to the zero-point energy. Show that the ground-state energy and wavefunction reduce to those for a box of width $L/2$.

3. If $|p_x\rangle$, $|p_y\rangle$, and $|p_z\rangle$ are the usual p orbitals, what are each of the following?

   (a) $\langle p_x | l_z | p_y \rangle$

   (b) $\langle p_x | l_x | p_y \rangle$

   (c) $\langle p_z | l_y | p_x \rangle$

   (d) $\langle p_z | l_x | p_y \rangle$

   (e) $\langle p_z | l_x | p_x \rangle$

4. Use the variational trial function $\psi(r, \theta, \phi) = Y_{10}(\theta, \phi) r e^{-cr^2}$ to estimate the hydrogen atom total energy. For what state does this energy provide an upper bound? What is the error in the estimate?

5. Consider a diatomic molecule, approximated as a rigid rotor, in a static electric field F that points in the z direction. In this case the rotor hamiltonian gets modified to

$$H = \frac{l^2}{2I} + Fd \cos \theta$$

where d is the dipole moment of the rotor.

(a) Assuming that the $Fd \cos \theta$ term is small, use first-order perturbation theory to determine how much the ground-state energy changes as a result of interaction with the field.

(b) Now consider the variational trial function $\phi = Y_{00} + \lambda Y_{10}$, where λ is a parameter. What is the optimum λ and the resulting energy for the ground state?

6. Consider the H_2O molecule in its ground vibrational state.

(a) Write down the hamiltonian governing vibrational motions using raising and lowering operators, one for each vibrational mode. Assume that the potential is harmonic.

(b) What are the ground-state energy and wavefunction associated with the hamiltonian in part (a)?

(c) If we add the cubic anharmonicity $V = C_{122}Q_1Q_2^2$ to the hamiltonian, what excited vibrational states mix with the ground state according to second-order perturbation theory?

7. Use the molecular orbital computer program GAUSSIAN (or an equivalent program) to study a triatomic molecule of your choice. Start with a HF/STO-3G calculation, optimizing the geometry, and then determine the vibrational frequencies. Draw a rough picture of the highest occupied and lowest unoccupied molecular orbitals (HOMO and LUMO).

2

Symmetry Considerations: Point Groups and Electronic Structure

2.1

Group Theory for Point Groups

The complete hamiltonian of a molecule [Eq. (1.65)] is usually invariant to several different kinds of symmetry operations. The most generally applicable symmetry operations are *permutation* of the coordinates of identical particles (either electrons or nuclei) and *inversion* of the coordinates of all particles through a coordinate origin (i.e., $x \rightarrow -x$, $y \rightarrow -y$, $z \rightarrow -z$ for all coordinates). In the Born–Oppenheimer approximation, one is interested in the symmetry properties of the electronic hamiltonian H [$H = T_e + V_{eN} + V_{ee}$ from Eq. (1.69)] for *fixed* nuclear positions. In this case H is invariant with respect to the symmetry operations that interrelate equivalent nuclei in a rigid molecule, such as reflection planes and rotations about specified axes. Except for linear molecules (which are considered in Chapter 3), the possible symmetry operations for a rigid molecule are always finite in number, and the collection of all such operations for a given molecule forms a *group* that is known as a *point group*.

In this chapter we consider the consequences of point group symmetry for the electronic wavefunctions of molecules. Symmetry plays an important role in providing labels for electronic wavefunctions, and a particular emphasis of this chapter is on determining these symmetry labels for molecular orbitals and for the complete many-electron wave-functions. Since symmetry applies equally well to very accurate and very approximate electronic wavefunctions, we use the latter in this chapter to keep the description as simple as possible.

2.1.1 Symmetry Operations

We begin by reviewing briefly some of the basic elements of group theory as applied to the point groups. Possible symmetry elements include:

E: identity operation
C_n: n-fold rotation (by an angle $2\pi/n$)
σ: symmetry plane reflection
S_n: n-fold improper rotation (C_n followed by σ perpendicular to rotation axis)
i: inversion

Exercise: Prove that $S_n^2 = C_n^2$, $S_n^n = \sigma$ (n odd), $S_n^n = E$ (n even) and $S_2^1 = i$.

If all of the symmetry elements of a given molecule are considered collectively, it is not difficult to show that they form a *group*. Specifically, this means that (1) any *products* of two symmetry elements are also elements of the group, (2) that there is always an *identity* symmetry element, (3) that every element has an inverse, and (4) that multiplication of symmetry elements is *associative* [$A(BC) = (AB)C$].

2.1.2 Representations

For any symmetry group, it is always possible to find sets of matrices which multiply in the same way that the group elements do. That is, if $AB = C$ for three group elements, then $M_A M_B = M_C$ for the matrices M corresponding to A, B, and C. We say then that these matrices form a *representation* of the group. For symmetry groups of molecules (point groups), an easy way to generate a representation is to consider the changes in the three cartesian coordinates x, y, z of a point brought about by the symmetry operations. In general, after such an operation, the point (x, y, z) is transformed to a new location (x', y', z'), and the relation between the two can be represented using a matrix. For the identity operation, for example, we have

$$\begin{pmatrix} x' \\ y' \\ z' \end{pmatrix} = \begin{pmatrix} 1 & 0 & 0 \\ 0 & 1 & 0 \\ 0 & 0 & 1 \end{pmatrix} \begin{pmatrix} x \\ y \\ z \end{pmatrix} \qquad (2.1)$$

so that the identity matrix forms a representation of E. Similarly, a C_n operation about the z axis would be represented by

$$\begin{pmatrix} \cos 2\pi/n & -\sin 2\pi/n & 0 \\ \sin 2\pi/n & \cos 2\pi/n & 0 \\ 0 & 0 & 1 \end{pmatrix} \qquad (2.2)$$

2.1.3 Similarity Transformations

Once we have a matrix representation of a group, it is not difficult to generate other representations via *similarity transformations*. For example, if Q is any nonsingular matrix and A_i is a matrix representing element i of a group, then $B_i = Q^{-1}A_iQ$ also forms a matrix representation of the group; this operation is called a similarity transformation of A_i.

Two important examples of similarity transformations are found when:

1. Q is an element of the $\{A_i\}$ group representation, and the B_i obtained is also an element of the $\{A_i\}$ representation. In this case we say that B_i and its counterpart A_i both belong to the same class.
2. For some Q we obtain B_i's that are all *block diagonal* in the same sense (looking like

$$\begin{pmatrix} x_{11} & x_{12} & 0 \\ x_{21} & x_{22} & 0 \\ 0 & 0 & x_{33} \end{pmatrix}$$

for example). In this case, each of the blocks forms a representation, and we say that the representation is *reducible*. If a representation cannot be reduced further, we say that it is *irreducible*.

2.1.4 Irreducible Representations

Let us now consider the properties of the irreducible representations of a group. Let $\Gamma_i(R)_{mn}$ be the (m, n)th matrix element of the ith irreducible representation of the group for the operation R. We also let h be the order of the group (the number of operations therein) and l_i be the dimension of the ith irreducible representation. We also define the *character* (or trace) $\chi_i(R)$ of a given irreducible representation i for a given operation R via

$$\chi_i(R) = \sum_n \Gamma_i(R)_{nn} \qquad (2.3)$$

Characters have a number of useful properties, one of which is that *characters are independent of a similarity transformation* (easily proved), from which it follows that *characters of elements in the same class are identical.*

One of the most useful properties of the irreducible representation is that the elements of the matrices therein, considered as a function of the group operations R, form a set of h mutually orthogonal vectors. This is a statement of the "great orthogonality theorem" (G.O.T.) (the proof of which is given in the book by Eyring, Walter, and Kimball). Mathematically stated, we have

$$\sum_R \Gamma_i(R)_{mn} \Gamma_j(R)^*_{m'n'} = \frac{h}{\sqrt{l_i l_j}} \delta_{ij} \delta_{mm'} \delta_{nn'} \qquad (2.4)$$

Immediate consequences of this theorem are (1) that $\Sigma_i \, l_i^2 = h$, (2) $\Sigma_R \chi_i(R)\chi_j^*(R) = h\delta_{ij}$, and (3) that the number of irreducible representations equals the number of classes.

2.1.5 Character Tables

Armed with this theorem it is possible to determine most of the characteristics of the irreducible representations of point groups. In particular, the possible characters $\chi_i(R)$ associated with each representation matrix can be determined. For a *one-dimensional* irreducible representation, these characters are of course identical to the matrices themselves, so the G.O.T. can be used to determine uniquely the one-dimensional representations. For *multidimensional* irreducible representations, the representation matrices are not unique (new ones can always be developed using similarity transformations), and the characters do not uniquely determine these representations. The characters are, however, invariant to a similarity transformation, and are useful for other purposes, as discussed below (Section 2.2.2).

The character tables for the point groups of molecules have been tabulated in many places (such as in Cotton; see the Bibliography) and we will not study them in depth. Let us illustrate the basic idea of character table construction by a simple example, the H_2O molecule. Referring to Fig. 2.1, we see that the symmetry elements associated with H_2O at equilibrium are E, C_2 (about z), $\sigma_v(xz)$, and $\sigma_v(yz)$. (The v refers to a vertical symmetry plane which contains the n-fold axis.) These four elements make up that C_{2v} symmetry group (in Schoenflies notation). It is not difficult to show that none of these symmetry elements belong to the same class, which implies that there are four classes, hence four irreducible representations. The constraint $\Sigma_i l_i^2 = h$ can then be used to argue

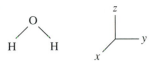

Figure 2.1. Coordinate system used for H_2O.

that all representations are one-dimensional $(1 + 1 + 1 + 1 = 4)$. From this, Table 2.1 can immediately be constructed.

Table 2.1

Character Table for C_{2v} Point Group

	E	C_2	$\sigma_v(xz)$	$\sigma_v(yz)$
A_1	1	1	1	1
A_2	1	1	-1	-1
B_1	1	-1	1	-1
B_2	1	-1	-1	1

Note that the symbols in the left-hand column are the so-called *Mulliken symbols*. They describe certain characteristics of each irreducible representation as follows:

1. *A* and *B* are used for one-dimensional representations, while *E* and *T* (or *F*) are used for two- and three-dimensional representations.
2. *A* is used for representations that are symmetric with respect to the C_n operations, while *B* is used for antisymmetric representations.
3. The subscripts 1 and 2 refer to representations that are either symmetric or antisymmetric with respect to σ_v or a C_2 perpendicular to the main *n*-fold axis.
4. The superscripts ' or " refer to representations that are symmetric or antisymmetric with respect to σ_h (reflection through a mirror plane that is perpendicular to the *n*-fold axis).
5. The subscripts *g* or *u* refer to representations that are symmetric or antisymmetric with respect to *i* (inversion).

Note that a representation having all characters equal to 1 is found for all groups and is called the *totally symmetric representation*.

2.1.6 Direct (or Tensor) Products

Earlier we noted that a representation of any symmetry group could be generated by considering the effect of the symmetry operations on the three cartesian coordinates (x, y, z). We would now like to consider what representation is generated when direct products of (x, y, z) are used as basis functions (such as $x^2, y^2, z^2, xy, yz, xz$), and how the representation thus generated is related to the one involving (x, y, z).

Let us introduce the notation $x_1 = x$, $x_2 = y$, $x_3 = z$, and so on, so that the transformation associated with the operator R can be written as

$$x_i' = Rx_i = \sum_j A_{ij}(R)x_j \tag{2.5}$$

Then the transformation associated with direct products such as $x_i x_j$ would be

$$Rx_i x_j = x_i' x_j' = \sum_k A_{ik}(R)x_k \sum_l A_{jl}(R)x_l = \sum_{kl} A_{ik}A_{jl}x_k x_l \tag{2.6}$$

Evidently, the forms of Eqs. (2.5) and (2.6) are identical provided that we consider the indices ij and kl as collective labels in the direct product vector space. Thus $C_{ij,kl}(R) = A_{ik}(R)A_{jl}(R)$ is a kind of supermatrix that forms a new representation of the symmetry group.

An important relation between the direct product representation and the representations that make it up is provided by considering the character of C:

$$\chi_C(R) = \sum_{ij} C_{ij,ij} = \sum_{ij} A_{ii}A_{jj} = \chi_A(R)\chi_A(R) \tag{2.7}$$

Equation (2.7) shows that the character of the C representation is simply the product of the characters of the representations that make it up (in this case, A^2).

2.1.7 Clebsch–Gordan Series

Often it is of interest to decompose a representation into irreducible representations. This is equivalent to block diagonalizing the matrices that define the representation, but it can be done with considerably less effort than that. It should be apparent that the characters associated with the reducible representation should equal the sums of the characters of the blocks into which it can be decomposed. Thus

$$\chi(R) = \sum_i a_i \chi_i(R) \tag{2.8}$$

where a_i is an integer coefficient that indicates how many times the irreducible representation i is contained in the reducible representation. Equation (2.8) is sometimes called the Clebsch–Gordan series. Multiplying Eq. (2.8) by $\chi_j^*(R)$, summing over R, and invoking consequence (2) to the G.O.T., we find that

$$a_j = \frac{1}{h} \sum_R \chi_j^*(R)\chi(R) \tag{2.9}$$

which provides a direct method for determining the coefficient a_j (and thereby decomposing the representation).

A direct consequence of Eqs. (2.7) and (2.9) is that the representation of the *direct product C* of two representations (say, A and B) will contain the totally symmetric representation only if A and B have at least one irreducible representation in common. To prove this, let's calculate the coefficient a_{TS} corresponding to the totally symmetric representation,

$$a_{TS} = \frac{1}{h} \sum_R \chi_C(R) = \frac{1}{h} \sum_R \chi_A(R)\chi_B(R) \tag{2.10}$$

If A and B are themselves irreducible, the G.O.T. indicates that $a_{TS} = \delta_{AB}$, which proves the theorem. If A and B are reducible, one must insert Eq. (2.8) for χ_A and χ_B. It then follows that a_{TS} will vanish unless A and B have an irreducible representation in common. (Note we have assumed here that the χ's are real. This is usually true, but if it is not, an analogous statement can still be made.)

2.2

Applications of Group Theory to Quantum Mechanics

2.2.1 Symmetry-Adapted Linear Combinations

The utility of group theory in electronic structure calculations arises from the fact that molecular electronic wave functions can always be constructed to belong to irreducible representations. This is because (1) if R is symmetry element of a molecule, R must commute with the electronic hamiltonian H of that molecule, and (2) simultaneous eigenfunctions of two operators may be constructed if those operators commute [as proved in the discussion of Eq. (1.12)].

We learned in Chapter 1 that solutions to the electronic Schrödinger equation are often constructed by diagonalizing the matrix representation of the secular equation

$$|H - ES| = 0 \tag{2.11}$$

In this equation the elements of H and S are $\langle \phi_i | H | \phi_j \rangle$ and $\langle \phi_i | \phi_j \rangle$, respectively, where ϕ_i is a basis function. If the ϕ_i's are constructed to belong to irreducible representations, $[H, R] = 0$ implies that the secular equation will automatically be block diagonalized. This is because the irreducible representation of the integrand in H_{ij} and S_{ij} must be totally symmetric if the resulting integral is to be nonzero (otherwise, the integrand will be an odd function of at least one variable). From our previous discussions we found that the representation of $\Gamma(S_{ij})$—namely, $\Gamma(\phi_i) \times \Gamma(\phi_j)$—would be totally symmetric only if ϕ_i and ϕ_j shared irreducible representations. Since that will not be the case if ϕ_i and ϕ_j belong to different irreducible representations, S_{ij} must be block diagonal. H_{ij} is also block diagonal because H is totally symmetric, so the representation generated by $\phi_i \times H \times \phi_j$ is the same as that of $\phi_i \times \phi_j$. Thus computational effort in solving the Schrödinger equation can be reduced considerably if we use symmetry-adapted linear combinations (SALCs) in setting it up.

2.2.2 Construction of SALCs

A very systematic method for constructing SALCs is to develop a *projection operator*, which will "project out" a function belonging to a specified irreducible representation when operating on an arbitrary function. To see how the projection operator is developed, let's first suppose that $\phi_t^i (t = 1 \ldots l_i)$ is a SALC belonging to the ith irreducible representation and the tth component. By definition of what we mean by irreducible representation, it must then follow that

$$R\phi_t^i = \sum_s \Gamma_i(R)_{ts} \phi_s^i \tag{2.12}$$

Now consider what happens when we multiply this equation by $\Gamma_j^*(R)_{t's'}$ and sum over R:

$$\sum_R \Gamma_j^*(R)_{t's'} R\phi_t^i = \sum_R \sum_s \Gamma_j^*(R)_{t's'} \Gamma_i(R)_{ts} \phi_s^i \tag{2.13}$$

Interchanging the two sums and invoking the G.O.T., we get

$$\sum_R \Gamma_j^*(R)_{t's'} R\phi_t^i = \sum_s \phi_s^i \frac{h}{\sqrt{l_i l_j}} \delta_{ij}\delta_{ss'}\delta_{tt'} = \frac{h}{l_j} \phi_{s'}^j \delta_{ij}\delta_{tt'} \tag{2.14}$$

Defining the operator $P_{t's'}^j$ as

$$P_{t's'}^j = \frac{l_j}{h} \sum_R \Gamma_j^*(R)_{t's'} R \tag{2.15}$$

we find that Eq. (2.14) becomes

$$P_{t's'}^j \phi_t^i = \phi_{s'}^j \delta_{ij}\delta_{tt'} \tag{2.16}$$

That is, $P_{t's'}^j$ operates on ϕ_t^i to give zero unless $i = j$ and $t = t'$, in which case $\phi_{s'}^j$ is obtained. For the special case $t' = s'$, we get

$$P^j_{t't'}\phi^i_t = \phi^j_{t'}\delta_{ij}\delta_{tt'} \qquad (2.17)$$

Now consider an arbitrary function ϕ. Although it will not in general be a symmetry-adapted linear combination, it may be decomposed into such functions using a Clebsch–Gordan series expression

$$\phi = \sum_{it} a_{it}\phi^i_t \qquad (2.18)$$

Applying $P^j_{t't'}$ to ϕ, we get

$$P^j_{t't'}\phi = \sum_{it} a_{it}P^j_{t't'}\phi^i_t = \sum_{it} a_{it}\phi^j_{t'}\delta_{ij}\delta_{tt'} = a_{jt'}\phi^j_{t'}$$

that is, out of the arbitrary function ϕ, $P^j_{t't'}$ has projected the symmetry eigenfunction $\phi^j_{t'}$. Note that if we had applied $P^j_{t't'}$ twice to ϕ, we get the same result as applying it once (i.e., $P^j_{t't'}$ is *idempotent*). It is then clear that $(P^j_{t't'})^2 = P^j_{t't'}$, which is sufficient to define $P^j_{t't'}$ as a projection operator.

Note that once we have generated a $\phi^j_{t'}$, the other components of the jth irreducible representation may be generated by using $P^j_{t's}$:

$$P^j_{t's}\phi^j_{t'} = \phi^j_s \qquad (2.19)$$

As an example of the use of a projection operator, consider the function $\phi = xz + yz$ in C_{2v} symmetry. Using the same coordinates and character tables as in Fig. 2.1 and Table 2.1 (Section 2.1.5), and realizing that for one-dimensional irreducible representations, the Γ's are identical to the characters, we find that the projector for the B_1 irreducible representation is

$$P^{B_1} = \frac{1}{4}\sum_R \chi_j(R)R = \frac{1}{4}[E - C_2 + \sigma_v(xz) - \sigma_v(yz)] \qquad (2.20)$$

Applying this to ϕ, we find that

$$P^{B_1}\phi = \frac{1}{4}(E - C_2 + \sigma_v(xz) - \sigma_v(yz))\phi$$

$$= \frac{1}{4}(xz + yz - (-xz - yz) + xz - yz - (-xz + yz)) = xz \quad (2.21)$$

Similarly, $P^{B_2}\phi = yz$ and $P^{A_1}\phi = P^{A_2}\phi = 0$. Thus we are guaranteed that xz is a SALC of the C_{2v} group, and simple inspection verifies that this is correct. Note that $P^{A_1}\phi = 0$. So ϕ is not sufficiently "arbitrary" to generate SALCs belonging to all irreducible representations.

For representations having $l_i > 1$, construction of $P^j_{t't'}$ requires knowledge of the $\Gamma_j(R)_{t't'}$, which are not always available. An alternative is to form a projection operator in terms of characters, although we must be forewarned that this does not provide all the desired $\phi_{t'}^{j}$'s. The character projection operator is defined as

$$P^j = \sum_{t'} P_{t't'}^j = \frac{l_j}{h} \sum_R \sum_{t'} \Gamma_j^*(R)_{t't'} R$$

or

$$P^j = \frac{l_j}{h} \sum_R \chi_j^*(R) R \tag{2.22}$$

Obviously, when we apply P^j to ϕ for j corresponding to a multidimensional representation, we get only one component of the $\phi_{t'}^{j}$'s. As will be demonstrated later, the other components can usually be obtained by auxiliary arguments.

2.2.3 Hückel Theory Applications

Let's now demonstrate the use of all of the group theory we have learned so far in applications to electronic structure problems by considering an application to the cyclopropenyl cation (Fig. 2.2), which has D_{3h} symmetry. Table 2.2 presents the character table. Let us number the symmetry operations according to Fig. 2.3 and assume that the molecule lies in the (x, y) plane.

In ordinary Hückel theory, we associate a p_z orbital with each carbon atom and then take linear combinations of these orbitals to form MOs. The secular equation is given by Eq. (2.11), where

$$H_{\mu\nu} = \begin{cases} \alpha & \text{if } \mu = \nu \\ \beta & \text{if } \mu, \nu \text{ adjacent} \\ 0 & \text{otherwise} \end{cases} \tag{2.23}$$

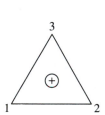

Figure 2.2. Cyclopropenyl cation.

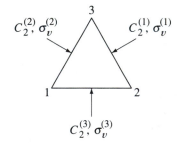

Figure 2.3. Symmetry operations for cyclopropenyl cation.

Table 2.2

Character Table for D_{3h} Point Group

D_{3h}	E	$2C_3$	$3C_2$	σ_h	$2S_3$	$3\sigma_v$		
A_1'	1	1	1	1	1	1		$x^2 + y^2, z^2$
A_2'	1	1	-1	1	1	-1	R_z	
E'	2	-1	0	2	-1	0	(x, y)	$x^2 - y^2, xy$
A_1''	1	1	1	-1	-1	-1		
A_2''	1	1	-1	-1	-1	1	z	
E''	2	-1	0	-2	1	0	(R_x, R_y)	xz, yz

For $C_3H_3^+$, this becomes

$$\begin{vmatrix} \alpha - E & \beta & \beta \\ \beta & \alpha - E & \beta \\ \beta & \beta & \alpha - E \end{vmatrix} = 0 \qquad (2.24)$$

This is only a 3×3 determinant, which is not difficult to evaluate explicitly. However, we can use group theory to simplify this, and although this is of minor consequence here, for larger molecules the reduction in effort can be substantial.

To use group theory to simplify, we must first determine the irreducible representations spanned by the AOs. To do this, we can construct 3×3 matrices analogous to those developed in Section 2.1.2. A simple set of rules for determining characters that may be derived from these matrices are:

1. If orbital is unchanged after applying symmetry operation, count $+1$ per orbital to the character.
2. If orbital changes sign, count -1.
3. If orbital changes location, count 0.

Applying these to the cyclopropenyl cation, we get, for the reducible representation,

D_{3h}	E	$2C_3$	$3C_2$	σ_h	$2S_3$	$3\sigma_v$
Γ	3	0	-1	-3	0	1

Now determine $a_i = (1/h) \sum_R \chi_i(R)\chi(R)$:

$$a_{A_1'} = \tfrac{1}{12}[3 - 1(3) - 3(1) + 3(1)] = 0$$

$$a_{A_2'} = \tfrac{1}{12}[3 - 1(3)(-1) - 3(1) + 1(3)(-1)] = 0$$

These indicate that wavefunctions cannot have A' symmetry

$$a_{A_1''} = \tfrac{1}{12}(3 - 3 + 3 - 3) = 0$$

$$a_{E'} = 0$$

$$\left.\begin{aligned} a_{A_2''} &= \tfrac{1}{12}(3 + 3 + 3 + 3) = 1 \\ a_{E''} &= \tfrac{1}{12}(6 + 6) = 1 \end{aligned}\right\} \quad \text{These indicate that } \Gamma = A_2'' + E'' \quad (2.25)$$

To construct SALCs, we first illustrate the laborious method based on projection operators. Previously, we had demonstrated that

$$P^j = \frac{l_j}{h} \sum_R \chi_j^*(R) R \qquad (2.26)$$

is a projection operator for the jth irreducible representation. Applying this to $j = A_2''$, we find that

$$P^{A_2''} = \tfrac{1}{12}(E + C_3 + C_3^2 - C_2^{(1)} - C_2^{(2)} - C_2^{(3)}$$
$$\qquad\qquad - \sigma_h - S_3 - S_3^2 + \sigma_v^{(1)} + \sigma_v^{(2)} + \sigma_v^{(3)}) \quad (2.27)$$

This can be applied to any of the three p_z orbitals to generate a SALC having A_2'' symmetry. Applying it to ϕ_1 gives us

$$P^{A_2''}\phi_1 = \tfrac{1}{12}(\phi_1 + \phi_2 + \phi_3 + \phi_1 + \phi_3 + \phi_2 + \phi_1$$
$$\qquad + \phi_2 + \phi_3 + \phi_1 + \phi_3 + \phi_2) = \tfrac{1}{3}(\phi_1 + \phi_2 + \phi_3) \quad (2.28)$$

Similarly,

$$P^{E''}\phi_1 = \tfrac{2}{12}(2\phi_1 - \phi_2 - \phi_3 + 2\phi_1 - \phi_2 - \phi_3)$$
$$\qquad = \tfrac{1}{3}(2\phi_1 - \phi_2 - \phi_3) \quad (2.29)$$

To generate the other component of the E'' representation, we take advantage of Eq. (2.12), which tells us that application of any group operation into one component of an irreducible representation generates linear combinations of all such components. If we normalize and require all components to be orthogonal, the remaining components can be projected out. Starting with

$$\psi_{E''}^{(1)} = \frac{1}{\sqrt{6}}(2\phi_1 - \phi_2 - \phi_3) \qquad \text{(normalized)} \qquad (2.30)$$

we find that

$$C_3\psi_{E''}^{(1)} = \frac{1}{\sqrt{6}}(2\phi_2 - \phi_3 - \phi_1) \qquad (2.31)$$

The overlap of this with $\psi_{E''}^{(1)}$ is

$$\langle \psi_{E''}^{(1)} \mid C_3\psi_{E''}^{(1)} \rangle = \frac{1}{\sqrt{36}}(-2 - 2 + 1) = -\frac{3}{\sqrt{36}} = -\frac{1}{2} \qquad (2.32)$$

so after subtracting this overlap from Eq. (2.31), we find that

$$C_3\psi_{E''}^{(1)} + \frac{1}{2}\psi_{E''}^{(1)} = \frac{\sqrt{3}}{2}\frac{1}{\sqrt{2}}(\phi_2 - \phi_3) \qquad (2.33)$$

Upon normalization, this second component is

$$\psi_{E''}^{(2)} = \frac{1}{\sqrt{2}}(\phi_2 - \phi_3) \qquad (2.34)$$

and Eq. (2.31) reduces to

$$C_3\psi_{E''}^{(1)} = -\frac{1}{2}\psi_{E''}^{(1)} + \frac{\sqrt{3}}{2}\psi_{E''}^{(2)} \qquad (2.35)$$

2.2.4 Shortcuts

Before proceeding further, it is useful to describe a much easier method for generating SALCs which is applicable to many systems. This involves using the C_n subgroup of D_{nh} to classify the irreducible representations. This can be used because the σ_h operation is already a symmetry operation of the p orbitals. C_n is easier to work with because all the irreducible representations are one-dimensional. The character table for C_3 is shown in Table 2.3.

Table 2.3

Character Table for C_3 Point Group

C_3	E	C_3	C_3^2
A	1	1	1
E	1	ε	ε^*
	1	ε^*	ε

where $\varepsilon = e^{2\pi i/3}$,

$$\tfrac{1}{2}(\varepsilon + \varepsilon^*) = \cos 120° = -\tfrac{1}{2},$$

$$\frac{\varepsilon - \varepsilon^*}{2i} = \sin 120° = \frac{\sqrt{3}}{2}.$$

Note that the second and third representations are collectively labeled E even though they are one-dimensional.

Applying the rules for characters, we find

	E	C_3	C_3^2
Γ	3	0	0

from which, by inspection,

$$\Gamma = A + E \tag{2.36}$$

Note that the correlation between representations in C_3 and D_{3h} is

C_3	D_{3h}
A	A_2''
E	E''

Whenever such a correlation cannot be established, another subgroup must be used. Since all the representations are one-dimensional, the generation of SALCs is easy. Almost by inspection (after normalization)

$$\psi_A = \frac{1}{\sqrt{3}} (\phi_1 + \phi_2 + \phi_3) \tag{2.37}$$

$$\psi_E^{(1)} = \frac{1}{\sqrt{3}} (\phi_1 + \varepsilon^*\phi_2 + \varepsilon\phi_3) \tag{2.38}$$

$$\psi_E^{(2)} = \frac{1}{\sqrt{3}} (\phi_1 + \varepsilon\phi_2 + \varepsilon^*\phi_3) \tag{2.39}$$

Obviously, ψ_A and $\psi_{A_2''}$ are the same, but the ψ_E's are not. Instead, they are linear combinations of one another:

$$\psi_E^{(1)} = \frac{1}{\sqrt{2}} \psi_{E''}^{(1)} - \frac{i}{\sqrt{2}} \psi_{E''}^{(2)} \tag{2.40}$$

$$\psi_E^{(2)} = \frac{1}{\sqrt{2}} \psi_{E''}^{(1)} + \frac{i}{\sqrt{2}} \psi_{E''}^{(2)} \tag{2.41}$$

2.2.5 Energy Eigenvalues

In the following, we use the SALCs:

$$\psi_1 = \psi_{A_2''} = \frac{1}{\sqrt{3}} (\phi_1 + \phi_2 + \phi_3) \tag{2.42}$$

$$\psi_2 = \psi_{E''}^{(1)} = \frac{1}{\sqrt{6}} (2\phi_1 - \phi_2 - \phi_3) \tag{2.43}$$

$$\psi_3 = \psi_{E''}^{(2)} = \frac{1}{\sqrt{2}} (\phi_2 - \phi_3) \tag{2.44}$$

To apply Hückel theory, we can automatically separate out the A_2'' subblock of the hamiltonian from the E'' block.

A_2'': The secular equation is just 1×1:

$$|\langle\psi_1|H|\psi_1\rangle - E\langle\psi_1|\psi_1\rangle| = 0 \tag{2.45}$$

with $\langle\psi_1|H|\psi_1\rangle = \alpha + 2\beta$, so this immediately gives

$$E = \alpha + 2\beta \tag{2.46}$$

E'': Here the secular equation is 2×2:

$$\begin{vmatrix} \langle\psi_2|H|\psi_2\rangle - E & \langle\psi_2|H|\psi_3\rangle \\ \langle\psi_3|H|\psi_2\rangle & \langle\psi_3|H|\psi_3\rangle - E \end{vmatrix} = 0 \tag{2.47}$$

with

$$\langle\psi_2|H|\psi_2\rangle = \tfrac{1}{6}(6\alpha - 6\beta) = \alpha - \beta \tag{2.48}$$

$$\langle\psi_3|H|\psi_3\rangle = \tfrac{1}{2}(2\alpha - 2\beta) = \alpha - \beta \tag{2.49}$$

and

$$\langle\psi_2|H|\psi_3\rangle = 0 = \langle\psi_3|H|\psi_2\rangle \tag{2.50}$$

Thus the secular equation is

$$\begin{vmatrix} \alpha - \beta - E & 0 \\ 0 & \alpha - \beta - E \end{vmatrix} = 0 \tag{2.51}$$

which gives

$$E = \alpha - \beta \text{ (twice)} \tag{2.52}$$

Note that the secular equation for the E'' block, (2.51), is diagonal, even though this is not required by symmetry. However, the eigenvalues of the E symmetry block will always be degenerate. Note also that since these SALCs diagonalize the secular equation, they are by themselves the eigenfunctions of the Schrödinger equation. This is *not* generally true.

Assuming that $\beta < 0$, we can generate the energy-level diagram in Fig. 2.4, with the two electrons occupying the a_2'' in the ground state. Note that in accordance with standard notation, we have used lowercase Mulliken symbols to label the one-electron orbitals in this diagram.

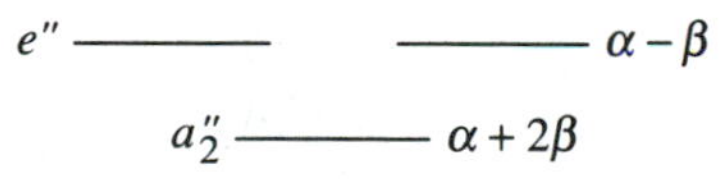

$$e'' \underline{\hspace{2cm}} \qquad \underline{\hspace{2cm}} \alpha - \beta$$

$$a_2'' \underline{\hspace{2cm}} \alpha + 2\beta$$

Figure 2.4. Energy-level diagram for cyclopropenyl cation.

Symmetry Properties of Many-Electron Wavefunctions

2.3.1 Many-Electron Configurations and Terms

Up to this point, only the one-electron orbital symmetries have been considered. Now in general the many-electron wavefunction for a given electron configuration can be written in terms of Slater determinants of the one-electron orbitals. These determinants involve products of orbitals, and if the orbitals are SALCs, the representations generated by the Slater determinant are determined by the direct product of the representations of the orbitals. In addition, simple models such as Hückel can be used to determine many-electron symmetries from the one-electron symmetries, as the symmetry properties of wavefunctions can be rigorously correct even for highly approximate wavefunctions.

2.3.2 Application to $C_3H_3^+$

For $C_3H_3^+$, we have a ground configuration $(a_2'')^2$. The representation generated by this is determined by applying the symmetry operations to the many-electron wavefunction, which in this case is just a Slater determinant. Since the Slater determinant involves a product of orbitals, the representation involves a direct product of the representations of the individual orbitals. For the ground state, this is $a_2'' \times a_2'' = A_1'$. Thus the ground term is $^1A'$ [uppercase Mulliken symbols are used to label many-electron states, with the spin multiplicity (singlet in this case) given as a superscript]. Quite generally, the representation associated with a totally filled shell is the *totally symmetric* one.

For the first excited configuration $(a_2'')^1(e'')^1$, we find that $a_2'' \times e'' = E'$. Here there is the possibility of either singlet or triplets. Since the electrons are in orbitals with different energies, there is no problem with the Pauli principle, and we get $^1E'$ and $^3E'$ as the allowed terms.

Now consider the configuration $(e'')^2$. Since there are two MOs, we can imagine several ways to arrange the electrons, but not all are consistent with the Pauli principle. Symmetry alone tells us that

$$e'' \times e'' = A_1' + A_2' + E' \qquad (2.53)$$

2.3.3 Incorporation of the Pauli Principle

An important question concerning Eq. (2.53) is: Which terms are singlets and which triplets? For the simple case of two electrons, the Slater determinant wavefunction can be written as a product of space and spin parts. This is not true for more than two electrons, but the two-electron result is important, so we will work it through in detail.

For this simple special case, then, we need to construct wavefunctions associated with A_1', A_2', and E' terms, then see which are symmetric and which antisymmetric with respect to interchange of spatial coordinates. Then the symmetric ones must be singlet, and the antisymmetric ones triplet to obey the Pauli principle.

To construct wavefunctions we can use projection operators. As before, it is easiest to use the C_3 group, although A_1' and A_2' are not distinguished thereby. Let

$$\chi_1 = \frac{1}{\sqrt{3}}(\phi_1 + \phi_2 + \phi_3) \tag{2.54}$$

$$\chi_2 = \frac{1}{\sqrt{3}}(\phi_1 + \varepsilon^*\phi_2 + \varepsilon\phi_3) \tag{2.55}$$

$$\chi_3 = \frac{1}{\sqrt{3}}(\phi_1 + \varepsilon\phi_2 + \varepsilon^*\phi_3) \tag{2.56}$$

Then

$$C_3\chi_1 = \chi_1 \tag{2.57}$$

$$C_3\chi_2 = \frac{1}{\sqrt{3}}(\phi_2 + \varepsilon^*\phi_3 + \varepsilon\phi_1) = \varepsilon\chi_2 \tag{2.58}$$

$$C_3\chi_3 = \varepsilon^*\chi_3 \tag{2.59}$$

Now consider the following two-electron spatial wavefunctions [these are all that can be imagined for $(e'')^2$]:

$$|1\rangle = \chi_2(1)\chi_2(2) \tag{2.60}$$

$$|2\rangle = \chi_3(1)\chi_3(2) \tag{2.61}$$

$$|3\rangle = \chi_2(1)\chi_3(2) \tag{2.62}$$

$$|4\rangle = \chi_3(1)\chi_2(2) \tag{2.63}$$

The characters generated from these are given by Table 2.4.

Table 2.4

Characters for $(e'')^2$ Configurations of $C_3H_3^+$

	E	C_3	C_3^2
$|1\rangle$	1	ε^*	ε
$|2\rangle$	1	ε	ε^*
$|3\rangle$	1	1	1
$|4\rangle$	1	1	1

Evidently, $|1\rangle$ and $|2\rangle$ form an E symmetry set, while $|3\rangle$ and $|4\rangle$ belong to A symmetry.

Since $|1\rangle$ and $|2\rangle$ are both symmetric spatially with respect to interchange of the two electrons, we immediately conclude that the E' symmetry term must be *singlet* ($^1E'$). For the A symmetry terms we must differentiate A_1' from A_2'. This is done most easily by determining symmetry with respect to $\sigma_v^{(1)}$. Since $\sigma_v^{(1)}\chi_1 = \chi_1, \sigma_v^{(1)}\chi_2 = \chi_3, \sigma_v^{(1)}\chi_3 = \chi_2$, we have

$$\sigma_v^{(1)}|3\rangle = |4\rangle \tag{2.64}$$

Thus neither $|3\rangle$ nor $|4\rangle$ belongs to an irreducible representation of D_{3h}, although both are of A symmetry for C_3. It is easy to see that the following linear combinations of $|3\rangle$ and $|4\rangle$ generate the desired D_{3h} irreducible representations (projection operators could be used if this were not obvious):

$$|5\rangle = \frac{1}{\sqrt{2}}(|3\rangle + |4\rangle) \tag{2.65}$$

$$|6\rangle = \frac{1}{\sqrt{2}}(|3\rangle - |4\rangle) \tag{2.66}$$

Clearly, $|5\rangle$ is symmetric with respect to σ_v and hence must be A_1' while $|6\rangle$ is A_2'. Furthermore, $|5\rangle$ is symmetric with respect to interchange of 1 and 2 and hence is singlet, while $|6\rangle$ is antisymmetric spatially and hence is triplet.

Overall, then, the allowed terms are $^1A_1'$, $^3A_2'$, and $^1E'$. These terms are all degenerate when electron repulsion is neglected, but when it is included, the degeneracy is removed. A generalization of Hund's rules that would be applicable to this case would be that:

1. The highest spin-multiplicity term has the lowest energy.
2. The highest spatial degeneracy term for a given multiplicity has the lowest energy.

The expected ordering is given in Fig. 2.5.

Figure 2.5. Energy-level diagram for $(e_2'')^2$ terms of $C_3H_3^+$.

2.3.4 Optical Spectrum: Dipole Selection Rules

In Chapter 5 we show that the probability of one-photon absorption or emission between states i and f for the electronic dipole mechanism is given by

$$I \propto \left| \int \Psi_i (\boldsymbol{\epsilon} \cdot \boldsymbol{\mu}) \Psi_f \, d\tau \right|^2 \tag{2.67}$$

where $\boldsymbol{\mu}$ is the dipole operator ($e\mathbf{r}$ for a one-electron transition) and $\boldsymbol{\epsilon}$ is the light polarization vector. There are many other mechanisms by which a transition can gain intensity, and these lead to different selection rules, but for now we consider just the simplest mechanism, the electronic dipole mechanism.

Now if our Ψ_i and Ψ_f belong to irreducible representations, we can determine if I is zero for a given polarization by examining the direct product for $\Psi_i \times (\boldsymbol{\epsilon} \cdot \mathbf{r}) \times \Psi_f$. This can be simplified usually by first determining $\Psi_i \times \Psi_f$, then decomposing and using the character table to see if some irreducible representation of $\Psi_i \times \Psi_f$ is the same as that of $(\boldsymbol{\epsilon} \cdot \mathbf{r})$.

Applying this to the ground-to-first excited-state transition in $C_3H_3^+$, we note that since Ψ_i is A_1', we need only compare the excited-state representations with those of x, y, or z.

Since E' is (x, y), we conclude that $A_1' \to E'$ is dipole allowed. Since the dipole matrix element is zero for terms of different spin symmetry, only singlet $\to$ singlet or triplet $\to$ triplet transitions are allowed. So the lowest allowed transition is predicted to be

$$^1A_1' \to {}^1E'$$

Question: What terms from the $(e'')^2$ configuration are optically connected to the ground state?

BIBLIOGRAPHY FOR CHAPTER 2

The "bible" of group theory as discussed in this chapter is:

Cotton, F. A., *Chemical Applications of Group Theory* (Wiley-Interscience, New York, 1971).

Other useful texts include:

Chesnut, D. B., *Finite Groups and Quantum Theory* (Wiley-Interscience, New York, 1974).

Eyring, H., J. Walter, and G. Kimball, *Quantum Chemistry* (Wiley, New York, 1944).

Pilar, F W., *Elementary Quantum Chemistry*, 2nd Ed. (McGraw-Hill, New York, 1990).

Tinkham, M., *Group Theory and Quantum Mechanics* (McGraw-Hill, New York, 1964).

The Hückel model of electronic structure is discussed in Cotton and in most of the texts mentioned in the Bibliography for Chapter 1.

PROBLEMS FOR CHAPTER 2

1. Apply Hückel theory to the p_z orbitals in cyclopentadienyl (assumed to be planar and symmetrical).

(a) What is the symmetry group?

(b) Give the secular equation using AOs.

(c) Find the irreducible representations spanned by the orbitals.

(d) Construct SALCs.

(e) Find the one-electron energies and wavefunctions.

(f) What are the many-electron symmetry and spin multiplicity of the ground state?

(g) What is the total electronic energy?

2. Consider the 2 + 2 addition reaction

$$H_2 + D_2 \rightarrow 2HD$$

Imagine the following reaction path:

$$
\begin{array}{cccccc}
\text{H} & \text{D} & & \text{H}\cdots\text{D} & & \text{H—D} \\
| & + \ | & \rightarrow & \vdots \quad \vdots & \rightarrow & + \\
\text{H} & \text{D} & & \text{H}\cdots\text{D} & & \text{H—D} \\
\end{array}
$$

$$
\begin{array}{cccc}
\text{(I)} & & \text{(II)} & \text{(III)}
\end{array}
$$

where the symmetry changes from D_{2h} to D_{4h} to D_{2h}.

(a) Apply the extended Hückel method to structures I, II, and III, making the Hückel-like assumptions:

$$H_{\mu\nu} = \begin{cases} \alpha & \mu = \nu \\ \beta & \mu \text{ adjacent to } \nu \\ 0 & \text{otherwise} \end{cases}$$

$$S_{\mu\nu} = \delta_{\mu\nu}$$

Obtain orbital energies and wavefunctions. Determine the ground-state electron configuration and many-electron symmetry and spin multiplicity.

(b) Now use symmetry to determine the correlation of the electronic orbitals as one progresses along the reaction coordinate. Draw a correlation diagram (energy level versus reaction coordinate). Show that this reaction is Woodward–Hoffmann forbidden (meaning that the ground state of the reagents correlates to an excited state of the products).

3. Consider the square planar complex $PtCl_4^{2-}$ (with Pt in a d^8 configuration). Let's describe σ bonding between metal and ligands using extended Hückel. The AO basis consists of p orbitals on each Cl^- directed toward the Pt, and the d orbitals on Pt. Assume that the only nonzero Hückel parameters are:

$$\langle d_j | H | d_j \rangle = \alpha_{Pt} \quad \text{(independent of } j\text{)}$$

$$\langle p_i | H | p_i \rangle = \alpha_{Cl} \quad \text{(independent of } i\text{)}$$

$$\langle p_i | H | d_j \rangle = \beta_{ij}$$

Choose $\beta_{ij} = 0$ for $j = d_{xy}$, d_{yz}, and d_{xz}. For the remainder, use the values in the following table:

$i =$	1	2	3	4
$j = d_{z^2}$	β_A	β_A	β_A	β_A
$j = d_{x^2-y^2}$	β_B	$-\beta_B$	β_B	$-\beta_B$

Assume that $\alpha_{Pt} > \alpha_{Cl}$ and $|\beta_B| > |\beta_A|$.

(a) What is the symmetry group?

(b) Give the secular equation using AOs.

(c) Find the irreducible representations spanned by the orbitals.

(d) Construct SALCs.

(e) Find the one-electron energies.

(f) Which orbitals are occupied in the ground state?

(g) What is the lowest-energy optically allowed transition?

4. Consider an application of extended Hückel theory to the H_3 molecule, using the $1s$ orbitals on each nucleus as a basis set. Let's imagine that we do not know the geometry of this molecule, so we will vary it to minimize the total

electronic energy. To do this, assume that the bend angle θ defined below is a variable to be determined, and that the hamiltonian matrix elements are as follows:

$$\langle i|H|i \rangle = \alpha \qquad (i = 1, 2, 3)$$

$$\langle 1|H|2 \rangle = \langle 2|H|3 \rangle = \beta$$

$$\langle 1|H|3 \rangle = 2\beta \left(1 - \sin \frac{\theta}{2} \right)$$

$$
\begin{array}{c}
H_2 \\
/ \;\theta\; \backslash \\
H_1 \qquad H_3
\end{array}
$$

(a) Construct SALCs and evaluate the one-electron energies for this basis.

(b) What energy levels do the electrons occupy? What is the total electronic energy? Considering the range $60° < \theta < 180°$, for which value of θ is the total energy minimized? If one considers the ion H_3^+, what is the total energy, and how does this energy vary with θ?

(c) Construct the molecular orbitals for arbitrary θ.

(d) Show how the symmetry labels of the orbitals correlate with θ between $\theta = 180$ and $60°$ (i.e., show how the linear molecule, isosceles, and equilateral triangle geometry point group labels interrelate).

(e) What are the many-electron term symbols for $\theta = 60°$ for H_3, H_3^+, and H_3^-?

3

Symmetry Considerations: Continuous Groups and Rotations

3.1

Introduction

There is a tremendous amount of literature on continuous groups, most of which is not used outside of physics and mathematics. Texts such as Hamermesh, Tinkham, Wigner, and others should be consulted for a more complete description. Here we describe some of the simplest chemically relevant continuous groups, and we also consider the closely related topic of angular momentum addition.

3.2

Continuous Groups; Electronic Structure of Linear Molecules

3.2.1 Two-Dimensional Rotation Group

The two-dimensional rotation group arises when considering the symmetry group of linear molecules like CO. Clearly, one element of symmetry of this molecule involves rotation about the molecular axis by an arbitrary angle ϕ. We label this rotation as $C(\phi)$. Now just as $(C_n)^m$, $m = 1, \ldots,$ n generates an n-dimensional abelian (multiplication is commutative) point group, the set $C(\phi)$, $\phi = [0, 2\pi]$ generates an infinite-dimensional abelian continuous group. ϕ in this case is a continuous parameter that labels the group elements.

Since the group is abelian, the representations of it are one-dimensional (only one-dimensional matrices commute in general). Furthermore, they must transform as

$$C(\phi_1)C(\phi_2) = C(\phi_1 + \phi_2) \tag{3.1}$$

since rotations about a single axis are additive. In other words, the matrices (or equivalently characters) of any representation must satisfy

$$\chi(\phi_1 + \phi_2) = \chi(\phi_1)\chi(\phi_2) \tag{3.2}$$

with $\chi(0) = 1$ as the identity character.

It is easy to verify that the function

$$\chi(\phi) = e^{im\phi} \tag{3.3}$$

satisfies the foregoing requirements. Now since $\chi(\phi + 2\pi) = \chi(\phi)$, the parameter m is constrained to be an integer ($m = 0, \pm1, \pm2, \ldots$). With these characters, the character table in Table 3.1 is constructed for the group [labeled C_∞ or $O(2)$]. This group is often called the *two-dimensional rotation group*. Note that $e^{im\phi}$ is the azimuthal part of the spherical harmonic Y_{lm}, so $\chi(\phi)$ is an eigenfunction of l_z, the angular momentum projection along the z axis, with eigenvalue $m\hbar$. The Mulliken symbols in this case (Σ, Π, Δ, etc.) are the uppercase Greek analogs of S, P, D, F, and so on.

Now we ask, what is the analog of the great orthogonality theorem?

Table 3.1

Character Table for Two-Dimensional Rotation Group

m	C_∞	E	$C(\phi)$
0	Σ	1	1
1	$\Pi \Big\{$	1	$e^{i\phi}$
-1		1	$e^{-i\phi}$
2	$\Delta \Big\{$	1	$e^{2i\phi}$
-2		1	$e^{-2i\phi}$

Well, with a little intuition we can imagine that $\Sigma_R \chi_i^*(R)\chi_j(R) = h\delta_{ij}$ becomes

$$\int_0^{2\pi} d\phi \, e^{-im\phi}e^{im'\phi} = 2\pi\delta_{mm'} \tag{3.4}$$

where the factor 2π is sometimes called the group "volume."

Decomposing a reducible representation is very much analogous to the situation for point groups. If $\chi(\phi)$ is the character for some representation, the Clebsch–Gordan series in this case is

$$\chi(\phi) = \sum_m a_m e^{im\phi} \tag{3.5}$$

with

$$a_m = \frac{1}{2\pi}\int_0^{2\pi} d\phi \, e^{-im\phi}\chi(\phi) \tag{3.6}$$

This of course is identical with the usual Fourier series. In addition, for a direct product representation such as $\chi = e^{im_1\phi}e^{im_2\phi}$, we obtain $a_m = \delta_{m,m_1+m_2}$, which defines a kind of Clebsch–Gordan coefficient for this group.

3.2.2 $C_{\infty v}$ Group

For linear molecules such as CO, there is actually an additional symmetry element, namely σ_v (for each ϕ). Adding this to the C_∞ group gives a group labeled $C_{\infty v}$. In this group the $C(\phi)$ and $C(-\phi)$ elements are related to each other by the similarity transformation $\sigma_v C(\phi)\sigma_v = C(-\phi)$. Thus $C(\phi)$ and $C(-\phi)$ belong to the same *class*, and representations may be two-dimensional. One can easily verify that the matrices

Table 3.2

Character Table for $C_{\infty v}$ Group

$C_{\infty v}$	E	$2C(\phi)$	σ_v
Σ^+	1	1	1
Σ^-	1	1	-1
Π	2	$2\cos\phi$	0
Δ	2	$2\cos 2\phi$	0
Φ	2	$2\cos 3\phi$	0

$$C(\phi) = \begin{bmatrix} e^{im\phi} & 0 \\ 0 & e^{-im\phi} \end{bmatrix} \quad and \quad \sigma_v = \begin{pmatrix} 0 & 1 \\ 1 & 0 \end{pmatrix} \qquad (3.7)$$

form a representation of the group for $m \neq 0$. For $m = 0$, we get two one-dimensional representations, depending on the sign of the character for σ_v (labeled Σ^+ and Σ^-). The character table is given in Table 3.2.

3.2.3 $D_{\infty h}$ Group

Molecules such as O_2, CO_2, and so on, also have a symmetry plane σ_h perpendicular to the molecular axis. If this symmetry element is added to the $C_{\infty v}$ group operations, one gets the $D_{\infty h}$ group. Note that $\sigma_h C(\phi) = S(\phi)$, so one can use $S(\phi)$ instead of σ_h to label the group elements, and customarily, this is done. [Note that $S(2\pi) = \sigma_h$ and $S(\pi) = i$.] The presence of the i operation gives the irreducible representations a definite inversion symmetry (splitting all the $C_{\infty v}$ representations into u and g). Table 3.3 shows the character table.

3.2.4 Simple Example: O_2

Now consider simple independent electron MO theory for homonuclear diatomics using the coordinates defined in Fig. 3.1. Labeling the nuclei "1" and "2," the following orbital combinations can be constructed belonging to the indicated irreducible representations:

Table 3.3

Character Table for $D_{\infty h}$ Group

$D_{\infty h}$	E	$2C(\phi)$	σ_v	i	$2S(\phi)$	C_2
Σ_g^+	1	1	1	1	1	1
Σ_g^-	1	1	-1	1	1	-1
Σ_u^+	1	1	1	-1	-1	-1
Σ_u^-	1	1	-1	-1	-1	1
Π_g	2	$2\cos\phi$	0	2	$-2\cos\phi$	0
Π_u	2	$2\cos\phi$	0	-2	$2\cos\phi$	0
Δ_g	2	$2\cos 2\phi$	0	2	$2\cos 2\phi$	0
Δ_u	2	$2\cos 2\phi$	0	-2	$-2\cos 2\phi$	0

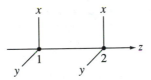

Figure 3.1. Coordinate system used to locate orbitals for homonuclear diatomic molecules.

$$1s_1 + 1s_2 = 1s\sigma_g$$
$$1s_1 - 1s_2 = 1s\sigma_u$$
$$2s_1 + 2s_2 = 2s\sigma_g$$
$$2s_1 - 2s_2 = 2s\sigma_u$$
$$2p_{z_1} - 2p_{z_2} = 2p\sigma_g$$
$$2p_{z_1} + 2p_{z_2} = 2p\sigma_u$$
$$2p_{x_1,y_1} + 2p_{x_2,y_2} = 2p\pi_u$$
$$2p_{x_1,y_1} - 2p_{x_2,y_2} = 2p\pi_g$$

For O_2, a typical energy-level diagram is given in Fig. 3.2.

To determine the terms generated by this configuration we decompose the direct product representation $2p\pi_g \times 2p\pi_g$. (The closed shells are totally symmetric and need not be considered.) The resulting product of characters gives

$D_{\infty h}$	E	$2C(\phi)$	σ_v	i	$2S(\phi)$	C_2
Γ	4	$4\cos^2\phi$	0	4	$4\cos^2\phi$	0

$$\underline{\qquad}\ 2p\sigma_u$$
$$\qquad\qquad \underline{+}\ 2p\pi_g$$
$$\qquad\qquad \underline{+\!\!+}\ 2p\pi_u$$
$$\underline{+\!\!+}\ 2p\sigma_g$$
$$\underline{+\!\!+}\ 2s\,\sigma_u$$
$$\underline{+\!\!+}\ 2s\,\sigma_g$$
$$\underline{+\!\!+}\ 1s\,\sigma_u$$
$$\underline{+\!\!+}\ 1s\,\sigma_g$$

Figure 3.2. Energy-level diagram for O_2, indicating electron occupations for the ground state.

This can be decomposed by inspection if we use the identity $\cos^2 \phi = \frac{1}{2}(1 + \cos 2\phi)$. From Table 3.3 we see that

$$\Gamma = \Delta_g + \Sigma_g^+ + \Sigma_g^-$$

The possible spin states are singlet or triplet, but because both unpaired electrons are in degenerate orbitals, the Pauli principle restricts which spatial states are associated with which spin states.

To see what the allowed combinations are, let's examine the spatial states of the electrons for the different terms. Since the symbols Σ, Δ, and so on, tell us about the projection of the electronic orbital angular momentum of the electrons about the z axis, it is convenient to use one-electron eigenfunctions of that angular momentum operator. Such eigenfunctions are given by

$$\pi_\pm = 2p\pi_{gx} \pm i2p\pi_{gy} = 2p_{x_1} \pm i2p_{y_1} - (2p_{x_2} \pm i2p_{y_2}) \quad (3.8)$$

Recalling from Chapter 1 that the combination of orbitals $p_x \pm ip_y \propto Y_{1\pm1}$, we see that the $\pi_\pm$ functions are eigenfunctions of l_z with eigenvalues ±1 in units of $\hbar$. Now consider the two-electron combinations that we can build up from $\pi_\pm$. Clearly, the following products are possible: $\pi_+(1)\pi_+(2)$, $\pi_-(1)\pi_-(2)$, $\pi_+(1)\pi_-(2)$, and $\pi_-(1)\pi_+(2)$. The first two combinations correspond to $m_1 = m_2 = +1$ or -1 and $M = m_1 + m_2 = \pm2$, and hence are components of the Δ_g representation. We can quickly verify this by finding the 2×2 matrix representation generated using the basis

$$\begin{pmatrix} \pi_+(1)\pi_+(2) \\ \pi_-(1)\pi_-(2) \end{pmatrix}$$

For example, $C(\phi)$ changes ϕ_1 to $\phi_1 + \phi$ and likewise for ϕ_2, so that $C(\phi)\pi_+(1)\pi_+(2) = e^{2i\phi}\pi_+(1)\pi_+(2)$. Overall, then,

$$C(\phi) \begin{pmatrix} \pi_+(1)\pi_+(2) \\ \pi_-(1)\pi_-(2) \end{pmatrix} = \begin{pmatrix} e^{2i\phi} & 0 \\ 0 & e^{-2i\phi} \end{pmatrix}\begin{pmatrix} \pi_+(1)\pi_+(2) \\ \pi_-(1)\pi_-(2) \end{pmatrix} \quad (3.9)$$

The character is $2 \cos 2\phi$. Similarly,

$$\sigma_v \begin{pmatrix} \pi_+(1)\pi_+(2) \\ \pi_-(1)\pi_-(2) \end{pmatrix} = \begin{pmatrix} \pi_-(1)\pi_-(2) \\ \pi_+(1)\pi_+(2) \end{pmatrix} = \begin{pmatrix} 0 & 1 \\ 1 & 0 \end{pmatrix}\begin{pmatrix} \pi_+(1)\pi_+(2) \\ \pi_-(1)\pi_-(2) \end{pmatrix} \quad (3.10)$$

$$i \begin{pmatrix} \pi_+(1)\pi_+(2) \\ \pi_-(1)\pi_-(2) \end{pmatrix} = \begin{pmatrix} 1 & 0 \\ 0 & 1 \end{pmatrix}\begin{pmatrix} \pi_+(1)\pi_+(2) \\ \pi_-(1)\pi_-(2) \end{pmatrix} \quad (3.11)$$

$$S(\phi)\begin{pmatrix} \pi_+(1)\pi_+(2) \\ \pi_-(1)\pi_-(2) \end{pmatrix} = \begin{pmatrix} e^{2i\phi} & 0 \\ 0 & e^{-2i\phi} \end{pmatrix}\begin{pmatrix} \pi_+(1)\pi_+(2) \\ \pi_-(1)\pi_-(2) \end{pmatrix} \quad (3.12)$$

$$C_2(\phi)\begin{pmatrix} \pi_+(1)\pi_+(2) \\ \pi_-(1)\pi_-(2) \end{pmatrix} = \begin{pmatrix} 0 & 1 \\ 1 & 0 \end{pmatrix}\begin{pmatrix} \pi_+(1)\pi_+(2) \\ \pi_-(1)\pi_-(2) \end{pmatrix} \quad (3.13)$$

One can easily verify that the characters generated by this representation are the same as Δ_g. Since both $\pi_+(1)\pi_+(2)$ and $\pi_-(1)\pi_-(2)$ are *symmetric* with respect to interchange of electrons 1 and 2, the only possible spin state is *singlet*. Thus $^1\Delta_g$ is one allowed term of O_2.

Similarly, one can show that both the $\pi_+(1)\pi_-(2)$ and $\pi_-(1)\pi_+(2)$ wavefunctions generate Σ representations, although neither function itself belongs to a single irreducible representation. There are two ways to generate functions that do belong. One is to reduce the two-dimensional representation generated by

$$\begin{pmatrix} \pi_+(1)\pi_-(2) \\ \pi_-(1)\pi_+(2) \end{pmatrix}$$

(it leads to $\Sigma_g^+ + \Sigma_g^-$, as expected), then find the desired SALCs. The other is to recognize that the SALCs thus formed must be either symmetric or antisymmetric with respect to interchange of the two electrons, and that the only two obvious linear combinations are

$$\psi_+ = \pi_+(1)\pi_-(2) + \pi_-(1)\pi_+(2) \tag{3.14}$$

$$\psi_- = \pi_+(1)\pi_-(2) - \pi_-(1)\pi_+(2) \tag{3.15}$$

These are indeed the desired SALCs, and by applying σ_v onto each, we find that $\sigma_v \psi_+ = \psi_+ (\Sigma_g^+$ symmetry) and $\sigma_v \psi_- = -\psi_- (\Sigma_g^-$ symmetry). Since ψ_+ is symmetric with respect to interchange of 1 and 2, we conclude that Σ_g^+ must be singlet. Similarly, Σ_g^- must be triplet.

Overall, then, the allowed terms generated by the ground configuration of O_2 are $^1\Delta_g$, $^3\Sigma_g^-$, and $^1\Sigma_g^+$. The observed ordering of these levels is shown in Fig. 3.3. It can be shown that transitions between the first two excited levels and the ground level are both spin and orbitally forbidden. This makes $^1\Delta_g$ and $^1\Sigma_g^+$ metastable, and enables their use as reactive intermediates.

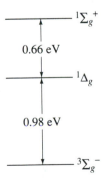

Figure 3.3. Energy levels generated from the ground configuration of O_2.

Three-Dimensional Rotation Group; Angular Momentum Addition

3.3.1 Angular Momentum Addition; Clebsch–Gordan Coefficients

Clebsch–Gordan coefficients arise in reducing the direct product representation of two electrons in atomic orbitals. We will not consider the group-theoretical properties of atomic orbitals extensively here, but suffice it to say that these orbitals form irreducible representations for the group of rotations in three dimensions [the so-called three-dimensional rotation group $O(3)$]. The spherical harmonics $Y_{lm}(\theta, \phi)$ are basis functions for the irreducible representations, with the l, m indices labeling these representations.

For many-electron atoms, just as for many-electron molecules, one often wants to construct electronic states belonging to specific irreducible representations by combining products of atomic orbitals for each electron. For molecules this can be done by inspection (or by constructing a fairly simple projection operator, as was done in Chapter 2). For atoms, it is more complicated (though conceptually the same), as we now discuss.

Consider, for example, a two-electron atom, having electrons with orbital quantum numbers $l_1 m_1$ and $l_2 m_2$. For a given l_1, l_2, we can determine the allowed many-electron term symbols by the vector addition

$$\mathbf{L} = \mathbf{l}_1 + \mathbf{l}_2 \tag{3.16}$$

$$L_z = l_{1z} + l_{2z} \tag{3.17}$$

$$M = m_1 + m_2 \tag{3.18}$$

This leads to the allowed L values,

$$L = l_1 + l_2, l_1 + l_2 - 1, \ldots, |l_1 - l_2| \tag{3.19}$$

Now if we examine the commutation properties of the operators L_x, L_y, L_z generated using Eq. (3.16), we find that they are identical to those for the corresponding l_1 and l_2 operators (see Section 1.4.2). This implies that the eigenfunctions of L^2 and L_z behave much like spherical harmonics:

$$L^2 y_{LM} = \hbar^2 L(L + 1) y_{LM} \tag{3.20}$$

$$L_z y_{LM} = \hbar M y_{LM} \tag{3.21}$$

In addition, since $[l_1^2, L^2] = [l_2^2, L^2] = [l_1^2, L_z] = [l_2^2, L_z] = 0$, the y_{LM}'s can be constructed as eigenfunctions of l_1^2, l_2^2. Now we ask, how are the y_{LM}'s related to $Y_{l_1 m_1}$ and $Y_{l_2 m_2}$? This is analogous to decomposing a direct product representation, and we write

$$y_{LM}^{l_1 l_2}(\Omega_1 \Omega_2) = \sum_{m_1 m_2} Y_{l_1 m_1}(\Omega_1) Y_{l_2 m_2}(\Omega_2) \langle l_1 m_1 l_2 m_2 | LM \rangle \quad (3.22)$$

where the coefficient $\langle l_1 m_1 l_2 m_2 | LM \rangle$ is called a Clebsch–Gordan (CG) or vector coupling coefficient. Using the orthonormality of the Y's, we easily show that

$$\langle l_1 m_1 l_2 m_2 | LM \rangle = \langle Y_{l_1 m_1} Y_{l_2 m_2} | y_{LM}^{l_1 l_1} \rangle \quad (3.23)$$

Our notation for the CG coefficients is identical to Zare. Other notations in common use are:

$$\langle l_1 m_1 l_2 m_2 | LM \rangle = \langle m_1 m_2 | LM \rangle \quad \text{(Schiff)}$$
$$= \langle l_1 l_2 m_1 m_2 | LM \rangle \quad \text{(Davydov)}$$
$$= \langle l_1 l_2 LM | l_1 l_2 m_1 m_2 \rangle \quad \text{(Baym)}$$

3.3.2 Properties of the Clebsch–Gordan Coefficients

Explicit values for all the CG coefficients can be determined using the following six properties:

1. *Angular momentum conservation constraints*
   (a) If we apply $L_z = l_{1z} + l_{2z}$ to Eq. (3.22) we find that

   $$My_{LM}^{l_1 l_2} = \sum_{m_1 m_2} Y_{l_1 m_1} Y_{l_2 m_2}(m_1 + m_2) \langle l_1 m_1 l_2 m_2 | LM \rangle \quad (3.24)$$

   Substituting Eq. (3.22) for $y_{LM}^{l_1 l_2}$ in this equation and rearranging, we find that

   $$\sum_{m_1 m_2} \langle l_1 m_1 l_2 m_2 | LM \rangle [M - m_1 - m_2] Y_{l_1 m_1} Y_{l_2 m_2} = 0 \quad (3.25)$$

   For this equation to be satisfied, we must either have $\langle \ \rangle = 0$ or $M = m_1 + m_2$. Thus all the $\langle \ \rangle$'s for which the conservation rule $M = m_1 + m_2$ does not hold are zero.
   (b) One can similarly show (by applying $L_\pm = l_{1\pm} + l_{2\pm}$) that $\langle \ \rangle = 0$ unless $|l_1 - l_2| \le L \le l_1 + l_2$.

2. *Orthogonality constraints.* If we choose the $y_{LM}^{l_1 l_2}$'s to be orthonormal, the integral of the product of $y_{LM}^{l_1 l_2}{}^*$ and $y_{L'M'}^{l_1 l_2}$ gives [using Eq. (3.22)]

   $$\delta_{LL'} \delta_{MM'} = \sum_{m_1 m_2} \langle l_1 m_1 l_2 m_2 | L'M' \rangle^* \langle l_1 m_1 l_2 m_2 | LM \rangle \quad (3.26)$$

Regarding the $\langle\ \rangle$'s as square matrices with row indices $m_1 m_2$ and column indices LM, that is,

$$\langle l_1 m_1 l_2 m_2 | LM \rangle = (\mathbf{C})^{LM}_{m_1 m_2} \tag{3.27}$$

one can rewrite Eq. 3.26 as

$$\mathbf{C}^{\dagger}\mathbf{C} = \mathbf{I} \tag{3.28}$$

which shows that the $\mathbf{C}$'s are unitary. Since the $\mathbf{C}$'s may also be chosen to be real (see property 5 below), it follows that $\mathbf{C}$ is orthogonal. Since the $\mathbf{C}$'s are square matrices, we can rewrite Eq. (3.28) as

$$\mathbf{C}\mathbf{C}^{\dagger} = \mathbf{I} \tag{3.29}$$

or

$$\sum_{LM} \langle l_1 m_1 l_2 m_2 | LM \rangle \langle l_1 m_1' l_2 m_2' | LM \rangle^* = \delta_{m_1 m_1'} \delta_{m_2 m_2'} \tag{3.30}$$

3. *Inverse of Eq. (3.22).* Multiplying Eq. (3.22) by $\langle l_1 m_1 l_2 m_2 | LM \rangle^*$, summing over LM and using Eq. (3.30), one can derive the inverse of Eq. (3.22):

$$Y_{l_1 m_1} Y_{l_2 m_2} = \sum_{LM} \langle l_1 m_1 l_2 m_2 | LM \rangle^* y^{l_1 l_2}_{LM} \tag{3.31}$$

4. *Recursion relations.* Applying the operator $L_{\pm} = l_{1\pm} + l_{2\pm}$ to Eq. (3.22), we get

$$L_{\pm} y^{l_1 l_2}_{LM} = \sqrt{L(L+1) - M(M \pm 1)}\, y^{l_1 l_2}_{LM \pm 1}$$

$$= \sum_{m_1 m_2} \{\sqrt{l_1(l_1 + 1) - m_1(m_1 \pm 1)}\ Y_{l_1 m_1 \pm 1} Y_{l_2 m_2}$$

$$+ \sqrt{l_2(l_2 + 1) - m_2(m_2 \pm 1)}\ Y_{l_1 m_1} Y_{l_2 m_2 \pm 1}\} \langle l_1 m_1 l_2 m_2 | LM \rangle \tag{3.32}$$

Now multiply this by $Y^*_{l_1 m_1'} Y^*_{l_2 m_2'}$ and integrate. Using the definition in Eq. (3.23) and the usual orthogonality relations among spherical harmonics, we easily find the following recursion relations:

$$\sqrt{L(L+1) - M(M \pm 1)}\ \langle l_1 m_1 l_2 m_2 | LM \pm 1 \rangle$$

$$= \{\sqrt{l_1(l_1 + 1) - m_1(m_1 \mp 1)}\ \langle l_1 m_1 \mp 1 l_2 m_2 | LM \rangle$$

$$+ \sqrt{l_2(l_2 + 1) - m_2(m_2 \mp 1)}\ \langle l_1 m_1 l_2 m_2 \mp 1 | LM \rangle\} \tag{3.33}$$

These are very useful for relating CG coefficients having different M, m_1, and m_2 values, as will be apparent in an example given below.

5. *Phase convention.* The phase of $y^{l_1 l_2}_{LM}$ is arbitrary, but it is customarily chosen such that

$$\langle l_1, m_1 = l_1, l_2, m_2 = L - l_1 | L, M = L \rangle = \text{real} \geq 0$$

Only the phase for these specific choices of m_1, m_2, and M need be chosen arbitrarily. The recursion relations determine the rest.

6. *Symmetry relations.* Using the recursion and other formulas, the following very useful symmetry relations may be derived:

$$\langle l_1 m_1 l_2 m_2 | LM \rangle = (-1)^{l_1 + l_2 - L} \langle l_2 m_2 l_1 m_1 | LM \rangle \tag{3.34}$$

$$= (-1)^{l_1 - L + m_2} \sqrt{\frac{2L + 1}{2l_1 + 1}} \langle LM l_2 - m_2 | l_1 m_1 \rangle \tag{3.35}$$

$$= (-1)^{l_1 + l_2 - L} \langle l_1 - m_1 l_2 - m_2 | L - M \rangle \tag{3.36}$$

3.3.3 Worked Examples

(a) *Evaluation for $l_1 = 1$, $l_2 = 0$.* Clearly, $L = 1$ and $M = m_1 = 1, 0$, and -1 are the only possible combinations that satisfy angular momentum conservation in this case. This implies that Eq. (3.22) is

$$y_{1M}^{10} = \langle 1 m_1 0 0 | 1 M \rangle Y_{1 m_1} Y_{00}$$

and using orthogonality, we infer that $|\langle 1 M 0 0 | 1 M \rangle| = 1$. The phase convention then tells us that $\langle 1 1 0 0 | 1 1 \rangle = +1$, and the lower recursion relation (3.33) that $\langle 1 0 0 0 | 1 0 \rangle = \langle 1 1 0 0 | 1 1 \rangle = 1$. Finally, the third symmetry relation [Eq. (3.36)] tells us that $\langle 1 -1 0 0 | 1 -1 \rangle = 1$.

(b) *Evaluation for $l_1 = 1$, $l_2 = 1$, $L = 2$.* Consider first the case $M = 2$. Equation (3.22), coupled with angular momentum conservation, tells us that

$$y_{22}^{11} = \langle 1 1 1 1 | 2 2 \rangle Y_{11} Y_{11}$$

Using the phase convention, together with orthogonality, we find that

$$\langle 1 1 1 1 | 2 2 \rangle = 1$$

To determine the CG coefficients for $M = 1$, we next use the lower recursion relation to show that

$$\sqrt{6 - 2} \langle 1 0 1 1 | 2 1 \rangle = \sqrt{2} \langle 1 1 1 1 | 2 2 \rangle$$

or

$$\langle 1 0 1 1 | 2 1 \rangle = \frac{1}{\sqrt{2}}$$

By symmetry, we then find that

$$\langle 1 1 1 0 | 2 1 \rangle = \frac{1}{\sqrt{2}}$$

For $M = 0$, again apply the lower recursion relation to obtain

$$\sqrt{6 - 0} \langle 1010|20 \rangle = \sqrt{2 - 0} \langle 1110|21 \rangle + \sqrt{2 - 0} \langle 1011|21 \rangle$$

So that

$$\langle 1010|20 \rangle = \frac{2}{\sqrt{6}}$$

Using the recursion relation again, we find that

$$\langle 111 - 1|20 \rangle = \frac{1}{\sqrt{6}}$$

$$\langle 1 - 111|20 \rangle = \frac{1}{\sqrt{6}}$$

The remaining coefficients for $M = -1, -2$ can be obtained by using the third symmetry relation. Table 3.4 summarizes all these results. Note that the resulting **C** matrix is orthogonal.

3.3.4 3-j and Higher Symbols

Another commonly used representation of the CG coefficients is provided by 3-j symbols. These are defined by

$$\begin{pmatrix} l_1 & l_2 & l_3 \\ m_1 & m_2 & m_3 \end{pmatrix} = \frac{(-1)^{l_1 - l_2 - m_3}}{\sqrt{2l_3 + 1}} \langle l_1 m_1 l_2 m_2 | l_3 - m_3 \rangle \qquad (3.37)$$

3-j symbols are useful because the symmetry properties of the CG coeffi-

Table 3.4

Clebsch–Gordan Coefficients for $l_1 = l_2 = 1$, $L = 2$

m_1	m_2	2	1	0	−1	−2
1	1	1	0	0	0	0
1	0	0	$1/\sqrt{2}$	0	0	0
0	1	0	$1/\sqrt{2}$	0	0	0
0	0	0	0	$2/\sqrt{6}$	0	0
1	−1	0	0	$1/\sqrt{6}$	0	0
−1	1	0	0	$1/\sqrt{6}$	0	0
0	−1	0	0	0	$1/\sqrt{2}$	0
−1	0	0	0	0	$1/\sqrt{2}$	0
−1	−1	0	0	0	0	1

The column header *M* spans the columns labeled 2, 1, 0, −1, −2.

cients are more transparent when expressed in terms of them. In particular, the relations in (3.22) and (3.23) become

(a) $\begin{pmatrix} l_1 & l_2 & l_3 \\ m_1 & m_2 & m_3 \end{pmatrix}$ $= 0$ unless $m_1 + m_2 + m_3 = 0$

$$|l_1 - l_2| \le l_3 \le l_1 + l_2$$

(b) $\begin{pmatrix} l_1 & l_2 & l_3 \\ m_1 & m_2 & m_3 \end{pmatrix}$ is unchanged by an even permutation of columns

(c) $\begin{pmatrix} l_1 & l_2 & l_3 \\ m_1 & m_2 & m_3 \end{pmatrix}$ is multiplied by $(-1)^{l_1+l_2+l_3}$ under an odd permutation of columns

One can also couple together three, four, or more angular momenta to obtain combined total angular momentum eigenstates. This leads to the use of more complicated vector coupling coefficients, such as 6-*j* and 9-*j* symbols, which will not be considered here (see the Bibliography).

BIBLIOGRAPHY FOR CHAPTER 3

Discussions relevant to continuous groups may be found in:

Hamermesh, M., *Group Theory and Its Application to Physical Problems* (Addison-Wesley, Reading, Mass., 1962).
Tinkham, M., *Group Theory and Quantum Mechanics* (McGraw-Hill, New York, 1964).
Wigner, E. P., *Group Theory and Its Application to the Quantum Mechanics of Atomic Spectra* (Academic Press, New York, 1959).

Basic texts that cover angular momentum addition and related topics include:

Baym, G., *Lectures on Quantum Mechanics* (Benjamin/Cummings, London, 1981).
Davydov, A. S., *Quantum Mechanics* (Pergamon Press, Oxford, 1976).
Schiff, L., *Quantum Mechanics* (McGraw-Hill, New York, 1968).
Zare, R. N., *Angular Momentum* (Wiley-Interscience, New York, 1988).

PROBLEMS FOR CHAPTER 3

1. Consider the metal–metal bond in a compound

(a) Ignoring the ligands for the moment, the M—M part of the molecule looks like a homonuclear diatomic. What combinations of the d-orbitals on each M will generate orbitals of σ, π, and δ symmetry? Assuming the energy ordering $\delta > \pi > \sigma$ for bonding orbitals and $\sigma > \pi > \delta$ for antibonding orbitals, what many electron terms (including spin) are generated when each metal atom is d^4?

(b) Suppose that the ligands reduce the overall symmetry of the molecule to D_{3h}. To what irreducible representations of this group do the generated orbitals in part (a) belong?

2. (a) Calculate all the $\langle l_1 m_1 l_2 m_2 | LM \rangle$ for $l_1 = l_2 = L = 1$.

(b) Use the results of part (a) to show that the $L = 1$ term arising from the $(p)^2$ configuration of an atom must be a spin triplet (i.e., ^3P).

3. If $\langle 1011|21 \rangle = 1/\sqrt{2}$, what is $\langle 11 - 22|11 \rangle$? (*Hint:* Use recursion and symmetry formulas to relate these two coefficients.)

4. Write down the ground and first excited molecular orbital configurations of the N_2 molecule. You should find two possible excited configurations, depending on whether the $2p\sigma_g$ energy is above or below $2p\pi_u$. Please include both possibilities. What terms arise from these configurations? What are the spin multiplicities of these terms? Which terms are connected by electric dipole transitions to the ground state?

5. The addition of three angular momenta (l_1, l_2, and l_3) is accomplished by first coupling any two (say, l_1, l_2) to generate states labeled by L_{12}, M_{12}, and then adding that to the third (l_3) to form states labeled by the total angular momentum L and projection M. Consider what happens when three electrons, each with spin s of $\frac{1}{2}$, are treated (so that $s_1 = s_2 = s_3 = \frac{1}{2}$). Find the wavefunctions (expressed in terms of α, β) associated with $S = \frac{3}{2}$ and $M = \frac{3}{2}, \frac{1}{2}, -\frac{1}{2}, -\frac{3}{2}$. (*Note:* You will need to determine Clebsch–Gordan coefficients for states with half-integral angular momenta. The formulas of this chapter can be applied to the half-integer case just as with the integer case.)

4

Time-Dependent Quantum Mechanics

4.1

Introduction

Up to this point, we have considered only the stationary (time-independent) solutions to the Schrödinger equation. However, much if not most of chemistry is concerned with how these solutions evolve in time. Important examples include (1) the response of an atom or molecule to *electromagnetic radiation* (light absorption, emission, scattering), (2) *collisions between atoms and molecules* (chemical reactions, energy transfer), and (3) *intramolecular energy transfer processes* (intramolecular electron transfer, radiationless transitions). This chapter presents the basic theory needed to describe these and other processes.

4.2

Time-Dependent Schrödinger Equation: Basis-Set Solution

In all or nearly all time-dependent problems, we can consider that initially the system is prepared in some stationary state, a time-dependent interaction is turned on and the state can undergo change. At a later time the interaction is turned off and a measurement of the final state is made. Letting H_0 be the time-independent initial hamiltonian, and $V(q, t)$ be the interaction potential (or hamiltonian), the time evolution of the system is determined by the time-dependent Schrödinger equation,

$$i\hbar \frac{\partial \psi(q, t)}{\partial t} = (H_0 + V(q, t))\psi(q, t) = H\psi(q, t) \tag{4.1}$$

This is a partial differential equation in terms of the coordinates q and time t. Normally, it is not at all easy to solve exactly, but often approximate solutions can be obtained which are quite accurate.

Suppose that the stationary states prior to turning on $V(q, t)$ are denoted as $\phi_n(q)$ ($n = 1, 2, \ldots$). These satisfy $H_0\phi_n = E_n\phi_n$ and have a time-dependent part $e^{-iE_n t/\hbar}$. We will assume that at the initial time (taken as $t = 0$), the system is in state ϕ_m [i.e., $\psi(q, 0) = \phi_m(q)$]. Since in general the ϕ_n's form a complete set, one can expand $\psi(q, t)$ for $t > 0$ in terms of them:

$$\psi(q, t) = \sum_n c_n(t)\phi_n(q)e^{-iE_n t/\hbar} \tag{4.2}$$

where the c_n's are coefficients. To determine the $c_n(t)$'s for $t > 0$, we substitute Eq. (4.2) into Eq. (4.1), obtaining

$$i\hbar \frac{\partial \psi}{\partial t} = i\hbar \sum_n \left[\dot{c}_n - \frac{i}{\hbar} E_n c_n\right] \phi_n(q)e^{-iE_n t/\hbar}$$

$$= \sum_n c_n(t)(H_0 + V(q, t))\phi_n e^{-iE_n t/\hbar}$$

$$= \sum_n c_n(t)(E_n + V(q, t))\phi_n e^{-iE_n t/\hbar} \tag{4.3}$$

where $\dot{c}_n \equiv dc_n/dt$. Notice that the terms containing E_n cancel in this equation. Multiplying by $\phi_k^*(q)$ and integrating, we find

$$i\hbar \sum_n \dot{c}_n e^{-iE_n t/\hbar}\langle k|n\rangle = \sum_n e^{-iE_n t/\hbar} c_n(t)\langle k|V|n\rangle \tag{4.4}$$

Since $\langle k|n\rangle = \delta_{kn}$, and defining $\langle k|V(q, t)|n\rangle = V_{kn}(t)$, this equation becomes

$$i\hbar\dot{c}_k = \sum_n e^{+i(E_k - E_n)t/\hbar} c_n(t)V_{kn}(t) \tag{4.5}$$

Now define $\omega_{kn} = (E_k - E_n)/\hbar$, and we obtain the following coupled ordinary differential equations for $c_k(t)$:

$$\dot{c}_k(t) = \frac{-i}{\hbar} \sum_n e^{i\omega_{kn}t}V_{kn}(t)c_n(t) \tag{4.6}$$

These, together with the boundary condition $c_k(t = 0) = \delta_{km}$, uniquely define an exact solution to the Schrödinger equation equivalent to solving Eq. (4.1). The form of Eq. (4.6) is, however, more amenable to numerical or approximate solution. Indeed, if the number of states known to be coupled in a given problem is small, and the frequency difference ω_{kn} and matrix elements $V_{kn}(t)$ are known, it is not difficult to solve Eq. (4.6) exactly on a computer. Unfortunately, it is rare that all of this information is available, and for this reason, the development of approximate analytic solutions is very useful.

4.3

Time-Dependent Perturbation Theory

4.3.1 First-Order Time-Dependent Perturbation Theory

Suppose that the interaction V_{kn} is small enough so that the change in $c_n(t)$ is small. Then, to a first approximation, $c_n(t)$ in the right-hand side of Eq. (4.6) is unchanged from its initial value δ_{nm}, and Eq. (4.6) is given by

$$\dot{c}_k^{(1)} = -\frac{i}{\hbar} V_{km}(t) e^{i\omega_{km}t} \tag{4.7}$$

This is trivially integrated to give

$$c_k^{(1)}(t) = c_k(t=0) - \frac{i}{\hbar} \int_0^t dt' \, V_{km}(t') e^{i\omega_{km}t'} \tag{4.8}$$

where $c_k(t=0) = \delta_{km}$.

Now from Eq. (4.2) we note that the projection of ψ onto $\phi_k(q)$ is $\langle \phi_k | \psi \rangle = c_k e^{-iE_k t/\hbar}$. The absolute square of this gives the probability P_k of finding the system in state k at time t. Thus, for $k \neq m$, we have, to first order,

$$P_k(t) = |c_k^{(1)}(t)|^2 = \frac{1}{\hbar^2} \left| \int_0^t dt' \, V_{km}(t') e^{i\omega_{km}t'} \right|^2 \tag{4.9}$$

4.3.2 Example: Collision-Induced Excitation of a Diatomic Molecule

Consider the atom–diatom collision system defined by Fig. 4.1. Often we are interested in calculating the probability that a molecule will change its internal state (vibrational, rotational, electronic, etc.) as a result of the

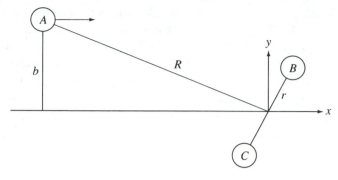

Figure 4.1. Coordinates used to describe the collision of an atom A with a diatomic molecule BC.

collision. This is very difficult to do exactly, but can be adequately approximated in some cases by assuming that the motion of A relative to BC can be obtained from a *straight-line classical trajectory*. This assumption is good for excitation that is determined by trajectories that have large impact parameters b.

In this case, the time dependence of the separation R between the atom and the diatomic center of mass is given by $R(t) = \sqrt{x^2 + y^2}$ and $y = b$, $x = v_0 t$, where v_0 is the initial velocity. Note that we consider $t = -\infty$ initially and $t = +\infty$ at end. This means that the lower limit of the integral in Eq. (4.9) should be $-\infty$.

The physically measurable quantity is the probability P_k evaluated at $t = +\infty$ since this defines the transition probability resulting from the collisional interaction after the collision is over. From Eq. (4.9) we see that this probability is

$$P_k(t = \infty) = \frac{1}{\hbar^2} \left| \int_{-\infty}^{\infty} dt'\ V_{km}(t') e^{i\omega_{km} t'} \right|^2 \tag{4.10}$$

which has the form of a Fourier transform of the interaction potential.

Now assume that the interaction potential V is given by $V(R, \mathbf{r}) = e^{-\beta R^2} U(\mathbf{r})$ (where β is a steepness parameter and U is some function of the molecular coordinates $\mathbf{r}$). Then the time dependence of V is given by

$$V(R, r) = e^{-\beta(b^2 + v_0^2 t^2)} U(r) \tag{4.11}$$

which has a gaussian dependence on time, with a width that depends inversely on the velocity v_0. In view of the Fourier relationship in Eq. (4.10), if the collision is slow, the lower frequency (i.e., lower ω_{km}) $V \to R$ (vibration to rotation) in the diatomic BC and $T \to R$ (translation to rotation) transitions will be favored while if the energy is high, the higher frequency $V \to T$ and $V \to E$ (vibration to electronic) transitions become more probable. Evaluating V_{km} for the assumed interaction potential, we find that

$$P_k = \frac{1}{\hbar^2} |U_{km}|^2 e^{-2\beta b^2} \left| \int_{-\infty}^{\infty} dt' \, e^{i\omega_{km}t'} e^{-\beta v_0^2 t'^2} \right|^2 \tag{4.12}$$

Using the result that

$$\int_{-\infty}^{\infty} dt \, e^{i\gamma t} e^{-\delta t^2} = \sqrt{\frac{\pi}{\delta}} \, e^{-\gamma^2/4\delta} \tag{4.13}$$

we find that

$$P_k = \frac{1}{\hbar^2} |U_{km}|^2 e^{-2\beta b^2} \exp\left(-\frac{\omega_{km}^2}{2\beta v_0^2}\right) \frac{\pi}{\beta v_0^2} \tag{4.14}$$

Converting from initial velocity v_0 to initial relative translational energy E_0 using $E_0 = \frac{1}{2}\mu v_0^2$, where μ is the translational reduced mass, we find that

$$P_k = \frac{\pi\mu}{2\beta\hbar^2} \frac{1}{E_0} e^{-2\beta b^2} \exp\left(-\frac{\mu\omega_{km}^2}{4\beta E_0}\right) |U_{km}|^2 \tag{4.15}$$

When plotted as a function of E_0 (Fig. 4.2), we see that $P_k \to 0$ as $E_0 \to 0$ and ∞, and that P_k peaks when

$$\frac{\partial P_k}{\partial E_0} = 0 = \text{const} \left(-\frac{1}{E_0^2} + \frac{1}{E_0} \frac{\omega_{km}^2 \mu}{4\beta E_0^2}\right) \exp\left(-\frac{\mu\omega_{km}^2}{4\beta E_0}\right) \tag{4.16}$$

which leads to

$$E_0 = \frac{\mu\omega_{km}^2}{4\beta} \tag{4.17}$$

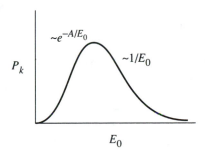

Figure 4.2. Dependence of transition probability P_k on collision energy E_0. The constant A is defined in Eq. 4.15.

Thus for a given ω_{km}, the E_0 required to give maximum P_k increases with increasing μ (heavier particles are less efficient at energy transfer), and decreasing β (flatter potentials are less efficient). Notice also that P_k decreases with increasing b (as physically makes sense), and that P_k is proportional to $|U_{km}|^2$.

The actual physical observable in a collision process is called the *cross section* σ_k, and for inelastic collisions it is related to P_k by integration over impact parameters

$$\sigma_k = 2\pi \int_0^\infty P_k(b)b \, db$$

$$= 2\pi \int_0^\infty |U_{km}|^2 \frac{\mu\pi}{2\beta\hbar^2} \frac{1}{E_0} e^{-2\beta b^2} \exp\left(-\frac{\mu\omega_{km}^2}{4\beta E_0}\right) b \, db$$

$$= \frac{\mu\pi^2}{4\beta^2\hbar^2} |U_{km}|^2 \frac{1}{E_0} \exp\left(-\frac{\mu\omega_{km}^2}{4\beta E_0}\right) \tag{4.18}$$

One can also calculate the *rate constant* for the $m \to k$ transition. This is done by multiplying σ_k by v_0, then averaging the result over a Boltzmann distribution of relative translational energies. Since $\sigma_k v_0[A]$ is the number of molecules of A that cause the $m \to k$ transition in BC per unit time for a specific velocity v_0, the total rate for all BC's is

$$\frac{d[BC]_{m\to k}}{dt} = -\sigma_k v_0[A][BC] \tag{4.19}$$

The coefficient $\sigma_k v_0$ in this expression is the rate constant k. After averaging over a Boltzmann distribution of relative velocities, we find that

$$k_{m\to k} = \frac{\mu^2}{4\beta^2\hbar E} |U_{km}|^2 \int_0^\infty \frac{1}{Q} e^{-E_0/kT} \exp\left(-\frac{\mu\omega_{km}^2}{4\beta E_0}\right) dE_0 \tag{4.20}$$

where Q is the translational partition function. This integral cannot be done exactly but by applying a *saddle-point method* (which is described in detail in Section 10.5), an approximation can be developed.

Consider the argument of the exponential in the integral $[(E_0/kT) + (\omega_{km}^2\mu)/(4\beta E_0)]$. This minimizes (and hence the exponential peaks) at $E_0^P = [(\omega_{km}^2\mu kT)/4\beta]^{1/2}$ and thus we can approximate this argument by the first two nonvanishing terms in the Taylor series expansion about E_0^P:

$$\frac{1}{2}\left(E_0 - \left(\frac{\omega_{km}^2\mu kT}{4\beta}\right)^{1/2}\right)^2 \frac{2}{kT}\left(\frac{4\beta}{\mu\omega_{km}^2 kT}\right)^{1/2} + 2\left(\frac{\omega_{km}^2\mu}{4\beta kT}\right)^{1/2} \tag{4.21}$$

The resulting integral is

$$k_{m\to k} = \frac{\sqrt{\mu} \, \pi^2 |U_{km}|^2}{\sqrt{2}\beta^2\hbar^2 kT} \left(\frac{\mu\omega_{km}^2 kT}{4\beta}\right)^{1/4} \exp\left(-\left(\frac{\omega_{km}^2\mu}{\beta kT}\right)^{1/2}\right) \tag{4.22}$$

where we have used

$$Q = \left(\frac{\mu kT}{2\pi\hbar^2}\right)^{3/2}$$

Notice here that the dominant temperature dependence of $k_{m\to k}$ is $\exp[-\text{const}|\omega_{km}|\mu^{1/2}/T^{1/2}]$. This means that plots of $\ln k_{m\to k}$ versus $1/T^{1/2}$ should be linear. In reality, realistic potentials are exponential rather than gaussian, and for that case, the same treatment gives $k \sim \exp(-\text{const}/T^{1/3})$. Also, k increases with decreasing energy gap $|\omega_{km}|$, with decreasing μ, and increasing β. All of these results are qualitatively correct and are much more general than this limited model would suggest.

4.3.3 Second-Order Perturbation Theory

Now let's substitute $c_n^{(1)}(t)$ into the right-hand side of the coupled equations [Eq. (4.6)] to get an improved (second-order) estimate of $c_k(t)$.

$$\frac{dc_k^{(2)}}{dt} = \frac{-i}{\hbar} \sum_n e^{i\omega_{kn}t} V_{kn}(t)[\delta_{nm} - \frac{i}{\hbar}\int_0^t dt'\ V_{nm}(t')e^{i\omega_{nm}t'}]$$

$$= \frac{-i}{\hbar} e^{i\omega_{km}t}V_{km}(t) + \left(\frac{-i}{\hbar}\right)^2 \sum_n e^{i\omega_{kn}t}V_{kn}(t)\int_0^t dt'\ V_{nm}(t')e^{i\omega_{nm}t'}$$

$$(4.23)$$

This is easily integrated to give

$$c_k^{(2)}(t) = \delta_{km} + \frac{-i}{\hbar}\int_0^t e^{i\omega_{km}t'}V_{km}(t')\ dt'$$

$$+ \left(\frac{-i}{\hbar}\right)^2 \sum_n \int_0^t dt'\ e^{i\omega_{kn}t'}V_{kn}(t')\int_0^{t'} dt''\ V_{nm}(t'')e^{i\omega_{nm}t''} \quad (4.24)$$

Notice here how the first two terms are just $c_k^{(1)}(t)$. The third term is then the second-order contribution to $c_k^{(2)}$ and involves matrix elements of the initial and final states with an intermediate state n. Thus even if $V_{km} = 0$ (i.e., the transition is forbidden in first order), the second-order coefficient may still be nonzero since there may be intermediate states n for which V_{kn} and V_{nm} are nonzero.

4.3.4 Simplifications and Extensions to Higher Order

Let's now proceed to generalize this second-order expression to get the Nth-order coefficient $c_n^{(N)}(t)$. To do this, we need to introduce a more compact notation. First note that

$$e^{i\omega_{kn}t}V_{kn}(t) = e^{iE_k t/\hbar}\langle k|V(t)|n\rangle e^{-iE_n t/\hbar} = \langle k|e^{iH_0 t/\hbar}V(t)e^{-iH_0 t/\hbar}|n\rangle \quad (4.25)$$

Here we *define* the exponential of an operator through its Taylor series expansion:

$$e^{-iH_0 t/\hbar}|n\rangle = \sum_l (l!)^{-1}\left(\frac{-iH_0 t}{\hbar}\right)^l |n\rangle$$

$$= \sum_l (l!)^{-1}\left(\frac{-iE_n t}{\hbar}\right)^l |n\rangle = e^{-iE_n t/\hbar}|n\rangle \quad (4.26)$$

Now define

$$V_I(t) = e^{iH_0 t/\hbar}V(t)e^{-iH_0 t/\hbar} \quad (4.27)$$

We will have more to say about the meaning of $V_I(t)$, but for now, simply regard it as an abbreviation.

In this notation, our expression for $c_k^{(2)}(t)$ becomes

$$c_k^{(2)}(t) = \langle k|m\rangle + \frac{-i}{\hbar}\int_0^t dt'\,\langle k|V_I(t')|m\rangle$$

$$+ \left(\frac{-i}{\hbar}\right)^2 \sum_n \int_0^t dt' \int_0^{t'} dt''\,\langle k|V_I(t')|n\rangle\langle n|V_I(t'')|m\rangle$$

$$= \langle k|\left\{1 - \frac{i}{\hbar}\int_0^t dt'\,V_I(t')\right.$$

$$\left. + \left(\frac{-i}{\hbar}\right)^2 \int_0^t dt' \int_0^{t'} dt''\,V_I(t')\sum_n |n\rangle\langle n|V_I(t'')\right\}|m\rangle \quad (4.28)$$

Now what is $\sum_n |n\rangle\langle n|$? Well clearly if the ϕ_n's form a complete set, we can expand any function in terms of them, and $|f\rangle = \sum_n |n\rangle\langle n|f\rangle$. This implies that $\sum_n |n\rangle\langle n| = 1$. With this, we can easily see how to generalize $c_k^{(2)}(t)$ to $c_k^N(t)$:

$$c_k^N(t) = \langle k|\left\{1 - \frac{i}{\hbar}\int_0^t dt'\,V_I(t') + \left(\frac{-i}{\hbar}\right)^2 \int_0^t dt' \int_0^{t'} dt''\,V_I(t')V_I(t'')\right.$$

$$\left. + \left(\frac{-i}{\hbar}\right)^3 \int_0^t dt' \int_0^{t'} dt'' \int_0^{t''} dt'''\,V_I(t')V_I(t'')V_I(t''') + \cdots\right\}|m\rangle$$

$$(4.29)$$

4.3.5 Time-Ordering Operators

A formally exact solution for $c_k^\infty(t)$ can now be developed by introducing the *Dyson chronological time-ordering operator* P, defined by

$$PA(t_1)B(t_2) = \begin{cases} A(t_1)B(t_2) & t_1 > t_2 \\ B(t_2)A(t_1) & t_1 < t_2 \end{cases} \tag{4.30}$$

P is an operator that acts on the operators A and B to order them so that the *earliest time is to the right*. The utility of this operator becomes apparent when we operate on the following expression:

$$P\left(\int_0^t dt'\, V_I(t')\right)^2 = \int_0^t dt' \int_0^t dt''\, PV_I(t')V_I(t'')$$

$$= \int_0^t dt' \int_0^{t'} dt''\, V_I(t')V_I(t'') + \int_0^t dt' \int_{t'}^t dt''\, V_I(t'')V_I(t') \tag{4.31}$$

Referring to Fig. 4.3, we note that the first integral on the right-hand side covers region I, while the second covers region II.

Now the second integral can be rewritten as $\int_0^t dt'' \int_0^{t''} dt'$, and upon interchanging t' and t'', it becomes equal to the first. Therefore,

$$P[\int_0^t dt'\, V_I(t')]^2 = 2 \int_0^t dt' \int_0^{t'} dt''\, V_I(t')V_I(t'') \tag{4.32}$$

In a similar way one can show that

$$P[\int_0^t dt'\, V_I(t')]^n = n! \int_0^t dt' \int_0^{t'} dt'' \cdots \int_0^{t^{n-1}} dt^n\, V_I(t') \cdots V_I(t^n) \tag{4.33}$$

With this result, one can easily show that

$$c_k^{\infty}(t) = \langle k| P \sum_{l=0}^{\infty} \frac{1}{l!} \left[\frac{-i}{\hbar} \int_0^t dt'\, V_I(t') \right]^l |m\rangle$$

$$= \langle k| P \exp \left\{ \frac{-i}{\hbar} \int_0^t dt'\, V_I(t') \right\} |m\rangle \tag{4.34}$$

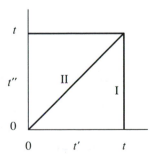

Figure 4.3. Time integration boundaries for Eq. (4.31).

where the *time-ordered exponential* is defined by its Taylor series representation. This expression for $c_k^{\infty}(t)$ is a formally exact solution to the Schrödinger equation. Computationally, it is not very useful, for its exact evaluation requires evaluating its series expansion. It is, however, a useful starting point for approximate evaluations, and also a number of formal properties of certain solutions to the time-dependent Schrödinger equation can be developed in terms of it.

4.4

Representations in Quantum Mechanics

4.4.1 Schrödinger Representation

At this point it is useful to introduce some new notation concerning the representation of time-dependent processes. Up to now our description has used the so-called *Schrödinger representation*, in which the time evolution of a system is described in terms of a time-dependent wavefunction $\psi_S(q, t)$. This is obtained by solving the time-dependent Schrödinger equation

$$i\hbar \frac{\partial \psi_S}{\partial t} = H\psi_S \qquad (4.35)$$

and for H independent of time, the solution is

$$\psi_S(q, t) = e^{-iHt/\hbar}\psi_S(q, 0) \qquad (4.36)$$

(Proof of this is obtained by differentiating the series representation of $e^{-iHt/\hbar}$.) For time-dependent H, we have [from (4.34)]

$$\psi_S(q, t) = \sum_n e^{-iE_n t/\hbar}|n\rangle c_n(t)$$

$$= e^{-iH_0 t/\hbar} \sum_n |n\rangle\langle n|P \exp\left[\frac{-i}{\hbar}\int_0^t dt'\ V_I(t')\right]|m\rangle$$

$$= e^{-iH_0 t/\hbar}P \exp\left[\frac{-i}{\hbar}\int_0^t dt'\ V_I(t')\right]\psi_S(q, 0) \qquad (4.37)$$

The replacement of $|m\rangle$ by $\psi_S(q, 0)$ follows from our assumptions of Section 4.1.2 concerning the choice of initial state.

Often, one rewrites these formal solutions as

$$\psi_S(q, t) = U(t)\psi_S(q, 0) \qquad (4.38)$$

where U is called the *propagator* or *time evolution operator*. For time-independent H, $U(t) = \exp[-iHt/\hbar]$.

In the Schrödinger representation one determines physical observables by taking the expectation value of the operator A_S corresponding to the observable. This expectation value is defined by

$$\langle A \rangle = \langle \psi_S(q, t) | A_S | \psi_S(q, t) \rangle \tag{4.39}$$

and the Schrödinger operator A_S is obtained using the postulates of quantum mechanics. Note that if A_S is time independent (as often occurs), the time dependence of $\langle A \rangle$ is determined by the time dependence of ψ_S.

4.4.2 Heisenberg Representation

Now let's introduce another way to describe things, known as the *Heisenberg representation*. In this, a time-*independent* wavefunction ψ_H is introduced which is ψ_S at $t = 0$ [i.e., $\psi_H = \psi_S(0)$]. Expectation values are invariant to this change of representation, so we get

$$\langle A(t) \rangle = \langle \psi_S(t) | A_S | \psi_S(t) \rangle = \langle U(t)\psi_S(0) | A_S | U(t)\psi_S(0) \rangle$$
$$= \langle \psi_H | U^+(t) A_S U(t) | \psi_H \rangle = \langle \psi_H | A_H | \psi_H \rangle \tag{4.40}$$

In this equation we have substituted from Eq. (4.38) and its adjoint. The last line indicates that an equation for $\langle A \rangle$ identical in form to Eq. (4.39) but expressed in the Heisenberg representation can be obtained provided that we define the Heisenberg operator as

$$A_H(t) \rightarrow U^+(t) A_S U(t) \tag{4.41}$$

For a time-independent hamiltonian, this can be rewritten as

$$A_H(t) = e^{iHt/\hbar} A_S e^{-iHt/\hbar} \tag{4.42}$$

From (4.42) it is immediately obvious that

$$H_H = H_S \tag{4.43}$$

With these definitions, we can now redo much of quantum mechanics using time-dependent operators $A_H(t)$ rather than time-dependent wavefunctions.

An equation for the time evolution of $A_H(t)$ can be obtained simply by differentiating Eq. (4.41):

$$\frac{dA_H}{dt} = \frac{dU^+}{dt} A_S U + U^+ A_S \frac{dU}{dt} + U^+ \frac{dA_S}{dt} U \tag{4.44}$$

Now dU/dt can be obtained by substituting Eq. (4.38) into the Schrödinger equation:

$$i\hbar \frac{dU}{dt} \psi_S(0) = HU\psi_S(0)$$

or

$$i\hbar \frac{dU}{dt} = HU \qquad (4.45)$$

Using this in Eq. (4.44) and defining

$$\frac{\partial A_H}{\partial t} = U^+ \frac{dA_S}{dt} U$$

we find that

$$\frac{dA_H}{dt} = \frac{i}{\hbar} U^+ H A_S U - \frac{i}{\hbar} U^+ A_S H U + \frac{\partial A_H}{\partial t}$$

or

$$\frac{dA_H}{dt} = \frac{i}{\hbar} [H_H, A_H] + \frac{\partial A_H}{\partial t} \qquad (4.46)$$

This is the *Heisenberg equation of motion*, an equation that possesses a profound analogy to the classical mechanical Hamilton's equation (when expressed in terms of Poisson brackets), and which occupies a position in the Heisenberg representation that is equally important as the Schrödinger equation in the Schrödinger representation. In this equation the third term appears only if the Schrödinger operators are explicit functions of time.

Let us now consider an example to illustrate solving the Heisenberg equation. Consider the motion of a free particle in one dimension. Since $H_S = P_S^2/2m$ is time independent, we can immediately assert that $H_S = H_H = P_H^2/2m$. Also, since $[P_S, X_S] = -i\hbar$ is time independent, $[P_H, X_H] = -i\hbar$. Substituting P_H for A_H in Eq. (4.46), we find that

$$\frac{dP_H}{dt} = 0 \qquad (4.47)$$

which implies that $P_H = $ constant $= P_S$.

The analogous equation for X_H gives

$$\frac{dX_H}{dt} = \frac{P_H}{m} = \text{constant} \qquad (4.48)$$

which implies that

$$X_H = \frac{P_H}{m} t + X_H(t = 0) = \frac{P_S}{m} t + X_S \qquad (4.49)$$

From this it follows that

$$\langle X(t) \rangle = \frac{\langle P \rangle t}{m} + \langle X(0) \rangle \tag{4.50}$$

which means that a quantum free-particle wavepacket on the average moves like its classical counterpart.

4.4.3 Interaction Representation

Whenever a partitioning of the hamiltonian of the type $H = H_0 + V(t)$ occurs, it is useful to work in a representation that combines certain aspects of both the Schrödinger and Heisenberg representations. In this so-called *interaction representation*, the wavefunction is defined via

$$\psi_I(t) = e^{iH_0 t/\hbar} \psi_S(t) \tag{4.51}$$

so that ψ_I would be ψ_H if $V(t) = 0$. The interaction representation Schrödinger equation is derived by differentiating Eq. (4.51) and substituting Eq. (4.35). This leads to

$$i\hbar \frac{\partial \psi_I}{\partial t} = H_0 \psi_I + e^{iH_0 t/\hbar} H \psi_S \tag{4.52}$$

Substituting $H = H_0 + V$ and using the inverse of Eq. (4.51), we find that

$$i\hbar \frac{\partial \psi_I}{\partial t} = e^{iH_0 t/\hbar} V(t) e^{-iH_0 t/\hbar} \psi_I \tag{4.53}$$

which can be reduced to

$$i\hbar \frac{\partial \psi_I}{\partial t} = V_I(t) \psi_I \tag{4.54}$$

where V_I was defined in Eq. (4.27). Note that $V_I(t)$ plays the role of an effective hamiltonian in this representation. From our previous exact formal solution for $\psi_S(t)$, we can immediately write the formal solution for ψ_I:

$$\psi_I = P \exp\left[\frac{-i}{\hbar} \int_0^t V_I(t') \, dt' \right] \psi_S(0) \tag{4.55}$$

To define operators in the interaction representation, we simply require that expectation values have the same form as in the Schrödinger or Heisenberg representations. Thus

$$\langle A \rangle = \langle \psi_S | A_S | \psi_S \rangle = \langle \psi_I | e^{iH_0 t/\hbar} A_S e^{-iH_0 t/\hbar} | \psi_I \rangle = \langle \psi_I | A_I | \psi_I \rangle \tag{4.56}$$

where

$$A_I = e^{iH_0 t/\hbar} A_S e^{-iH_0 t/\hbar} \tag{4.57}$$

An equation of motion for A_I analogous to the Heisenberg equation is easily generated and is given as follows:

$$\frac{dA_I(t)}{dt} = \frac{i}{\hbar} [H_0, A_I(t)] + \frac{\partial A_I}{\partial t} \tag{4.58}$$

4.5

Transition Probabilities per Unit Time

Up to this point, we have left the time dependence of $V(t)$ arbitrary. If, however, $V(t)$ is either a constant or a periodic function (after being turned on), then some additional development is possible, leading to simplified expressions for transition rates. Since this often happens in chemical problems, we now consider it in detail.

4.5.1 Perturbation Theory for a Constant Interaction Potential

Consider first the case where V = constant. We assume that V is turned on at $t = 0$, then off at $t = \tau$, and we would like to determine the probability of transition to a final state that is not the same as the initial state. An example where such a situation occurs in nature is *radiationless transitions*. In that case, light absorption to an electronically excited state (which is then coupled by nonadiabatic or spin-orbit coupling to another state) initiates the interaction. This coupling is not an explicit function of time (it depends on the coordinates of the nuclei), so it can be thought of as being constant in applying time-dependent perturbation theory.

If we use first-order perturbation theory to calculate $c_k(t)$ for $t \geq \tau$, we find from Eq. (4.8) (for $k \neq m$)

$$c_k = \frac{-i}{\hbar} \int_0^\tau dt' \; V_{km} e^{i\omega_{km}t'} \tag{4.59}$$

where V_{km} is the interaction potential. Since V_{km} = constant, this can be integrated to give

$$c_k = \frac{-i}{\hbar} V_{km} \frac{e^{i\omega_{km}\tau} - 1}{i\omega_{km}} = -V_{km} \frac{e^{i\omega_{km}\tau} - 1}{\hbar\omega_{km}} \tag{4.60}$$

The probability of being in state k is thus

$$P_k^m = |c_k|^2 = |V_{km}|^2 \frac{2 - 2\cos\omega_{km}\tau}{(\hbar\omega_{km})^2} = |V_{km}|^2 \frac{\sin^2[(E_k - E_m)\tau/2\hbar]}{[(E_k - E_m)/2]^2} \tag{4.61}$$

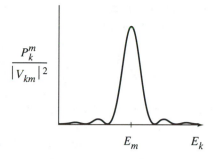

Figure 4.4. E_k dependence of $P_k^m/|V_{km}|^2$.

If we assume that $|V_{km}|^2$ is a slowly varying function of k, the k dependence of P_k^m is determined by the term which follows it in Eq. (4.61). Figure 4.4 shows how this varies with E_k. Note that for large enough τ $[\tau \gtrsim (2\pi\hbar)/E_k]$ a sharply peaked function is obtained, with the height of the peak growing as $\tau^2/\hbar^2$ and width decreasing as $2\pi\hbar/\tau$. The approximate area of the peak is

$$\frac{2\pi\hbar}{\tau}\frac{\tau^2}{\hbar^2} = \frac{2\pi\tau}{\hbar}$$

which grows linearly with τ. In addition, note that we have equal probability for upward and downward transitions.

Considering the radiationless transitions example again, this result indicates that a time τ after the initial electronic excitation, transitions to the second state have generated a distribution of those states such as is pictured in Fig. 4.5. Notice that the energy width (or energy uncertainty)

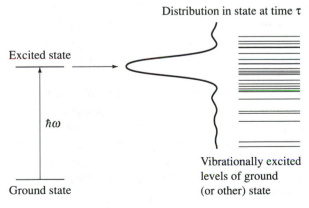

Figure 4.5. Schematic diagram of the state distribution at time τ induced via radiationless transition from a photoexcited state.

of the distribution after a time τ is $\Delta E \approx [(2\pi\hbar)/\tau]$ so that $\tau\,\Delta E \approx 2\pi\hbar > \hbar$, in accord with the usual uncertainty principle.

4.5.2 Fermi's Golden Rule

Now suppose (1) that the final states are so closely spaced in energy that they form a near continuum with density of states $\rho(E_k)$ (the density of states is the number of states per unit energy); (2) that we are only interested in the long-time behavior of the system (i.e., $\tau \gg [(2\pi\hbar)/(E_k - E_m)]$ for typical k); (3) that V_{km} and $\rho(E_k)$ have a weak dependence on k; and (4) that first-order perturbation theory is still valid under these assumptions. Then the total probability of transition is

$$P_T = \sum_k P_k^m = \sum_k |V_{km}|^2 \frac{\sin^2[(E_k - E_m)\tau/2\hbar]}{[(E_k - E_m)/2]^2} \qquad (4.62a)$$

Now replace the sum by an integral and change the integration variable from state index to energy as follows:

$$\approx \int dk\, |V_{km}|^2 \frac{\sin^2(\cdot)\tau/\hbar}{(\cdot)^2} = \int dE_k\, \rho(E_k)|V_{km}|^2 \frac{\sin^2(\cdot)\tau/\hbar}{(\cdot)^2}$$

$$\approx \rho(E_m)|V_{km}|^2 \int_{-\infty}^{\infty} dE_k\, \frac{\sin^2(E_k - E_m)\tau/2\hbar}{(E_k - E_m)^2/4} \qquad (4.62b)$$

The last integral is just $2\pi\tau/\hbar$, so we get

$$P_T = \frac{2\pi\tau}{\hbar}\, \rho(E_m)|V_{km}|^2$$

Thus in this limit, the total probability of a transition is a linear function of time, and we can usefully define a *rate* of transition via

$$w_T = \frac{P_T}{\tau} = \frac{2\pi}{\hbar}\, \rho(E_m)|V_{km}|^2 \qquad (4.63)$$

This is called *Fermi's golden rule* and it represents a very useful result of time-dependent perturbation theory for many problems. In fact there are usually very few problems for which the assumptions above are not satisfied when light absorption and emission are concerned. For intramolecular dynamics and molecular collisions, usually, assumption 4 is not satisfied, so one needs to go to a higher-order theory. This still leads to an expression like Eq. (4.63), but the matrix element $|V_{km}|^2$ is replaced by a more complicated matrix element, examples of which will be given later.

4.5.3 State-to-State Form of Fermi's Golden Rule

An *alternative form* of Fermi's golden rule (the "state-to-state" form) arises in considering the behavior of P_k^m in the same limit as discussed above. In particular, note that

$$F(E_k - E_m) = \frac{\sin^2(E_k - E_m)\tau/2\hbar}{\tau[(E_k - E_m)/2]^2} = \frac{P_k^m}{\tau|V_{km}|^2}$$

has the property that

$$\int_{-\infty}^{\infty} F \, dE_k = \frac{2\pi}{\hbar} \tag{4.64}$$

while $F(0) \to \infty$ for $\tau \to \infty$. Thus for large enough τ, F looks like a big spike, as indicated in Fig. 4.6. This is quite similar to the behavior of a delta function (see Appendix A). The latter satisfies

$$\delta(0) = \infty$$
$$\int \delta(x) \, dx = 1$$

From this it follows that

$$\lim_{\tau \to \infty} F(E_k - E_m) = \frac{2\pi}{\hbar} \delta(E_k - E_m) \tag{4.65}$$

Thus, starting from P_k^m, the following expression for the state-to-state rate can be given:

$$w_{km} = \lim_{\tau \to \infty} \frac{P_k^m}{\tau} = \frac{2\pi}{\hbar} \delta(E_k - E_m)|V_{km}|^2 \tag{4.66}$$

The $\delta(E_k - E_m)$ expresses the result that in the $\tau \to \infty$ limit, only transi-

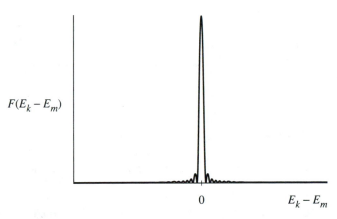

Figure 4.6. Appearance of $F(E_k - E_m)$ for large τ.

tions which obey energy conservation can be caused by a secular (time-independent) interaction. Note that

$$\int \rho(E_k) w_{km} \, dE_k = \frac{2\pi}{\hbar} |V_{km}|^2 \rho(E_m)$$

as should be the case.

4.5.4 Treatment of Periodic Interactions

The generalization of the golden rule to the treatment of periodic interactions is quite simple. Supposing that $V(t) = U e^{\pm i\omega t}$ (where U is independent of time but can be a function of coordinate and momentum operators), the equation analogous to Eq. (4.59) is

$$c_k(t) = \frac{-i}{\hbar} \int_0^\tau dt' \, U_{km} e^{i(\omega_{km} \pm \omega)t'} = -U_{km} \frac{e^{i(\omega_{km} \pm \omega)\tau} - 1}{\hbar(\omega_{km} \pm \omega)} \quad (4.67)$$

This is the same as Eq. (4.60) but with V replaced by U and ω_{km} by $\omega_{km} \pm \omega$. Thus the derivation of the golden rule works the same way except that $E_k - E_m$ is replaced by $E_k - E_m \pm \hbar\omega$. Evidently, then, the peak in P_k^m will occur at $E_k = E_m \mp \hbar\omega$ and Fermi's rule becomes

$$w_T = \frac{2\pi}{\hbar} |U_{km}|^2 \, \rho(E_m \mp \hbar\omega) \quad (4.68)$$

and

$$w_{km} = \frac{2\pi}{\hbar} |U_{km}|^2 \delta(E_k - E_m \pm \hbar\omega) \quad (4.69)$$

Note that $e^{+i\omega t}$ causes transitions for which the final energy E_k is $E_m - \hbar\omega$ (which would happen in the *emission* of light), while $e^{-i\omega t}$ leads to $E_k = E_m + \hbar\omega$ (characteristic of *absorption*). In the next chapter we examine the application of this form of Fermi's golden rule to the interaction of radiation and matter. The function U, which has so far been left as arbitrary, will then become a specific function that is determined by this interaction.

BIBLIOGRAPHY FOR CHAPTER 4

Time-dependent quantum mechanics is treated at an elementary level in the following texts:

Atkins, P. W., *Molecular Quantum Mechanics*, 2nd Ed. (Oxford University Press, New York, 1983).

Flygare, W. H., *Molecular Structure and Dynamics* (Prentice Hall, Englewood Cliffs, N.J., 1978).

More rigorous development of time-dependent quantum mechanics may be found in:

Baym, G., *Lectures on Quantum Mechanics* (Benjamin/Cummings, London, 1981).

Cohen-Tannoudji, C., B. Diu, and F. Laloe, *Quantum Mechanics* (Wiley, New York, 1976).

Davydov, A. S., *Quantum Mechanics* (Pergamon Press, Oxford, 1976).

Eyring, H., J. Walter, and G. Kimball, *Quantum Chemistry* (Wiley, New York, 1944).

Merzbacher, E., *Quantum Mechanics* (Wiley, New York, 1970).

Rapp, D., *Quantum Mechanics* (Holt, Rinehart and Winston, New York, 1971).

Schiff, L. I., *Quantum Mechanics* (McGraw-Hill, New York, 1968).

PROBLEMS FOR CHAPTER 4

1. Consider the collision of an ion such as H^+ with H_2. At large separations R between the ion and molecule (where the straight-line trajectory model is accurate), the ion–molecule interaction potential has the general form

$$V(R, \theta) = \frac{-e^2}{R^4} [V_0 + V_2 P_2(\cos \theta)]$$

where

$$V_0 = \frac{\alpha_\parallel + 2\alpha_\perp}{3} \qquad V_2 = \frac{\alpha_\parallel - \alpha_\perp}{3} \qquad P_2(\cos \theta) = \frac{3 \cos^2 \theta - 1}{2}$$

In this formula, e is the electronic charge, and $\alpha_\parallel$ and $\alpha_\perp$ are the parallel and perpendicular static polarizabilities of the H_2 molecule. Since $V(R, \theta)$ depends on orientation angle θ between R and the diatomic axis r, the ion–molecule coupling will cause rotational excitation and deexcitation of H_2.

(a) Assuming the straight-line trajectory approximation, derive an expression for the transition probability from

$$H_2(v = 0, j = 0, m_j = 0) \quad \text{to} \quad H_2(v = 0, j = 2, m_j = 0, \pm 1, \pm 2).$$

Evaluate this probability explicitly for $b = 2, 10, 50$ Å, and $E_0 = 10$ eV using $\alpha_\parallel = 1.0$ Å^3 and $\alpha_\perp = 0.63$ Å^3. Since the rotational spacing is small, you can assume when evaluating the transition probability that $\omega_{km} = 0$.

(b) To integrate the result of part (a) over impact parameters to get a cross section, it is necessary to truncate the integration at a minimum impact parameter b_0. b_0 corresponds to the "radius" of H_2 and represents the distance of closest approach between H_2 and H^+ in a head-on collision. Assuming that $b_0 = 2$ Å, what are the cross sections for the above-mentioned transitions at 10 eV?

2. Consider the time evolution of the spin state of an electron in a magnetic field. Let the unperturbed hamiltonian be the Zeeman hamiltonian (with static magnetic field B_0 taken in the z direction): $H_0 = \gamma B_0 S_z$, where γ is the gyromagnetic ratio and S_z is the usual spin operator. Let the coupling hamiltonian be a time-varying Zeeman interaction in the x direction, $V = \gamma B_1 S_x \cos \omega t$, where B_1 is the perturbing magnetic field and ω is the frequency.

(a) If the electron is initially ($t = 0$) in state α, what is the probability of ending up in state β as a function of time? Take $\omega = \gamma B_0$, and use first-order perturbation theory.

(b) How does the answer to part (a) change when second-order perturbation theory is used?

(c) Now let's solve the time-dependent Schrödinger equation exactly for the same problem. To do this, first write down the coupled equations for the coefficients C_α and C_β associated with the states α and β, respectively. Now take the limit $\omega = \gamma B_0$ in these equations. Carefully consider the time dependence of each term and neglect all terms that vary as $e^{\pm 2i\omega t}$ in the resulting differential equations. Show that the solutions are the same as those obtained from perturbation theory (taking $\omega = \gamma B_0$) in the limit of small B_1.

3. Consider a collision between two atoms in which two electronic potential curves cross to cause a change of state (as in $Na + I \rightarrow Na^+ + I^-$). To a first approximation, the transition probability may be calculated by first-order perturbation theory, with the motions of the nuclei treated classically.

(a) Suppose that the interaction matrix element V_{12} between the two states 1 and 2 is a constant independent of time, while the energies E_1 and E_2 of each state are linear functions of time (i.e., $E_1 = \alpha_1 t$ and $E_2 = \alpha_2 t$, where α_1 and α_2 are constants). (Here $t = 0$ is taken as the moment when the two curves intersect.) What is the time dependence of each state in the absence of V_{12}? What is the perturbation theory expression for the transition probability P_{12} between states 1 and 2? [*Hint:* Eq. (4.9) has to be rederived for the case where the zero-order energies are time dependent.] Also, the following integral will prove useful.

$$\int_0^\infty \sin x^2 \, dx = \int_0^\infty \cos x^2 \, dx = \left(\frac{\pi}{8}\right)^{1/2}$$

(b) One way to rationalize the linear dependence of $E_1 - E_2$ on time is to assume that the nuclei move with a constant velocity v in the vicinity of the crossing point. Show that under these circumstances P_{12} is identical in the limit of small V_{12} with the *Landau–Zener expression* for the transition probability,

$$P_{12} = 1 - e^{-2\pi\gamma}$$

where $\gamma = V_{12}^2 / \hbar v |s_1 - s_2|$ and $s_1 - s_2$ is the difference between the slopes of the potential curves at their point of intersection.

4. (a) Solve Heisenberg's equations of motion for the time evolution of the raising and lowering operators b_H^+ and b_H for a harmonic oscillator. Assume that the hamiltonian

$$h = \frac{p^2}{2m} + \frac{1}{2} m\omega^2 x^2$$

has been simplified using $P = p/(\hbar\omega m)^{1/2}$, $Q = x(m\omega/\hbar)^{1/2}$, $H = h/\hbar\omega$ into

$$H = \frac{1}{2}(P^2 + Q^2)$$

so that

$$b = \frac{Q + iP}{\sqrt{2}} \qquad b^+ = \frac{Q - iP}{\sqrt{2}}$$

(b) Show that

$$Q_H(t) = Q \cos \omega t + P \sin \omega t$$

where Q and P are the Schrödinger operators defined in part (a).

5. Two spin-$\frac{1}{2}$ particles S_1 and S_2 interact in the absence of a magnetic field via a coupling $V = \lambda S_1 \cdot S_2$, where λ is constant. If the spin quantum numbers are $m_1 = \frac{1}{2}$, $m_2 = -\frac{1}{2}$ at $t = 0$, what is the probability of $m_1 = -\frac{1}{2}$, $m_2 = \frac{1}{2}$ at time t? Use first-order perturbation theory.

5

Interaction of Radiation with Matter

5.1

Introduction

This chapter begins with a brief review of the properties of classical electromagnetic fields. The rest of the chapter concerns the interaction of these fields with matter. Throughout we consider that the fields are classical functions of coordinates and time while the matter is quantum mechanical. This *semiclassical* treatment is not strictly correct, for in reality both field and matter are quantum mechanical—as we discuss in Chapter 6. The major defect of this treatment is the omission of *spontaneous emission,* but as we shall see, it is not difficult to include it in an *ad hoc* manner.

5.2

Electromagnetic Fields

5.2.1 Vector Potentials and Wave Equations

To get started, we have to know how to describe an electromagnetic field mathematically, and how particles interact with electromagnetic fields. Our first goal will be to determine the electric field **E** and the magnetic field **B** (i.e., the magnetic induction—*not* **H**) for an electromagnetic wave moving in free space. To be rigorously correct, we should start from *Maxwell's equations* (partial differential equations which determine **E** and **B**), but here we use a simpler description.

First, let's express **E** and **B** in terms of scalar and vector potentials. From electrostatics we have $\mathbf{E} = -\nabla\phi$, where ϕ is the scalar potential. If the field is time dependent, however, this must be generalized to

$$\mathbf{E} = -\nabla\phi - \frac{1}{c}\frac{\partial\mathbf{A}}{\partial t} \tag{5.1}$$

and

$$\mathbf{B} = \nabla \times \mathbf{A} \tag{5.2}$$

where **A** is called the *vector* potential and $\nabla \times \mathbf{A}$ denotes the *curl* of **A** and is defined via

$$\nabla \times \mathbf{A} = \begin{vmatrix} \mathbf{i} & \mathbf{j} & \mathbf{k} \\ \partial/\partial x & \partial/\partial y & \partial/\partial z \\ A_x & A_y & A_z \end{vmatrix}$$

Note that because **A** is a vector function, each component of **A** can separately depend on x, y, and z.

A major reason for introducing these potentials is that they reduce the number of field components needed to define the electromagnetic wave from 6 (the components of **E** and **B**) to 4 (ϕ and the components of **A**). Also, two of the four Maxwell equations are solved automatically by determining **E** and **B** from ϕ and **A**, and the other two equations can then be used to determine ϕ and **A** (see below). These equations do not, however, determine ϕ and **A** uniquely. Instead, one must impose additional conditions to define them. Called choosing a *gauge*, this arises because the transformations $\phi \to \phi + f(t)$; $\mathbf{A} \to \mathbf{A} + \nabla g(\mathbf{r})$ for arbitrary $f(t)$ and $g(\mathbf{r})$ do not alter the physical observables **E** and **B**. Although the choice of gauge is arbitrary, a customary choice will be used here, the *Coulomb gauge*. In this we simply choose $\nabla \cdot \mathbf{A} = 0$. As we shall see, this one additional constraint will define ϕ and **A** uniquely to within an additive constant.

From Maxwell's equations, one can derive the following equations for ϕ and **A** for an electromagnetic wave moving through empty space:

$$\nabla^2\phi = 0 \qquad \text{(Laplace's equation)} \tag{5.3}$$

$$\nabla^2\mathbf{A} = \frac{1}{c}\nabla\left(\frac{\partial\phi}{\partial t}\right) + \frac{1}{c^2}\frac{\partial^2\mathbf{A}}{\partial t^2} \tag{5.4}$$

Since space is isotropic, with no charges, the only allowable solution to Eq. (5.3) is ϕ = constant, so Eq. (5.4) becomes

$$\nabla^2 \mathbf{A} = \frac{1}{c^2} \frac{\partial^2 \mathbf{A}}{\partial t^2} \tag{5.5}$$

which is known as the classical *wave equation*.

5.2.2 Plane Waves

It is not difficult to show that Eq. (5.5) is solved by any function of $\mathbf{k} \cdot \mathbf{r} - \omega t$ since $\nabla^2 f(\mathbf{k} \cdot \mathbf{r} - \omega t) = k^2 f''$ and

$$\frac{1}{c^2} \frac{\partial^2 f}{\partial t^2} = \frac{\omega^2}{c^2} f''$$

Thus as long as $k^2 = \omega^2/c^2$, f is a solution. The actual choice of f depends on the imposition of *boundary conditions*, and for our purposes, it is convenient to choose f to describe a *freely propagating plane wave*. Thus we take

$$\mathbf{A}(r, t) = \mathbf{A}_0'(e^{i(\mathbf{k} \cdot \mathbf{r} - \omega t)} + e^{-i(\mathbf{k} \cdot \mathbf{r} - \omega t)}) \tag{5.6}$$

It is not difficult to show that the $e^{i\mathbf{k} \cdot \mathbf{r}}$ represents a plane wave moving in the $\mathbf{k}$ direction with wavelength $2\pi/|\mathbf{k}|$ while $e^{-i\mathbf{k} \cdot \mathbf{r}}$ is a plane wave moving in the $-\mathbf{k}$ direction. Applying the gauge condition $\nabla \cdot \mathbf{A} = 0$ to Eq. (5.6), we find that

$$\nabla \cdot \mathbf{A} = -2\mathbf{k} \cdot \mathbf{A}_0' \sin(\mathbf{k} \cdot \mathbf{r} - \omega t) = 0 \tag{5.7}$$

This can equal zero everywhere only if $\mathbf{A}_0'$ and $\mathbf{k}$ are perpendicular. Since there are two vectors which can be simultaneously perpendicular to $\mathbf{k}$ and to each other it follows that there are two possible *polarizations* of light. If we let $\boldsymbol{\varepsilon}$ be a unit vector that points in the direction of $\mathbf{A}_0'$ (the direction of polarization), then $\boldsymbol{\varepsilon} \cdot \mathbf{k} = 0$. If we define $|\mathbf{A}_0'| = \frac{1}{2}A_0$, then

$$\mathbf{A}(\mathbf{r}, t) = A_0 \boldsymbol{\varepsilon} \cos(\mathbf{k} \cdot \mathbf{r} - \omega t) \tag{5.8}$$

which is an expression for the vector potential that will be used extensively below. Note that the choice for the phase of $\mathbf{A}$ (i.e., replacing $\mathbf{k} \cdot \mathbf{r} - \omega t$ by $\mathbf{k} \cdot \mathbf{r} - \omega t$ + constant) is arbitrary. It will not influence our results.

Given the expression above for $\mathbf{A}$ it is easy to derive expressions for $\mathbf{E}$ and $\mathbf{B}$ as follows:

$$\mathbf{E} = -\frac{1}{c} \frac{\partial \mathbf{A}}{\partial t} = \frac{-\omega}{c} A_0 \boldsymbol{\varepsilon} \sin(\mathbf{k} \cdot \mathbf{r} - \omega t) \tag{5.9}$$

$$\mathbf{B} = \nabla \times \mathbf{A} = -A_0(\mathbf{k} \times \boldsymbol{\varepsilon}) \sin(\mathbf{k} \cdot \mathbf{r} - \omega t) \tag{5.10}$$

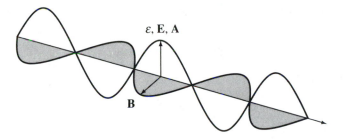

Figure 5.1. E, A, and **B** fields for a plane-polarized electromagnetic wave.

One can now depict the mutually perpendicular **E** and **B** fields as shown in Fig. 5.1.

5.2.3 Energy and Photon Number Density

The quantity A_0 is related to the *energy density* in the field. Classically, this is a continuous quantity, but if the field were quantized, we would find that the number of photons N at any frequency ω must be integral. The relation between energy density and photon number is determined by imagining that our field is contained in a volume V and that the time-averaged energy E associated with this field equals $N\hbar\omega$, where $\hbar\omega$ is the energy per photon. Then

$$E = N\hbar\omega = \int d^3r \; \overline{\text{energy density}} \tag{5.11}$$

where the bar refers to a time average, and the energy density is related to **E** and **B** via the expression

$$\text{energy density} = \frac{\mathbf{E}^2 + \mathbf{B}^2}{8\pi} \tag{5.12}$$

Substituting for **E** and **B** from above, we find that

$$N\hbar\omega = \int d^3r \left(\frac{2\pi}{\omega}\right)^{-1} \int_0^{2\pi/\omega} dt \; \frac{\mathbf{E}^2 + \mathbf{B}^2}{8\pi}$$

$$= \int d^3r \left(\frac{\omega}{2\pi}\right) \int_0^{2\pi/\omega} dt \; \frac{\omega^2 A_0^2}{4\pi c^2} \overline{\sin^2(\mathbf{k}\cdot\mathbf{r} - \omega t)} = V\frac{\omega^2 A_0^2}{8\pi c^2} \tag{5.13}$$

To obtain this, we have used the relations $|\mathbf{k}\times\boldsymbol{\varepsilon}|^2 = \omega^2/c^2$ and

$$\overline{\sin^2(\mathbf{k}\cdot\mathbf{r} - \omega t)} = \frac{1}{2}$$

From this it follows that

$$A_0 = 2c \left(\frac{2\pi\hbar N}{\omega V}\right)^{1/2} \tag{5.14}$$

which relates A_0 to the number of photons per unit volume (N/V).

5.3

Interaction between Matter and Field .

5.3.1 Classical Theory

Classically, if a particle with charge e interacts with an electromagnetic field, that particle experiences a force $\mathbf{F}$ given by

$$\mathbf{F} = e\left(\mathbf{E} + \frac{\mathbf{v} \times \mathbf{B}}{c}\right) \tag{5.15}$$

The development of a hamiltonian that reproduces this force when substituted into Hamilton's equations is rather difficult. Here we just give the result, which is

$$H = \frac{1}{2m}\left(\mathbf{p} - \frac{e}{c}\mathbf{A}\right)^2 \tag{5.16}$$

Thus the vector potential $\mathbf{A}$ field does not act like an ordinary potential, but rather, it changes the effective momentum of the charged particle. The next section shows that these formulas are consistent.

5.3.2 Derivation of Classical Hamiltonian

From classical mechanics it can be shown that the equations of motion for the coordinate and momentum are related to the hamiltonian via Hamilton's equations as follows (we consider the x coordinate here):

$$\dot{p}_x = \frac{-\partial H}{\partial x} \tag{5.17a}$$

$$\dot{x} = \frac{\partial H}{\partial p_x} \tag{5.17b}$$

Using Eq. (5.16), these equations become

$$\dot{p}_x = \frac{e}{mc}\left(\mathbf{p} - \frac{e}{c}\mathbf{A}\right) \cdot \frac{\partial \mathbf{A}}{\partial x} \tag{5.18a}$$

$$\dot{x} = \frac{1}{m}\left(p_x - \frac{e}{c}A_x\right) \qquad (5.18b)$$

Then

$$m\ddot{x} = F_x = \dot{p}_x - \frac{e}{c}\dot{A}_x$$

$$= \frac{e}{mc}\left(\mathbf{p} - \frac{e}{c}\mathbf{A}\right)\cdot\frac{\partial\mathbf{A}}{\partial x} - \frac{e}{c}\frac{dA_x}{dt} = \frac{e}{c}\left(\dot{\mathbf{r}}\cdot\frac{\partial\mathbf{A}}{\partial x} - \frac{dA_x}{dt}\right) \qquad (5.19)$$

Since

$$\frac{dA_x}{dt} = \frac{\partial A_x}{\partial t} + \frac{\partial A_x}{\partial x}\dot{x} + \frac{\partial A_x}{\partial y}\dot{y} + \frac{\partial A_x}{\partial z}\dot{z} \qquad (5.20)$$

we get

$$F_x = \frac{e}{c}\left[\dot{x}\frac{\partial A_x}{\partial x} + \dot{y}\frac{\partial A_y}{\partial x} + \dot{z}\frac{\partial A_z}{\partial x}\right]$$

$$- \frac{e}{c}\left[\frac{\partial A_x}{\partial t} + \dot{x}\frac{\partial A_x}{\partial x} + \dot{y}\frac{\partial A_x}{\partial y} + \dot{z}\frac{\partial A_x}{\partial z}\right]$$

$$= -\frac{e}{c}\frac{\partial A_x}{\partial t} + \frac{e}{c}\left[\dot{y}\left(\frac{\partial A_y}{\partial x} - \frac{\partial A_x}{\partial y}\right) + \dot{z}\left(\frac{\partial A_z}{\partial x} - \frac{\partial A_x}{\partial z}\right)\right] \qquad (5.21)$$

or

$$\mathbf{F} = -\frac{e}{c}\frac{\partial\mathbf{A}}{\partial t} + \frac{e}{c}[\mathbf{v}\times(\boldsymbol{\nabla}\times\mathbf{A})] = e\left[\mathbf{E} + \frac{1}{c}\mathbf{v}\times\mathbf{B}\right] \qquad (5.22)$$

5.3.3 Quantum Hamiltonian for a Particle in an Electromagnetic Field

Now let's convert H to its quantum counterpart. This is trivially done by replacing $\mathbf{p}$ by $-i\hbar\boldsymbol{\nabla}$, yielding

$$H = \frac{1}{2m}\left[-i\hbar\boldsymbol{\nabla} - \frac{e}{c}\mathbf{A}\right]^2 \qquad (5.23)$$

If the charged particle also experiences a static potential V_s (such as would happen for electrons in a molecule, for example), this potential is simply added to H. Thus we have

$$H = -\frac{\hbar^2}{2m}\nabla^2 + V_s + \frac{i\hbar e}{2mc}(\boldsymbol{\nabla}\cdot\mathbf{A} + \mathbf{A}\cdot\boldsymbol{\nabla}) + \frac{e^2}{2mc^2}\mathbf{A}\cdot\mathbf{A} = H_0 + V$$

$$(5.24)$$

where H_0 is the hamiltonian in the absence of the field and V is the matter–field interaction hamiltonian:

$$V = \frac{i\hbar e}{2mc} (\nabla \cdot \mathbf{A} + \mathbf{A} \cdot \nabla) + \frac{e^2}{2mc^2} \mathbf{A} \cdot \mathbf{A} \qquad (5.25)$$

This is the desired $V(t)$ for use in studying light absorption and emission, but before we consider such applications, let us first make a few generalizations and simplifications with it.

1. First, for the case of the interaction with *many charged particles* (a molecule, for instance), V becomes

$$V = \sum_i \left\{ \frac{i\hbar e_i}{2m_i c} (\nabla_i \cdot \mathbf{A}(\mathbf{r}_i) + \mathbf{A}(\mathbf{r}_i) \cdot \nabla_i) + \frac{e_i^2}{2m_i c^2} \mathbf{A}(\mathbf{r}_i) \cdot \mathbf{A}(\mathbf{r}_i) \right\} \quad (5.26)$$

In other words, we simply sum the interaction over the particles, taking into account the charge e_i and mass m_i of each particle. For a molecule undergoing electronic excitation, we can often ignore the sum over nuclei, since their interactions with a field are much smaller, due to the $1/m_i$ term in the sum.
2. The first term in V can be rewritten as

$$\nabla \cdot \mathbf{A} + \mathbf{A} \cdot \nabla = 2\mathbf{A} \cdot \nabla \qquad (5.27)$$

since $(\nabla \cdot \mathbf{A}\psi) = (\nabla \cdot \mathbf{A})\psi + (\mathbf{A} \cdot \nabla)\psi = (\mathbf{A} \cdot \nabla)\psi$ by virtue of the gauge condition $\nabla \cdot \mathbf{A} = 0$. Thus we can write

$$V = -\frac{e}{mc} (\mathbf{A} \cdot \mathbf{p}) + \frac{e^2}{2mc^2} \mathbf{A} \cdot \mathbf{A} \qquad (5.28)$$

3. Finally, for many applications, it is a very good approximation to neglect the $(e^2/2mc^2)\mathbf{A} \cdot \mathbf{A}$ term. This is known as the *weak-field approximation,* but in fact it applies to fields that we might think of as being quite strong.

The following estimate will indicate that only for extremely large fields does the A^2 term approach the $\mathbf{A} \cdot \mathbf{p}$ term in magnitude. Consider the ratio

$$\frac{2|p|}{\frac{e}{c}|A|}$$

(roughly the ratio of these two terms). We would like to determine how big a field is needed to make this ratio equal to unity. Consider an electron in a ground-state hydrogen atom. We can estimate p using the Bohr model as follows. For circular orbits the angular momentum is

$$l \cong \hbar = pr \qquad (5.29)$$

which implies that $p \cong \hbar/r \cong \hbar/a_0$. Also, using (5.14) and assuming that the energy E is related to photon number N by $E = N\hbar\omega$,

$$A \simeq A_0 = 2c \left(\frac{2\pi\hbar\omega N}{\omega^2 V}\right)^{1/2} = 2c\hbar \left(\frac{2\pi E}{(\hbar\omega)^2 V}\right)^{1/2} \qquad (5.30)$$

So if the ratio $2|p|/(e/c)|A|$ is to equal unity, then

$$1 = \frac{2p}{\frac{e}{c}A} \simeq \frac{2\hbar/a_0}{\frac{e}{c}2c\hbar \left(\frac{2\pi E}{(\hbar\omega)^2 V}\right)^{1/2}} = \frac{\hbar\omega}{ea_0 \left(\frac{2\pi E}{V}\right)^{1/2}} \qquad (5.31)$$

Now imagine that we have a laser beam with a given power per unit area P/A. Then

$$\frac{P}{A} = \frac{E/t}{A} = \frac{E}{At} = \frac{Ec}{V} \qquad (5.32)$$

where t is time. Thus

$$\frac{P}{A} = \frac{Ec}{V} = \frac{c}{2\pi}\left(\frac{\hbar\omega}{ea_0}\right)^2 \qquad (5.33)$$

Substituting, we find that (taking $\hbar\omega = 13.6$ eV, the energy needed to ionize the hydrogen atom)

$$\frac{P}{A} = \frac{3 \times 10^{10} \text{ cm/s}}{2\pi}\left(\frac{13.6 \times 1.6022 \times 10^{-12} \text{ erg}}{4.8 \times 10^{-10} \text{ esu} \times 0.529 \times 10^{-8} \text{ cm}}\right)^2$$

$$= 3.5 \times 10^{23} \frac{\text{erg}}{\text{cm}^2 \text{ s}} = 3.5 \times 10^{16} \frac{\text{W}}{\text{cm}^2} \qquad (5.34)$$

This is an enormous laser flux that can be achieved only with very short pulse lasers.

It should also be noted that this same calculation tells us the strength of the electron kinetic energy relative to the $(e/mc)A \cdot p$ term; that is,

$$\frac{\dfrac{p^2}{2m}}{\dfrac{e}{mc}\mathbf{A}\cdot\mathbf{p}} \approx \frac{p}{\dfrac{e}{c}A}$$

This indicates that the coupling of matter to electromagnetic fields is typically very weak, and hence that perturbation theory is quite appropriate.

An alternative way to the same conclusion is simply to notice that the atomic unit of field gradient is just (hartree)/(bohr)(electron charge), which is 27.21 V/0.529 $\times$ 10^{-8} cm $\approx 10^9$ V/cm. Physically, this is the field gradient at the position of the first Bohr orbit of the H atom. This is far higher than ordinary laboratory electrostatic or magnetostatic field gradients, and much higher than those afforded by ordinary flashlamps or lasers. For such experiments, then, the weak field approximation is wholly appropriate. *Very* intense laser fields can approach or even exceed 10^9 V/cm; for such situations, the weak-field approximation will fail.

5.4

Absorption and Emission of Light

5.4.1 Application of Fermi's Golden Rule

We now wish to examine the molecular transitions that can be induced in first order by the $(e/mc)\mathbf{A}\cdot\mathbf{p}$ interaction. Rewriting, we have

$$V = -\frac{e}{mc}\mathbf{A}\cdot\mathbf{p} = -\frac{e}{mc}A_0\cos(\mathbf{k}\cdot\mathbf{r} - \omega t)\boldsymbol{\varepsilon}\cdot\mathbf{p}$$

$$= \frac{-e}{2mc}A_0(e^{i(\mathbf{k}\cdot\mathbf{r}-\omega t)} + e^{-i(\mathbf{k}\cdot\mathbf{r}-\omega t)})\boldsymbol{\varepsilon}\cdot\mathbf{p} = U(\mathbf{k})e^{-i\omega t} + U(-\mathbf{k})e^{i\omega t} \quad (5.35)$$

The last line is in a form that enables us to use Fermi's golden rule for periodic interactions (Chapter 4) to calculate the rate of transitions induced by V. We noted previously (Section 4.5.4) that for $V = Ue^{\pm i\omega t}$, the rate expression is

$$w_{km} = \frac{2\pi}{\hbar}|U_{km}|^2\,\delta(E_k - E_m \pm \hbar\omega)$$

Thus the $Ue^{i\omega t}$ term causes $E_k = E_m - \hbar\omega$ and hence leads to *stimulated emission*, while the $Ue^{-i\omega t}$ term causes $E_k = E_m + \hbar\omega$ and hence *stimulated absorption* (or just absorption). An electromagnetic field can therefore cause transitions in both directions. Note also that the delta function ensures that only states having $E_k = E_m \pm \hbar\omega$ can be reached; that is, energy must be conserved.

Using Eq. (5.35), the total rates of emission and absorption are given by

$$w_{abs}(m \rightarrow k) = \frac{2\pi}{\hbar}|U_{km}(\mathbf{k})|^2\rho(E_m + \hbar\omega) \quad (5.36)$$

$$w_{em}(m \rightarrow k) = \frac{2\pi}{\hbar}|U_{km}(-\mathbf{k})|^2\rho(E_m - \hbar\omega) \quad (5.37)$$

where

$$U_{km}(\mathbf{k}) = -\frac{eA_0}{2mc}\langle k|e^{i\mathbf{k}\cdot\mathbf{r}}\boldsymbol{\varepsilon}\cdot\mathbf{p}|m\rangle \quad (5.38)$$

5.4.2 Dipole Approximation

The expression for U_{km} can be considerably simplified by noting that for visible light, $e^{i\mathbf{k}\cdot\mathbf{r}} \sim 1$ for those r's considered in the matrix elements. This is because the range of r's is typically molecular dimensions (10 to 100 Å), while $k = 2\pi/\lambda$ is much smaller [i.e., for $r = 50$ and $\lambda = 5000$ Å, $kr = (2\pi/5000)50 = 2\pi/100 = 0.06$]. This is the *long-wavelength approximation*, and although it is not perfect, it does describe the most intense optical transitions.

Introducing this approximation into U_{km}, we find that

$$U_{km} = -\frac{eA_0}{2mc} \langle k|\boldsymbol{\varepsilon}\cdot\mathbf{p}|m\rangle \tag{5.39}$$

Now let's rewrite this expression. First note that for any single-particle hamiltonian H_0,

$$\mathbf{p} = \frac{im}{\hbar} [H_0, \mathbf{r}]$$

Proof: In one dimension, this easily follows from the expression

$$\frac{im}{\hbar}\left[\frac{p^2}{2m}, r\right] = \frac{ip}{\hbar}[p, r] = p$$

Substituting into (5.39), we find that

$$U_{km} = -\frac{eA_0}{2mc}\frac{im}{\hbar}\,\boldsymbol{\varepsilon}\cdot\langle k|[H_0, \mathbf{r}]|m\rangle = \frac{-ieA_0}{2c\hbar}(E_k - E_m)\langle k|\boldsymbol{\varepsilon}\cdot\mathbf{r}|m\rangle \tag{5.40}$$

The second expression follows from $H_0|m\rangle = E_m|m\rangle$. In addition, $E_k - E_m = \hbar\omega_{km} = \pm\hbar\omega$ for resonant transitions; thus

$$U_{km} = \frac{\mp ieA_0}{2c}\,\omega\boldsymbol{\varepsilon}\cdot\langle k|\mathbf{r}|m\rangle \tag{5.41}$$

Now $e\mathbf{r}$ is just the dipole operator for a single electron. Thus

$$e\langle k|\mathbf{r}|m\rangle = \text{dipole matrix element} \tag{5.42a}$$

For many electrons, this generalizes to

$$\boldsymbol{\mu}_{km} = e\Big\langle k\Big| \sum_i \mathbf{r}_i \Big|m\Big\rangle \tag{5.42b}$$

Thus we can write

$$U_{km} = \frac{\mp iA_0\omega}{2c}\,\boldsymbol{\varepsilon}\cdot\boldsymbol{\mu}_{km} \tag{5.43}$$

Since U_{km} is proportional to $\boldsymbol{\mu}_{km}$, this expression is called the *electric dipole* approximation for U_{km}.

5.4.3 Photon Density of States

To complete the evaluation, we need to determine the density of states. Now typically we would need to consider both photon and matter states. However, if we consider emission and absorption between discrete states of molecules, the matter states need not be summed over. Rather, just a sum over photon states need be done. Since the same photon states appear in both emission and absorption, the same density will do for both. To calculate the photon density, imagine that the field is in a cube of length L, with $V = L^3$. Let dN be the number of states between N and $N + dN$. For a one-dimensional box, if l_x labels the states, then $dN = dl_x$. But in three dimensions,

$$dN = dl_x \, dl_y \, dl_z \tag{5.44}$$

For a particle in a three-dimensional box, each set of x, y, and z state labels is determined by applying periodic boundary conditions: $e^{ik_x x} = e^{ik_x(x+L)}$, so that $k_x L = 2l_x \pi$, $l_x = 0, \pm 1, \pm 2, \ldots$. Thus

$$dl_x = \frac{L}{2\pi} dk_x \tag{5.45}$$

and

$$dN = \left(\frac{L}{2\pi}\right)^3 dk_x \, dk_y \, dk_z \tag{5.46}$$

Now transform this to polar coordinates, converting $dk_x \, dk_y \, dk_z$ into $k^2 \, dk \, d\Omega$, where Ω is the angle specifying the orientation of $\mathbf{k}$. Since $k = \omega/c$, we have

$$dN = \frac{V}{(2\pi)^3} \frac{1}{c^3} \omega^2 \, d\omega \, d\Omega = \rho(E) \, dE \tag{5.47}$$

Then the density of states is simply dN/dE, and we have

$$\rho(E) = \frac{V}{(2\pi)^3} \frac{\omega^2}{c^3} \frac{d\omega}{dE} \, d\Omega \tag{5.48}$$

and since $E = \hbar\omega$,

$$\rho(E) = \frac{V}{(2\pi c)^3} \frac{\omega^2}{\hbar} \, d\Omega \tag{5.49}$$

5.4.4 Emission Rate

With this result we can now write down an explicit expression for the rate of emission (now considered to be the differential rate into the solid angle $d\Omega$) as follows:

$$dw_{em}(m \rightarrow k) = \frac{2\pi}{\hbar} |U_{km}|^2 \, \rho(E_m - \hbar\omega)$$

$$= \frac{2\pi}{\hbar} \frac{A_0^2 \omega^2}{4c^2} |\boldsymbol{\varepsilon} \cdot \boldsymbol{\mu}_{km}|^2 \frac{V}{(2\pi c)^3} \frac{\omega^2}{\hbar} \, d\Omega \qquad (5.50)$$

Substituting for A_0 from Eq. (5.14), we find that

$$dw_{em}(m \rightarrow k) = \frac{N\omega^3}{2\pi\hbar c^3} |\boldsymbol{\varepsilon} \cdot \boldsymbol{\mu}_{km}|^2 \, d\Omega \qquad (5.51)$$

The polar angles θ and ϕ contained in $d\Omega$ here represent the angles of the stimulating field wave vector relative to a molecule fixed frame, which we define such that $\boldsymbol{\mu}_{km}$ is along the z axis (see Fig. 5.2). Since $\boldsymbol{\varepsilon}$ is perpendicular to $\mathbf{k}$, $\boldsymbol{\varepsilon} \cdot \boldsymbol{\mu}_{km} = |\boldsymbol{\mu}_{km}| \sin \theta$ (assuming, for simplicity, that $\boldsymbol{\varepsilon}$, $\mathbf{k}$, and $\boldsymbol{\mu}_{km}$ are coplanar). Thus

$$\frac{dw_{em}(m \rightarrow k)}{d\Omega} = \frac{N\omega^3}{2\pi\hbar c^3} |\boldsymbol{\mu}_{km}|^2 \sin^2 \theta \qquad (5.52)$$

which means that the efficiency of stimulated emission varies as the $\sin^2$ of the angle between the photon beam direction and $|\boldsymbol{\mu}_{km}|^2$. Also, $dw/d\Omega$ is proportional to N (i.e., intensity varies linearly in the photon energy or intensity). This implies that when $N \rightarrow 0$, the rate vanishes. This means that an atom in an excited state cannot emit if there is no stimulating external field.

In reality, this is not the case, as *spontaneous* emission can also occur. It turns out that we can account for the influence of spontaneous emission in an *ad hoc* way by replacing N by $N + 1$ in emission but not absorption. (See Section 6.2.5 for the correct treatment.) The 1 factor gives us a residual emission that occurs in the absence of a stimulating field and is called spontaneous emission. We will include this factor in what follows so that all expressions are correct.

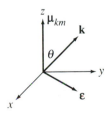

Figure 5.2. Orientation of $\mathbf{k}$ and $\boldsymbol{\varepsilon}$ relative to coordinate system where z axis is along $\boldsymbol{\mu}_{km}$.

Integrating $dw/d\Omega$ over all angles gives the total rate as follows:

$$w_{em}(m \rightarrow k) = \frac{4}{3\hbar} (N + 1) \left(\frac{\omega}{c}\right)^3 |\mu_{km}|^2 \qquad (5.53)$$

Notice how this emission rate is proportional to $|\mu_{km}|^2$, the square of the dipole matrix element.

For spontaneous emission ($N = 0$), the sum of $w_{em}(m \rightarrow k)$ over all possible final states (i.e., those with an energy lower than E_m) gives the total radiative decay rate—just the inverse of the radiative lifetime τ_{rad}. Thus

$$\frac{1}{\tau_{rad}} = \sum_{E_k < E_m} \frac{4}{3\hbar} \left(\frac{|\omega_{km}|}{c}\right)^3 |\mu_{km}|^2 \qquad (5.54)$$

5.4.5 Absorption Rate

The rate expression for *absorption* of light under the assumption of this treatment is the same as for emission, since $|U_{km}|^2$ is the same in the dipole approximation and the density of field states at a given frequency has nothing to do with whether it is emission or absorption. Thus

$$w_{abs}(m \rightarrow k) = N \frac{4}{3\hbar} \left(\frac{\omega}{c}\right)^3 |\mu_{km}|^2 \qquad (5.55)$$

where in this case, we do *not* replace N by $N + 1$ since there is no spontaneous absorption. Note that if a sum over matter states had been included, the rate expressions for absorption and emission would in general be different, as the density of final matter states is different.

5.4.6 Einstein A and B Coefficients

A commonly used procedure for representing absorption and emission rates involves the use of the Einstein A and B coefficients. These represent the transition rates in such a way that the field-dependent effects (proportional to N) are separated from the field-independent effects. Specifically, for *stimulated* absorption and emission, the Einstein B coefficient for the $m \rightarrow k$ transition is defined via

$$w_{stim}(m \rightarrow k) = B(m \rightarrow k)\bar{\rho} \qquad (5.56)$$

where $\bar{\rho}$ is the total energy in the field per unit volume between frequencies ν and $\nu + d\nu$ (the so-called "radiation field density"). $\bar{\rho}$ is the field energy density appropriate to a field inside a blackbody cavity and is given by

$$\bar{\rho}(\nu) \, d\nu = \frac{2N\hbar\omega}{V} \rho(E) \, dE \qquad (5.57)$$

where ρ is the density of states defined earlier. In other words, we take the density of states, multiply by the energy per state ($N\hbar\omega$), then by 2 (to sum over the two possible polarizations for each state), then divide by the volume (so that $\bar{\rho}$ is the energy per unit volume per unit frequency). Substituting for $\rho(E)$ from (5.49), we find that

$$\bar{\rho}(\nu) = \frac{2N\omega^3}{(2\pi c)^3} \frac{dE}{d\nu} \int d\Omega \qquad (5.58)$$

In this expression, we are only interested in the angle-integrated field density, so we take $\int d\Omega = 4\pi$. Since $dE/d\nu = h = 2\pi\hbar$, we find that

$$\bar{\rho}(\nu) = \frac{2N\omega^3}{\pi c^3} \hbar \qquad (5.59)$$

Thus, from our expression for $w_{em}(m \rightarrow k)$, and the definition of $B(m \rightarrow k)$, we have

$$B(m \rightarrow k) = \frac{\dfrac{4N}{3\hbar} \left(\dfrac{\omega}{c}\right)^3 |\boldsymbol{\mu}_{km}|^2}{\dfrac{2N}{\pi} \dfrac{\omega^3}{c^3} \hbar} = \frac{2\pi}{3\hbar^2} |\boldsymbol{\mu}_{km}|^2 \qquad (5.60)$$

which shows us that $B(m \rightarrow k)$ is simply proportional to $|\boldsymbol{\mu}_{km}|^2$.

Note that if we consider the B coefficients for the $m \rightarrow k$ and $k \rightarrow m$ processes, we find that since $|\boldsymbol{\mu}_{km}|^2 = |\boldsymbol{\mu}_{mk}|^2$,

$$B(m \rightarrow k) = B(k \rightarrow m) \qquad (5.61)$$

Thus the rates of absorption and stimulated emission between the two states are the same.

For *spontaneous* emission, the A coefficient is simply defined to be

$$w_{\text{spont}}(m \rightarrow k) = A(m \rightarrow k) \qquad (5.62)$$

so that

$$A(m \rightarrow k) = \frac{4}{3\hbar} \left(\frac{\omega}{c}\right)^3 |\boldsymbol{\mu}_{km}|^2 \qquad (5.63)$$

From this it follows that

$$\frac{A(m \rightarrow k)}{B(m \rightarrow k)} = \frac{2\hbar}{\pi} \left(\frac{\omega}{c}\right)^3 \qquad (5.64)$$

That is, the ratio of spontaneous to stimulated emission varies as the cube of the emitted photon frequency. This means that spontaneous emission is more important in the ultraviolet (UV) region than in the infrared (IR). Of course, the relative *rates* depend on $\bar{\rho}$, but for typically encountered light sources, one often finds that spontaneous emission dominates the emission process in the UV and above, while stimulated emission dominates in the far infrared and below, with both processes commonly seen in between.

It is interesting to note that Einstein actually worked out the relation between A and B in 1917, long before the quantum theory was developed that could derive this from first principles. Rather, he used Planck's radiation law together with simple arguments based on microscopic reversibility to derive this.

5.4.7 Oscillator Strengths

Another way to represent emission rates is in terms of *oscillator strengths* f_{km}. These are defined as

$$f_{km} = \frac{2m\omega}{3\hbar e^2} |\boldsymbol{\mu}_{km}|^2 \tag{5.65}$$

where m is the electron mass. f_{km} is a dimensionless quantity that would be unity for a harmonically bound electron. f_{km} thus measures the intensity of a transition relative to this harmonic model.

From Eqs. (5.60) and (5.65) it is easy to see that

$$f_{km} = \frac{\hbar m\omega}{\pi e^2} B(m \rightarrow k) \tag{5.66}$$

5.4.8 Electric Quadrupole, Magnetic Dipole Mechanisms

Recall that in developing our expression for the rate of absorption and emission of light we made the long-wavelength approximation of replacing $e^{i\mathbf{k}\cdot\mathbf{r}}$ by 1. This gave us an expression involving an electric dipole matrix element for the transition rate, and transitions for which this is nonzero are called *electric-dipole-allowed* transitions. Often, even when this matrix element is zero, weak transitions are still seen. Although there are a number of reasons why this might occur, one often encountered reason is that the terms neglected in the long-wavelength approximation may be contributing. To see what is possible, let's expand $e^{i\mathbf{k}\cdot\mathbf{r}}$ in a Taylor series and keep the $i\mathbf{k}\cdot\mathbf{r}$ term.

$$e^{i\mathbf{k}\cdot\mathbf{r}} = 1 + i\mathbf{k}\cdot\mathbf{r} + \cdots \tag{5.67}$$

Recall that the transition rate expression involves the matrix element U_{km}, and in this case it is given by

$$U_{km} = -\langle k|\frac{eA_0}{2mc}(1 + i\mathbf{k}\cdot\mathbf{r})(\boldsymbol{\varepsilon}\cdot\mathbf{p})|m\rangle \qquad (5.68)$$

Assuming that the electric dipole matrix element is zero, we have

$$U_{km} = -i\frac{eA_0}{2mc}\langle k|(\mathbf{k}\cdot\mathbf{r})(\boldsymbol{\varepsilon}\cdot\mathbf{p})|m\rangle \qquad (5.69)$$

Now

$$(\mathbf{k}\cdot\mathbf{r})(\boldsymbol{\varepsilon}\cdot\mathbf{p}) = \frac{1}{2}[(\mathbf{k}\cdot\mathbf{r})(\boldsymbol{\varepsilon}\cdot\mathbf{p}) - (\boldsymbol{\varepsilon}\cdot\mathbf{r})(\mathbf{k}\cdot\mathbf{p})] + \frac{1}{2}[(\mathbf{k}\cdot\mathbf{r})(\boldsymbol{\varepsilon}\cdot\mathbf{p}) + (\boldsymbol{\varepsilon}\cdot\mathbf{r})(\mathbf{k}\cdot\mathbf{p})]$$
$$(5.70)$$

Using the identity

$$(\mathbf{a}\cdot\mathbf{c})(\mathbf{b}\cdot\mathbf{d}) - (\mathbf{b}\cdot\mathbf{c})(\mathbf{a}\cdot\mathbf{d}) = (\mathbf{a}\times\mathbf{b})\cdot(\mathbf{c}\times\mathbf{d}) \qquad (5.71)$$

in the first bracket in this expression, we get

$$(\mathbf{k}\cdot\mathbf{r})(\boldsymbol{\varepsilon}\cdot\mathbf{p}) = \frac{1}{2}(\mathbf{k}\times\boldsymbol{\varepsilon})(\mathbf{r}\times\mathbf{p}) + \frac{1}{2}[(\mathbf{k}\cdot\mathbf{r})(\boldsymbol{\varepsilon}\cdot\mathbf{p}) + (\boldsymbol{\varepsilon}\cdot\mathbf{r})(\mathbf{k}\cdot\mathbf{p})] \quad (5.72)$$

Considering just the first term on the right-hand side of this expression, we notice that it contains the orbital angular momentum $\mathbf{L} = \mathbf{r}\times\mathbf{p}$ of the electron being considered. Thus the matrix element of the first term is

$$-\frac{ieA_0}{2mc}\left(\frac{1}{2}\right)(\mathbf{k}\times\boldsymbol{\varepsilon})\cdot\langle k|\mathbf{L}|m\rangle = \frac{-ieA_0}{2mc}\frac{mc}{e}(\mathbf{k}\times\boldsymbol{\varepsilon})\cdot\langle k|\mathbf{M}|m\rangle \quad (5.73)$$

where $\mathbf{M} = e/(2mc)\mathbf{L}$ is the magnetic dipole moment operator for the electron. To show this, note that for a Bohr model atom, the magnetic moment equals the area enclosed by an orbit times the current associated with that orbit, divided by c.

$$M = \frac{1}{c}Aj = \frac{1}{c}(\pi r^2)\left(e\frac{\omega}{2\pi}\right) \qquad (5.74)$$

Since $mvr = L$ and $\omega = v/r$,

$$\mathbf{M} = \frac{e}{2mc}\mathbf{L} \qquad (5.75)$$

Then

$$U_{km} = \frac{-iA_0}{2}(\mathbf{k}\times\boldsymbol{\varepsilon})\cdot\mathbf{M}_{km} \qquad (5.76)$$

which is the *magnetic dipole* contribution to the rate.

Now let's evaluate the second term from above. Writing down the vector dot products in terms of components, we have

$$\frac{1}{2} \left[(\mathbf{k} \cdot \mathbf{r})(\boldsymbol{\varepsilon} \cdot \mathbf{p}) + (\boldsymbol{\varepsilon} \cdot \mathbf{r})(\mathbf{k} \cdot \mathbf{p}) \right] = \frac{1}{2} \sum_{ij} (k_i r_i \varepsilon_j p_j + \varepsilon_j r_j k_i p_i)$$

$$= \frac{1}{2} \sum_{ij} k_i \varepsilon_j (r_i p_j + r_j p_i) \qquad (5.77)$$

Now $p_j = (im)/\hbar \, [H_0, r_j]$ and $r_i p_j = p_j r_i + [r_i, p_j] = p_j r_i + i\hbar \delta_{ij}$. Thus the second term is

$$\frac{1}{2} \sum_{ij} k_i \varepsilon_j [p_j r_i + r_j p_i] + \frac{1}{2} i\hbar \sum_{ij} k_i \varepsilon_j \delta_{ij} \qquad (5.78)$$

The last term in this expression is simply

$$\frac{1}{2} i\hbar \sum_i k_i \varepsilon_i = \frac{1}{2} i\hbar \, \mathbf{k} \cdot \boldsymbol{\varepsilon} = 0$$

since $\mathbf{k}$ and $\boldsymbol{\varepsilon}$ are perpendicular. Thus we get

$$\frac{1}{2} \sum_{ij} k_i \varepsilon_j [p_j r_i + r_j p_i] = \frac{im}{\hbar} \frac{1}{2} \sum_{ij} k_i \varepsilon_j [H_0 r_j r_i - r_j H_0 r_i + r_j H_0 r_i - r_j r_i H_0]$$

$$= \frac{im}{2\hbar} \sum_{ij} k_i \varepsilon_j [H_0, r_j r_i] \qquad (5.79)$$

and omitting the magnetic dipole term, we have

$$U_{km} = \frac{-ieA_0}{2mc} \frac{im}{2\hbar} \sum_{ij} k_i \varepsilon_j \langle k | [H_0, r_j r_i] | m \rangle \qquad (5.80)$$

The matrix element in the sum can then be evaluated to be

$$\sum_{ij} \varepsilon_j \langle k | (E_k - E_m) r_j r_i | m \rangle k_i = \frac{\hbar \omega_{km}}{e^2} \sum_{ij} \varepsilon_j (Q_{km})_{ji} k_i = \frac{\hbar \omega_{km}}{e^2} \boldsymbol{\varepsilon} \cdot Q_{km} \cdot \mathbf{k}$$

$$(5.81)$$

where

$$(Q_{km})_{ij} = \langle k | (er_i)(er_j) | m \rangle \qquad (5.82)$$

is the (i, j)th matrix element of the electric quadrupole moment tensor. Thus

$$U_{km} = \frac{A_0 \omega_{km}}{4ce} \boldsymbol{\varepsilon} \cdot Q_{km} \cdot \mathbf{k} \qquad (5.83)$$

Combining this electric quadrupole term with the magnetic dipole term, we have

$$U_{km} = \frac{-iA_0}{2} (\mathbf{k} \times \boldsymbol{\varepsilon}) \cdot \mathbf{M}_{km} + \frac{A_0 \omega_{km}}{4ce} \boldsymbol{\varepsilon} \cdot Q_{km} \cdot \mathbf{k} \qquad (5.84)$$

This expression can now be substituted into the rate formula for emission and absorption. Often one finds that *selection rules* make one or the other of the two terms therein be zero. For example, $\mathbf{M}$ behaves like rotations about x, y, or z in its symmetry properties, and thus belongs to the same irreducible representation as R_x, R_y, and R_z. Q_{km}, on the other hand, depends on the products x^2, xy, y^2, and so on. Both of these terms can in general generate selection rules different from the electric dipole result and can therefore give nonzero transition probabilities even when the transition is electric dipole forbidden. In these cases the transition is said to be magnetic dipole allowed or electric quadrupole allowed, respectively.

5.4.9 Molecular Transitions: Franck–Condon Factors

For electronic transitions in molecules, one needs to consider both the electronic and nuclear degrees of freedom in evaluating the dipole (or quadrupole) matrix elements. Thus if we consider the dipole matrix element $\boldsymbol{\mu}_{km}$, the states k and m refer to combined electronic plus nuclear states. If the Born–Oppenheimer approximation is invoked, we write these states as products:

$$\psi_k = \chi^k_{\nu_k}(q_N)\phi_k(q_e; q_N) \qquad (5.85)$$

and similarly for state m. Here q_e represents the electronic coordinates and q_N the nuclear coordinates. ϕ_k is the electronic wavefunction for state k and $\chi^k_{\nu_k}$ the nuclear wavefunction for nuclear quantum numbers ν_k. The dipole matrix element in this notation is

$$\boldsymbol{\mu}_{km} \rightarrow \boldsymbol{\mu}_{k\nu_k, m\nu_m} = \int dq_N \int dq_e \, \chi^k_{\nu_k} \phi_k \boldsymbol{\mu} \chi^m_{\nu_m} \phi_m \qquad (5.86)$$

where $\boldsymbol{\mu}$ is the dipole operator appropriate for the system of interest. By rearranging this expression, we can write

$$\boldsymbol{\mu}_{k\nu_k, m\nu_m} = \int dq_N \, \chi^k_{\nu_k} \mathbf{M}_{km}(q_N) \chi^m_{\nu_m} \qquad (5.87)$$

where we have defined the *electronic matrix element* $\mathbf{M}_{km}$ (not to be confused with the magnetic dipole matrix element of Section 5.4.8) as

$$\mathbf{M}_{km}(q_N) = \int dq_e \, \phi_k(q_e; q_N)\boldsymbol{\mu}\phi_m(q_e; q_N) \qquad (5.88)$$

Note that if the Born–Oppenheimer approximation is accurate, the ϕ's must be slowly varying functions of q_N, and one could argue that this would imply that $\mathbf{M}_{km}$ is also slowly varying (admittedly, with less rigor). If so, then $\mathbf{M}_{km}$ can be expanded about some q_N which we call q_N°.

Preferably, q_N° is the point of maximum nuclear overlap, but often the equilibrium position in the mth or kth state is used. In either case we have

$$\mathbf{M}_{km}(q_N) = \mathbf{M}_{km}(q_N^\circ) + \frac{\partial \mathbf{M}_{km}}{\partial q_N}\bigg|_{q_N^\circ}(q_N - q_N^\circ) + \cdots \qquad (5.89)$$

Keeping only the constant term, we find that

$$\boldsymbol{\mu}_{k\nu_k, m\nu_m} = \mathbf{M}_{km}(q_N^\circ) \int \chi_{\nu_k}^k \chi_{\nu_m}^m \, dq_N \qquad (5.90)$$

The nuclear overlap integral is called the *Franck–Condon overlap*. This provides an easily evaluated expression for determining the relative final-state probabilities (i.e., the intensity as a function of ν_m). The electronic matrix element is needed to get absolute probabilities, but often only the relative ones are needed. Note that in the evaluation of the Franck–Condon overlap integral, symmetry may not be useful since the symmetries of the molecule in electronic states m and k need not be the same. If the symmetry groups are the same, the states ν_k and ν_m must belong to the same irreducible representation if the integral is to be nonzero.

For diatomics, the dependence of the overlap on ν_k can easily be determined graphically using the electronic potential energy curves. Figure 5.3 shows a typical situation, in which the excited-state equilibrium geometry is located at larger internuclear distances than the ground

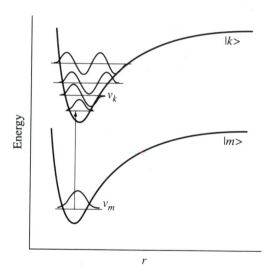

Figure 5.3. Electronic potential energy curves for two states in a diatomic molecule.

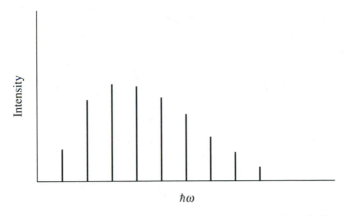

Figure 5.4. Vibrational bands determined by overlap of vibrational wavefunctions for the potential curves in Fig. 5.3.

state. For moderately displaced minima, evaluation of the Franck–Condon overlap leads to the vibrational distribution depicted in Fig. 5.4. Of course, each line here should really be a band, since there will be many rotational transitions of similar energy for a given vibrational transition. Often, there is appreciable overlap between an initial bound state and a final dissociative continuum state, in which case a portion of the band spectrum will be continuous. Figure 5.5 shows the net result of adding rotational lines and a vibrational continuum to the spectrum.

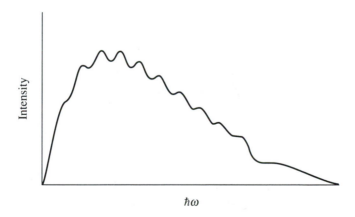

Figure 5.5. Portion of an electronic spectrum, showing smoothed-out vibrational bands and a dissociative continuum.

The presence of centrifugal or other barriers in the excited state can lead to *predissociation*, as illustrated in Fig. 5.6, wherein there is a significant probability of transition to a quasibound vibrational state. This results in broadened lines. Note that the transition probability is proportional to the square of the overlap of initial and final states. It is possible that a specific transition may have a nearly zero overlap, even though it is symmetry allowed. In such cases we say that the transition is *Franck–Condon forbidden*. This typically occurs when the two potential curves have very different equilibrium geometries. Indeed, the Franck–Condon overlap integral typically constrains transitions to be *vertical* (i.e., no change in nuclear coordinate), as one might expect given that the nuclei have little time to move during the time scale of the electronic transition.

If the equilibrium dipole matrix element $\mathbf{M}_{km}(q_N^\circ)$ is zero, often the $\partial \mathbf{M}_{km}/\partial q_N$ term in the expansion of $\mathbf{M}_{km}(q)$ is nonzero. This term (the first non-Condon term) is

$$\boldsymbol{\mu}_{k\nu_k,m\nu_m} = \frac{\partial \mathbf{M}_{km}}{\partial q_N} \Big|_{q_N^\circ} \int dq_N \, \chi_{\nu_k}^k (q_N - q_N^\circ) \chi_{\nu_m}^m \tag{5.91}$$

Note that if the equilibrium positions and frequencies of the m and k states are identical and the vibrational states are harmonic oscillator eigenfunctions, this matrix element requires a change in the vibrational quantum number to be nonzero. This gives rise to *vibronic transitions*, wherein a change in both the electronic and nuclear states occurs. Equation (5.90) can also cause vibronic transitions, but it cannot cause changes in overall vibrational state symmetry, whereas Eq. (5.91) can.

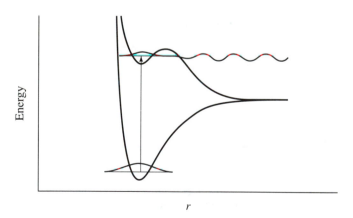

Figure 5.6. Potential curves and wavefunctions associated with a predissociative final state.

5.5

Light Scattering

5.5.1 Qualitative Description of Light Scattering

We now turn our attention to the scattering of electromagnetic radiation by atoms and molecules. First consider the phenomenology of the light-scattering process. If an electric field $\mathbf{E}$ interacts with any atom or molecule, it induces a dipole moment $\boldsymbol{\mu}^{ind}$ in it which is related to $\mathbf{E}$ to a first approximation via

$$\boldsymbol{\mu}^{ind} = \alpha\mathbf{E} \tag{5.92}$$

where α is the *polarizability tensor* (a 3×3 matrix). For very large fields, there will also be terms proportional to E^2 that will be important, but we will defer discussion of these to Problem 4 and Chapter 11.

If $\mathbf{E}$ is an oscillatory function of t (such as in an electromagnetic field), then so will be $\boldsymbol{\mu}^{ind}$. Classically, oscillating dipoles act as antennas and radiate. This radiated light is what we think of as scattered light.

Now α in general depends on the molecular coordinates q_N [i.e., $\alpha = \alpha(q_N)$], and to the approximation that the displacements from equilibrium q_N° are small, it can be expanded:

$$\alpha(q_N) = \alpha(q_N^\circ) + \frac{\partial \alpha}{\partial q_N}(q_N - q_N^\circ) + \cdots \tag{5.93}$$

The first term is the rigid molecule polarizability, which does not depend on time. Upon multiplying by $\mathbf{E}(t)$, we get a $\boldsymbol{\mu}$ that oscillates at the same frequency as $\mathbf{E}$. The radiation emitted by $\boldsymbol{\mu}$ will thus be at the same frequency ω. This is called *Rayleigh scattering*. The $\partial \alpha/\partial q_N \times (q_N - q_N^\circ)$ term classically oscillates at a vibrational frequency ω_v, so when it multiplies $\mathbf{E}(t)$, we get a $\boldsymbol{\mu}$ that oscillates with a mixture of the two frequencies $\omega \pm \omega_v$. The scattered light at frequency $\omega - \omega_v$ is called *Stokes–Raman scattering*, and that at $\omega + \omega_v$ is called *anti-Stokes–Raman scattering*.

5.5.2 Simplification of Electric Dipole Interaction

To get a time-dependent quantum description of light scattering we need to be able to write down an interaction potential between $\boldsymbol{\mu}$ and the electromagnetic field $\mathbf{E}$. To simplify this analysis, we now show that for any dipole $\boldsymbol{\mu}$, in the long-wavelength approximation the interaction is of the form $-\boldsymbol{\mu} \cdot \mathbf{E}$. We demonstrate this by showing that it works for absorption and emission. In the long-wavelength limit, we have, from Eq. (5.9),

$$\mathbf{E} = \frac{-\omega A_0}{c} \boldsymbol{\varepsilon} \sin(\mathbf{k} \cdot \mathbf{r} - \omega t) \approx \frac{\omega A_0}{c} \boldsymbol{\varepsilon} \sin \omega t \qquad (5.94)$$

So if

$$V(t) = -\boldsymbol{\mu} \cdot \mathbf{E} = -\boldsymbol{\mu} \cdot \boldsymbol{\varepsilon} \frac{\omega A_0}{c} \sin \omega t = -\boldsymbol{\mu} \cdot \boldsymbol{\varepsilon} \frac{\omega A_0}{2ic} (e^{i\omega t} - e^{-i\omega t})$$

$$= U e^{-i\omega t} + U^* e^{+i\omega t} \qquad (5.95)$$

where

$$U = \frac{(\boldsymbol{\mu} \cdot \boldsymbol{\varepsilon}) \omega A_0}{2ic} \qquad (5.96)$$

the k, m matrix element of U is

$$U_{km} = \frac{\omega A_0}{2ic} \boldsymbol{\varepsilon} \cdot \boldsymbol{\mu}_{km} \qquad (5.97)$$

which is identical to the U_{km} obtained from the $(e/mc) \mathbf{A} \cdot \mathbf{p}$ interaction [Eq. (5.43)]. Thus as a shortcut to the dipole approximation we can use a $-\boldsymbol{\mu} \cdot \mathbf{E}$ interaction rather than the more laborious $(e/mc) \mathbf{A} \cdot \mathbf{p}$.

5.5.3 Interaction between Field and Induced Dipole; Two-Photon Process

Now we consider the interaction potential between two fields $\mathbf{E}_\omega(t)$ and $\mathbf{E}_{\omega'}(t)$ due to their mutual interaction with a molecule. This will lead to a description of several two-photon processes, of which scattering is one example. Suppose that these two fields have frequencies ω and ω', and in the long-wavelength approximation are given by

$$\mathbf{E}_\omega = \frac{A_0 \omega}{c} \boldsymbol{\varepsilon} \sin \omega t \qquad (5.98a)$$

$$\mathbf{E}_{\omega'} = \frac{A_0' \omega'}{c} \boldsymbol{\varepsilon}' \sin \omega' t \qquad (5.98b)$$

Then if the field $\mathbf{E}_\omega$ interacts with a molecule to induce a dipole $\boldsymbol{\mu}^{\text{ind}} = \alpha \cdot \mathbf{E}_\omega$, the interaction of this dipole with the field $\mathbf{E}_{\omega'}$ is

$$V(t) = -\mathbf{E}_{\omega'}(t) \cdot \boldsymbol{\mu}^{\text{ind}} = -\mathbf{E}_{\omega'}(t) \cdot \alpha \cdot \mathbf{E}_\omega(t) \qquad (5.99)$$

Substituting and multiplying, we get

$$V(t) = \frac{-\omega A_0}{c}\frac{\omega' A_0'}{c}(\boldsymbol{\varepsilon}'\cdot\boldsymbol{\alpha}\cdot\boldsymbol{\varepsilon})\frac{1}{2i}(e^{i\omega t} - e^{-i\omega t}) \times \frac{1}{2i}(e^{i\omega' t} - e^{-i\omega' t})$$

$$= \frac{1}{4}\frac{\omega A_0}{c}\frac{\omega' A_0'}{c}(\boldsymbol{\varepsilon}'\cdot\boldsymbol{\alpha}\cdot\boldsymbol{\varepsilon})(e^{i(\omega+\omega')t} + e^{-i(\omega+\omega')t}$$

$$- e^{i(\omega-\omega')t} - e^{-i(\omega-\omega')t})$$

$$= U(e^{i(\omega+\omega')t} + e^{-i(\omega+\omega')t} - e^{i(\omega-\omega')t} - e^{-i(\omega-\omega')t}) \qquad (5.100)$$

where

$$U = \frac{1}{4}\frac{\omega A_0}{c}\frac{\omega' A_0'}{c}(\boldsymbol{\varepsilon}'\cdot\boldsymbol{\alpha}\cdot\boldsymbol{\varepsilon}) \qquad (5.101)$$

The transitions that are induced by the time-dependent interaction are as follows:

1. $e^{-i(\omega+\omega')t}$ causes absorption of one photon of frequency ω and one of frequency ω'. This is two-photon absorption, a coherent process whose rate is not the same as the rate of sequential absorption of two photons by a molecule.

   Notice that both orderings depicted in Fig. 5.7 of the two photons are possible. Generally, both orderings will contribute to the absorption rate, although occasionally (say, when a real intermediate state at one of the photon energies exists) the amplitude for one process can dominate over the other. This will become more apparent after an expression for α is developed.
2. $e^{i(\omega+\omega')t}$ causes two-photon emission, as in Fig. 5.8.
3. $e^{i(\omega-\omega')t}$ causes the emission of a photon of frequency ω and absorption of one at frequency ω'. Suppose that $\omega' > \omega$. Then the two possible transition schemes are as depicted in Fig. 5.9. For either scheme, the net effect is called *Stokes–Raman scattering*. If $|m\rangle$ is the ground

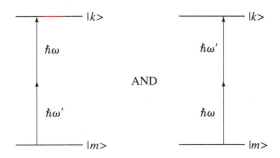

Figure 5.7. Two-photon absorption.

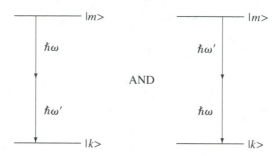

Figure 5.8. Two-photon emission.

state of a molecule, typically the first process pictured in Fig. 5.9 contributes predominantly to the Raman amplitude.

4. $e^{-i(\omega-\omega')t}$ causes emission of $\hbar\omega'$ and absorption of $\hbar\omega$. For $\omega' > \omega$, we get the transitions in Fig. 5.10. This corresponds to *anti-Stokes–Raman scattering*.

We see then that the types of processes caused by two fields interacting with an atom or molecule are all two-photon processes. For three fields, we would get three-photon processes, and so on. We also would get three-photon processes from two fields provided that one couples nonlinearly to the molecule [i.e., adding terms dependent on E^2, E^3, and so on, to Eq. (5.92)].

5.5.4 Raman Scattering

Now let's calculate the rate of Stokes–Raman scattering. The appropriate interaction term is

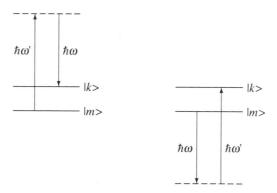

Figure 5.9. Stokes–Raman scattering.

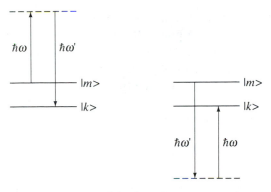

Figure 5.10. Anti-Stokes–Raman scattering.

$$V(t) = U e^{i(\omega - \omega')t} \tag{5.102}$$

and since this is in a form to which Fermi's rule is immediately applicable, we get

$$w_{km} = \frac{2\pi}{\hbar} \rho |U_{km}|^2 = \frac{2\pi}{\hbar} \rho \frac{1}{16} \frac{\omega^2 A_0^2}{c^2} \frac{\omega'^2 A_0'^2}{c^2} |\boldsymbol{\varepsilon}' \cdot \boldsymbol{\alpha}_{km} \cdot \boldsymbol{\varepsilon}|^2 \tag{5.103}$$

Substituting

$$A_0 = 2c \left(\frac{2\pi\hbar N}{\omega V} \right)^{1/2} \tag{5.14}$$

we get

$$w_{km} = \frac{2\pi}{\hbar} \rho \frac{2\pi\hbar N}{V} \omega \frac{2\pi\hbar N'}{V} \omega' |\boldsymbol{\varepsilon}' \cdot \boldsymbol{\alpha}_{km} \cdot \boldsymbol{\varepsilon}|^2 \tag{5.104}$$

At this point we can correct for spontaneous emission, much as we did previously by replacing N by $N + 1$ for the emitted field E_ω. This gives

$$w_{km} = \frac{2\pi}{\hbar} \rho \omega \omega' \left(\frac{2\pi\hbar}{V} \right)^2 N'(N + 1) |\boldsymbol{\varepsilon}' \cdot \boldsymbol{\alpha}_{km} \cdot \boldsymbol{\varepsilon}|^2 \tag{5.105}$$

It is customary to express Raman intensities in terms of *cross sections*. The differential cross section $d\sigma$ is defined as

$$d\sigma = \frac{\text{rate of scattering into solid angle } d\Omega}{\text{incident photon flux in beam being absorbed}} \tag{5.106}$$

The rate of scattering into $d\Omega$ is just w_{km} evaluated using the emitted field density of states:

$$\rho = \frac{V}{(2\pi c)^3} \frac{\omega^2}{\hbar} \, d\Omega \tag{5.107}$$

The incident photon flux is simply the number of photons per unit time per unit area that cross a given point. If our volume V consists of a photon beam of area A and length L, then in a time $t = L/c$ the number of photons crossing is simply the number of photons in V. Recalling that the number in the absorbing beam is N', we see that

$$\text{number per unit area per unit time} = \frac{N'}{A(L/c)} = \frac{N'c}{V} \quad (5.108)$$

Thus

$$d\sigma = \frac{2\pi}{\hbar} \frac{V}{(2\pi c)^3} \frac{\omega^2}{\hbar} \, d\Omega \, \omega\omega' \left(\frac{2\pi\hbar}{V}\right)^2 \frac{N'(N+1)|\boldsymbol{\varepsilon}' \cdot \boldsymbol{\alpha}_{km} \cdot \boldsymbol{\varepsilon}|^2}{N'c/V}$$

$$= \frac{\omega^3\omega'(N+1)}{c^4} |\boldsymbol{\varepsilon}' \cdot \boldsymbol{\alpha}_{km} \cdot \boldsymbol{\varepsilon}|^2 \, d\Omega \quad (5.109)$$

so that the differential cross section is

$$\frac{d\sigma}{d\Omega} = (N+1) \frac{\omega^3\omega'}{c^4} |\boldsymbol{\varepsilon}' \cdot \boldsymbol{\alpha}_{km} \cdot \boldsymbol{\varepsilon}|^2 \quad (5.110)$$

For *Rayleigh scattering*, $\omega' = \omega$ and

$$\frac{d\sigma}{d\Omega} = (N+1) \frac{\omega^4}{c^4} |\boldsymbol{\varepsilon}' \cdot \boldsymbol{\alpha}_{km} \cdot \boldsymbol{\varepsilon}|^2 \quad (5.111)$$

Note that if α_{km} has weak frequency dependence (which is often the case), then $d\sigma/d\Omega$ is proportional to ω^4, which explains why the sky is blue (blue light scatters more effectively than red). Also, if $\boldsymbol{\varepsilon}' = \boldsymbol{\varepsilon}$, we call the scattering *polarized*, whereas if $\boldsymbol{\varepsilon}' \perp \boldsymbol{\varepsilon}$, it is *depolarized*.

5.5.5 Evaluation of α_{km}: Kramers–Heisenberg Formula

Previously, we had defined α via $\boldsymbol{\mu}^{ind} = \alpha \cdot \mathbf{E}$. That is, α is the proportionality constant between the applied field $\mathbf{E}$ and the induced dipole moment $\boldsymbol{\mu}^{ind}$. $\boldsymbol{\mu}^{ind}$ is calculated by determining the expectation value of the dipole operator between molecular wavefunctions which have been perturbed by the applied field E. Thus

$$\boldsymbol{\mu}_{km}(t) = \langle \psi_k(t)|\boldsymbol{\mu}|\psi_m(t)\rangle \quad (5.112)$$

but it should be apparent that $\boldsymbol{\mu}_{km}(t)$ will include contributions from both the permanent dipole moment and the induced one. In fact, if there is no applied field, then $\psi_m(t) = \phi_m e^{-iE_m t/\hbar}$, and we get

$$\boldsymbol{\mu}_{km}(t) = e^{i\omega_{km}t}\boldsymbol{\mu}_{km} \quad (5.113)$$

where $\boldsymbol{\mu}_{km}$ is the usual dipole matrix element we use in Fermi's golden rule. For a general $\psi_m(t)$, we expect to get

$$\boldsymbol{\mu}_{km}(t) = e^{i\omega_{km}t}(\boldsymbol{\mu}_{km}^{\text{perm}} + \boldsymbol{\mu}_{km}^{\text{ind}})$$

where $\boldsymbol{\mu}_{km}^{\text{perm}}$ is the usual dipole matrix element and $\boldsymbol{\mu}_{km}^{\text{ind}}$ is the induced one, from which we can get α_{km}. To get $\psi_m(t)$ and $\psi_k(t)$ we use first-order perturbation theory:

$$\psi_j(t) = \sum_n c_n^j(t)e^{-iE_nt/\hbar}\phi_n \qquad j = k,\, m \qquad (5.114)$$

with

$$c_n^j = \delta_{nj} - \frac{i}{\hbar}\int_0^t dt'\, e^{i\omega_{nj}t'} V_{nj}(t') \qquad (5.115)$$

For the interaction of a single field with matter, we have previously seen that

$$V(t) = Ue^{-i\omega t} + U^*e^{i\omega t} \qquad (5.96)$$

where

$$U_{km} = \frac{\omega A_0}{2ic}\,\boldsymbol{\varepsilon}\cdot\boldsymbol{\mu}_{km} \qquad (5.97)$$

in the long-wavelength approximation. Then for $j = k,\, m$, we have

$$\begin{aligned}
c_n^j(t) &= \delta_{nj} - \frac{i}{\hbar}\int_0^t dt'(U_{nj}e^{-i\omega t'} + U_{nj}^*e^{i\omega t'})e^{i\omega_{nj}t'} \\
&= \delta_{nj} - \frac{i}{\hbar}\left[U_{nj}\frac{e^{-i(\omega-\omega_{nj})t} - 1}{-i(\omega - \omega_{nj})} + U_{nj}^*\frac{e^{+i(\omega+\omega_{nj})t} - 1}{i(\omega + \omega_{nj})}\right] \qquad (5.116)
\end{aligned}$$

Substituting this into the expressions for ψ_m and ψ_k, we get

$$\begin{aligned}
\psi_m(t) &= e^{-iE_mt/\hbar}\phi_m - \frac{i}{\hbar}\sum_n e^{-iE_nt/\hbar}\phi_n\left[\frac{U_{nm}(e^{-i(\omega-\omega_{nm})t} - 1)}{-i(\omega - \omega_{nm})}\right.\\
&\qquad\left. + U_{nm}^*\frac{(e^{i(\omega+\omega_{nm})t} - 1)}{i(\omega + \omega_{nm})}\right]\\
&= e^{-iE_mt/\hbar}\left\{\phi_m + \frac{1}{\hbar}\sum_n\phi_n\left[\frac{U_{nm}(e^{-i\omega t} - e^{i\omega_{mn}t})}{\omega - \omega_{nm}}\right.\right.\\
&\qquad\left.\left. - \frac{U_{nm}^*(e^{i\omega t} - e^{i\omega_{mn}t})}{\omega + \omega_{nm}}\right]\right\} \qquad (5.117)
\end{aligned}$$

Now at this point we notice that part of the perturbation-induced change in the wavefunction oscillates at frequency ω and part at frequency ω_{mn}. Only the first part is of interest here since that will be in phase with $\mathbf{E}(t)$.

The second part will be out of phase and cannot effectively cause transitions. Neglecting it is sometimes called the *rotating wave approximation*.

This leads us to

$$\psi_m = e^{-iE_m t/\hbar} \left\{ \phi_m + \frac{1}{\hbar} \sum_n \phi_n \left[\frac{U_{nm} e^{-i\omega t}}{\omega - \omega_{nm}} - \frac{U_{nm}^* e^{i\omega t}}{\omega + \omega_{nm}} \right] \right\} \quad (5.118a)$$

$$\psi_k = e^{-iE_k t/\hbar} \left\{ \phi_k + \frac{1}{\hbar} \sum_n \phi_n \left[\frac{U_{nke} e^{-i\omega t}}{\omega - \omega_{nk}} - \frac{U_{nk}^* e^{i\omega t}}{\omega + \omega_{nk}} \right] \right\} \quad (5.118b)$$

Now let's evaluate $\boldsymbol{\mu}_{km}(t)$, to first order in U:

$$\boldsymbol{\mu}_{km}(t) = e^{i\omega_{km} t} \left\{ \boldsymbol{\mu}_{km}^{\text{perm}} + \frac{1}{\hbar} \sum_n \langle k | \boldsymbol{\mu} | n \rangle \left[\frac{U_{nm} e^{-i\omega t}}{\omega - \omega_{nm}} - \frac{U_{nm}^* e^{i\omega t}}{\omega + \omega_{nm}} \right] \right.$$

$$\left. + \frac{1}{\hbar} \sum_n \langle n | \boldsymbol{\mu} | m \rangle \left[\frac{U_{nk}^* e^{i\omega t}}{\omega - \omega_{nk}} - \frac{U_{nke} e^{-i\omega t}}{\omega + \omega_{nk}} \right] \right\}$$

$$= e^{i\omega_{km} t} \left\{ \boldsymbol{\mu}_{km}^{\text{perm}} - \frac{e^{i\omega t}}{\hbar} \sum_n \left[\frac{\boldsymbol{\mu}_{kn} U_{nm}^*}{\omega + \omega_{nm}} - \frac{\boldsymbol{\mu}_{nm} U_{nk}^*}{\omega - \omega_{nk}} \right] \right.$$

$$\left. + \frac{e^{-i\omega t}}{\hbar} \sum_n \left[\frac{\boldsymbol{\mu}_{kn} U_{nm}}{\omega - \omega_{nm}} - \frac{\boldsymbol{\mu}_{nm} U_{nk}}{\omega + \omega_{nk}} \right] \right\}$$

$$= e^{i\omega_{km} t} \{ \boldsymbol{\mu}_{km}^{\text{perm}} + \boldsymbol{\mu}_{km}^{\text{ind}} \} \quad (5.119)$$

where

$$\boldsymbol{\mu}_{km}^{\text{ind}} = \frac{\omega A_0}{2ic} \frac{1}{\hbar} \left\{ e^{i\omega t} \sum_n \left[\frac{\boldsymbol{\mu}_{kn} \boldsymbol{\mu}_{nm}^*}{\omega + \omega_{nm}} - \frac{\boldsymbol{\mu}_{nm} \boldsymbol{\mu}_{nk}^*}{\omega - \omega_{nk}} \right] \right.$$

$$\left. + e^{-i\omega t} \sum_n \left[\frac{\boldsymbol{\mu}_{kn} \boldsymbol{\mu}_{nm}}{\omega - \omega_{nm}} - \frac{\boldsymbol{\mu}_{nm} \boldsymbol{\mu}_{nk}}{\omega + \omega_{nk}} \right] \right\} \cdot \boldsymbol{\varepsilon} \quad (5.120)$$

Note that the dot product in Eq. (5.120) involves the second dipole matrix element in each term.

Now we want to reexpress this in the form $\boldsymbol{\mu}_{km}^{\text{ind}} = \alpha_{km} \cdot \mathbf{E}$, where

$$\mathbf{E} = \frac{\omega A_0}{c} \boldsymbol{\varepsilon} \sin \omega t \quad (5.121)$$

To do this, consider the (i,j)th component of the (k,m)th matrix element of the first set of square brackets in Eq. (5.120). Taking μ_{km} to be hermitian, and defining

$$(\alpha_{km})_{ij} = \frac{1}{\hbar} \sum_n \left\{ \frac{\langle k | \mu_i | n \rangle \langle m | \mu_j | n \rangle}{\omega + \omega_{nm}} - \frac{\langle n | \mu_i | m \rangle \langle k | \mu_j | n \rangle}{\omega - \omega_{nk}} \right\} \quad (5.122)$$

Eq. (5.120) becomes

$$\boldsymbol{\mu}_{km}^{\text{ind}} = \frac{\omega A_0}{2ic} \{ e^{i\omega t} \alpha_{km} \cdot \boldsymbol{\varepsilon} - e^{-i\omega t} \alpha_{mk}^* \cdot \boldsymbol{\varepsilon} \} \quad (5.123)$$

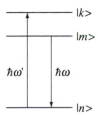

Figure 5.11. Transitions associated with $\omega = -\omega_{nm}$.

Since α is hermitian, $\alpha^*_{mk} = \alpha_{km}$, so we can write

$$\mu^{ind}_{km} = \alpha_{km} \cdot \frac{\omega A_0}{c} \, \varepsilon \sin \omega t = \alpha_{km} \cdot \mathbf{E} \qquad (5.124)$$

Thus Eq. (5.122) is the famous Kramers–Heisenberg dispersion relation.

Note that it consists of two terms, one of which would be resonant if $\omega = -\omega_{nm}$ and the other if $\omega = \omega_{nk}$. The first term corresponds to an intermediate state with energy $E_n = E_m - \hbar\omega$ and comes from the diagram depicted in Fig. 5.11. The second involves a state with energy $E_n = E_k + \hbar\omega$ and involves the diagram in Fig. 5.12. Thus we see that there are two contributions to the Raman amplitude, corresponding to the two possible orderings of the photon absorption and emission.

The *Rayleigh intensity* corresponds to α_{mm}, and from the expression above we find that

$$\alpha_{mm} = \frac{1}{\hbar} \sum_n \mu_{mn} \mu_{nm} \frac{2\omega_{nm}}{\omega^2_{nm} - \omega^2} \qquad (5.125)$$

Note that because of the infinite sum in Eqs. (5.122) and (5.125), the expressions for α_{mm} and α_{km} can be inconvenient in actual calculations of scattered intensities. One alternative that is useful for zero-frequency polarizabilities is to do a *finite field* calculation wherein α is obtained by evaluating the following expression (numerically evaluated by finite difference):

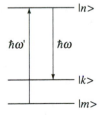

Figure 5.12. Transitions associated with $\omega = \omega_{nk}$.

$$(\alpha_{mm})_{ij} = \frac{\partial^2 \langle H \rangle}{\partial E_i \, \partial E_j} \tag{5.126}$$

where $\langle H \rangle$ is the expectation value of the molecular hamiltonian, including the molecule–field interaction.

BIBLIOGRAPHY FOR CHAPTER 5

Atkins, P. W., *Molecular Quantum Mechanics*, 2nd Ed. (Oxford University Press, New York, 1983).

Baym, G., *Lectures on Quantum Mechanics* (Benjamin/Cummings, London, 1981).

Davydov, A. S., *Quantum Mechanics* (Pergamon Press, Oxford, 1976).

Eyring, H., L. Walter, and J. Kimball, *Quantum Chemistry* (Wiley, New York, 1944).

Flygare, W. H. *Molecular Structure and Dynamics* (Prentice Hall, Englewood Cliffs, N.J., 1978).

Schiff, L. I., *Quantum Mechanics* (McGraw-Hill, New York, 1988).

PROBLEMS FOR CHAPTER 5

1. What is the radiative lifetime of the hydrogen atom in the $2p_z$ state? The answer is roughly 10^{-9} s.

2. (a) What is the density of states of a free particle (e.g., an electron)? The derivation works the same way as for photons except that the relation between energy and wave vector is $E = \hbar^2 k^2 / 2m$.

(b) What is the density of states of a harmonic oscillator?

3. What is the total cross section σ for photoionization of a ground-state hydrogen atom using 20-eV photons? To solve this problem, first derive an expression for the differential cross section $d\sigma/d\Omega$ for absorption of light by a ground-state hydrogen atom. The differential cross section is simply the rate of absorption divided by the flux of incident photons. This flux is just $N\rho c / V$ where ρ is the density of photon states. In determining the rate expression, the relevant state density should be taken to be the product of photon and electron state densities. (For the latter, see Problem 2.) After ionization, the electron is a free particle, with wavefunction $\psi = \exp(i\mathbf{k}' \cdot \mathbf{r})/V^{1/2}$, where $\mathbf{k}'$ is the final electron wavevector. k' is related to the electron translational energy via $E' = \hbar^2 k'^2/2m$, where E' is related to the initial energy (20 eV) by energy conservation. What is E' for the $\hbar\omega = 20$ eV? [*Hint:* After you have gotten to the point of writing down an expression for the differential cross section, there is a nasty integral to work out. The following formula will help:

$$\int_0^\infty r^2\,dr \int_0^{2\pi} d\phi \int_0^\pi \sin\theta\,d\theta\, e^{-i\mathbf{k}'\cdot\mathbf{r}} r\cos\theta\, e^{-r/a_0} = \frac{32\pi k'}{a_0 i(a_0^{-2} + k'^2)^3}$$

Once you have $d\sigma/d\Omega$, you can determine σ by straightforward integration over scattering angles.] Evaluate σ at $\hbar\omega = 20$ eV, and show how σ varies with E'.

4. When high-intensity radiation interacts with matter, the induced dipole μ^{ind} is related to the field E_ω by an expression

$$\mu^{\text{ind}} = \alpha E_\omega + \beta E_\omega^2 + \gamma E_\omega^3 + \cdots$$

which includes terms that depend nonlinearly on E_ω. The first nonlinear term is proportional to the hyperpolarizability β. Allowing the induced dipole moment from this term to interact with another radiation field $E_{\omega'}$ leads to the interaction

$$V_{\text{int}} = -E_{\omega'}\cdot\beta\cdot E_\omega^2$$

(a) Using the semiclassical theory of radiation and the long-wavelength approximation, find the term in V_{int} that causes *Stokes–hyper-Raman scattering* (i.e., absorption of two ω photons, emission of one ω' photon with $\omega' < 2\omega$).

(b) What other transitions does V_{int} cause?

(c) Develop an expression for the rate of hyper-Raman scattering. How does $d\sigma/d\Omega$ depend on ω and ω' if β is frequency independent?

5. Consider that an electron is harmonically bound to a molecule. This means that the electronic hamiltonian is

$$H_0 = \frac{p^2}{2m} + \frac{1}{2} m\omega_0^2 x^2$$

where x is the electronic coordinate and ω_0 is the frequency of oscillation of the electron. The electronic wavefunctions in this case are simply harmonic oscillator states, and the energy levels are $E_n = \hbar\omega_0(n + \frac{1}{2})$, where n is the electronic quantum number.

Show that the Rayleigh polarizability for this system is

$$\alpha = \frac{e^2/m}{\omega_0^2 - \omega^2}$$

6. Consider the $(e^2/2mc^2)\,A^2$ term in the interaction of radiation and matter. In the long-wavelength approximation and for the vector potential $\mathbf{A}$ defined in the text, what types of spectroscopic processes can this term induce (i.e., one- or two-photon emission, Raman scattering, etc.)?

7. For a molecule with D_{3h} symmetry in both the initial and final electronic states, if the initial state belongs to the A_1' irreducible representation and the final to A_2', to what irreducible representations must the vibrational coordinate q_N belong to make the non-Condon term nonzero?

8. (a) Apply extended Hückel theory to the linear H_3 molecule ($D_{\infty h}$ symmetry). Assume that the Hückel rules apply in determining hamiltonian matrix ele-

ments and overlaps. Determine energy levels and symmetries of the lowest three states.

(b) What is the lowest-energy transition that is electric dipole allowed? Electric quadrupole allowed? Magnetic dipole allowed?

9. Using the identity

$$[[x, H], x] = 2xHx - Hx^2 - x^2H$$

and assuming that the potential is independent of momentum, prove that

$$\frac{2m}{\hbar^2} \sum_k (E_k - E_s)|x_{sk}|^2 = 1$$

Here $x_{sk} = \langle s|x|k \rangle$. Similar expressions hold for the y- and z-coordinate matrix elements. Then replacing x by the dipole operator for N electrons, $\boldsymbol{\mu} = e\Sigma_i \mathbf{r}_i$, and substituting the definition of oscillator strengths, prove the Thomas–Reiche–Kuhn sum rule,

$$\sum_k f_{ks} = N$$

Note that the result is independent of s—what does that mean?

6

Occupation Number Representations

Introduction

In Chapter 2 we discussed symmetry aspects of molecular electronic structure. One very important concept underlying our discussion there was that of a basis set. Although such operators as H, x, p, and so on, are defined as functions of coordinates and derivatives, the discussion of molecular electronic structure with which most chemists are familiar involves discussion not in arbitrary spatial terms, but in basis sets. Thus organic chemists talk familiarly of $2p\pi$ orbitals on carbon atoms in aromatic molecules, silicon chemists discuss $p\pi$–$d\pi$ bonding, transition metal chemists invoke $d\pi$–$p\pi$ back donation, and materials scientists are interested in electron trapping in narrowband semiconductors. It is therefore convenient to develop a theoretical description of electronic structure that can directly describe the electronic wavefunctions and states in terms of basis sets. A convenient language for doing this involves the use of the occupation-number representation, sometimes called second quantization. This method exhibits some very convenient properties:

1. It greatly simplifies notation for observables and wavefunctions, as well as certain manipulations.

2. It automatically handles antisymmetry (Pauli principle) requirements in the treatment of electronic structure problems.
3. It conveniently describes processes such as photoionization or electron capture, in which the actual number of electrons changes.
4. It can be used to describe boson systems such as photons or vibrational quanta as well as fermion (electron) systems.
5. It can easily be used in the study of time-dependent problems, especially in the Heisenberg representation.
6. Most important, it facilitates visualizing and thinking about the behavior of vibrations, phonons, and electrons in solids and in molecules.

All of these advantages come from one simple idea: We replace abstract operators with their basis-set representations. These, in turn, are described in terms of occupation numbers (filled, partly full, or empty orbitals). Before discussing occupation number methods for electrons (fermions), it is useful to consider the simpler situation of vibrational normal modes and quantized radiation fields that use occupation number representations appropriate for bosons.

6.2

Occupation Number Representations for Harmonic Molecular Vibrations and Quantized Radiation Fields

6.2.1 Single Harmonic Oscillator

The occupation number representation was developed by Dirac to describe the harmonic oscillator. His presentation is reviewed in Section 1.4.1, and a key feature for the present discussion is that it involves defining raising and lowering operators b^+ and b (also known as creation and annihilation operators) such that for an eigenstate $|n\rangle$, $b^+|n\rangle$ is proportional to $|n + 1\rangle$ while $b|n\rangle$ is proportional to $|n - 1\rangle$. In addition, the operator b^+b defines what we call a *number operator,* whose eigenvalues specify the number of quanta in each eigenstate (i.e., $b^+b|n\rangle = n|n\rangle$).

The utility of using raising and lowering operators to solve harmonic oscillator problems may be illustrated by considering the mixing of states that arises from the cubic anharmonic coupling term

$$V_{anh} = \frac{1}{6} \frac{d^3V}{dx^3}\Big|_0 x^3 \tag{6.1}$$

By solving Eqs (1.37a,b) for x in terms of b and b^+, that is,

$$x = \left(\frac{\hbar}{2\mu\omega}\right)^{1/2} (b + b^+) \tag{6.2}$$

one can rewrite (6.1) as

$$V_{\text{anh}} = \frac{1}{6}\frac{d^3V}{dx^3}\left(\frac{\hbar}{2\mu\omega}\right)^{3/2} (b + b^+)^3 \tag{6.3}$$

The evaluation of matrix elements of V_{anh} can now be readily accomplished using the relations (1.45a,b), leading to

$$
\langle m|V_{\text{anh}}|n\rangle = \frac{1}{6}\frac{d^3V}{dx^3}\left(\frac{\hbar}{2\mu\omega}\right)^{3/2} \langle m|(b + b^+)^3|n\rangle = \frac{1}{6}\frac{d^3V}{dx^3}\left(\frac{\hbar}{2\mu\omega}\right)^{3/2}
$$
$$
\left\{ \delta_{m,n+3}[(n + 1)(n + 2)(n + 3)]^{1/2} \right.
$$
$$
+ \ \delta_{m,n+1}3(n + 1)^{3/2}
$$
$$
+ \ \delta_{m,n-1}3n^{3/2}
$$
$$
\left. + \ \delta_{m,n-3}[(n)(n - 1)(n - 2)]^{1/2} \right\} \tag{6.4}
$$

so that this perturbation can add or subtract one or three quanta.

6.2.2 Normal Modes

For polyatomic molecules, or for solids, the vibrational modes can be described approximately in terms of normal coordinates. Within this description, the hamiltonian for the molecule, without translational or rotational terms, is written

$$H = \sum_q \left(\frac{p_q^2}{2\mu_q} + \frac{1}{2} k_q x_q^2\right) \tag{6.5}$$

in perfect analogy to (1.36). The sum runs over $3N - 6$ internal coordinates ($3N - 5$ for linear species) where N is the number of atoms, and anharmonic terms in the potential have been neglected.

The harmonic hamiltonian (6.5) is separable, so that the total energy is a sum of the single-mode energies and the wavefunction is a product of single-mode wavefunctions. Quantization may now be introduced by defining raising and lowering operators b_q^+ and b_q:

$$b_q = \frac{1}{\sqrt{2}}\left\{\left(\frac{\mu_q\omega_q}{\hbar}\right)^{1/2} x_q + \frac{i}{\sqrt{\hbar\omega_q\mu_q}}\,p_q\right\} \tag{6.6a}$$

$$b_q^+ = \frac{1}{\sqrt{2}}\left\{\left(\frac{\mu_q\omega_q}{\hbar}\right)^{1/2} x_q - \frac{i}{\sqrt{\hbar\omega_q\mu_q}}\,p_q\right\} \tag{6.6b}$$

and assuming the commutation relations

$$[b_q, b_{q'}^+] = \delta_{qq'} \tag{6.6c}$$

$$[b_q, b_{q'}] = 0 \tag{6.6d}$$

The hamiltonian (6.5) then becomes

$$H = \sum_q (b_q^+ b_q + \tfrac{1}{2})\hbar\omega_q \tag{6.6e}$$

The commutation relations (6.6c,d) are particularly important: They hold for particles obeying Bose–Einstein statistics (photons or deuterons, as well as vibrations) as long as the states are orthogonal. The eigenstates and energies can be written

$$E(n_1, n_2, n_3, \ldots) = \sum_q (n_q + \tfrac{1}{2})\hbar\omega_q \tag{6.7a}$$

$$
\begin{aligned}
|n_1 n_2 n_3 \cdots\rangle &= \frac{(b_1^+)^{n_1}(b_2^+)^{n_2} \cdots |000\ldots\rangle}{(n_1!\, n_2!\, n_3!\cdots)^{1/2}} \\
&= \prod_q \frac{(b_q^+)^{n_q}}{(n_q!)^{1/2}} |000\ldots\rangle
\end{aligned}
\tag{6.7b}
$$

Just as for the single harmonic oscillator, any operator that can be written as a function of the p_q and x_q variables can equivalently be written in terms of the b_q, b_q^+. For example, intramolecular vibrational energy transfer can be so described. The Taylor expansion of the overall, multimode vibrational potential will include so-called diagonal cubic anharmonicities such as that in (6.1) and off-diagonal anharmonicities, such as

$$V_{12}^{(3)} = \frac{1}{3} \frac{\partial^3 V}{\partial x_1\, \partial x_2^2}\Big|_{r_q = r_q^o} x_1 x_2^2 \tag{6.8a}$$

This can be reexpressed as

$$V_{12}^{(3)} = A_{12}^{(3)}(b_1^+ + b_1)(b_2^+ + b_2)^2 \tag{6.8b}$$

where the constant $A_{12}^{(3)}$ includes the derivative of (6.8a) and the constants from Eqs. (6.6a,b). Then using golden rule arguments, to lowest order the only vibrational energy exchange mediated by $V_{12}^{(3)}$ will involve the two processes $(n_1 n_2 \rightarrow n_1 + 1, n_2 - 2)$, $(n_1 n_2 \rightarrow n_1 - 1, n_2 + 2)$, since the others permitted by (6.8b) cannot conserve vibrational energy; even these terms will give a small contribution unless modes 1 and 2 have a $2:1$ resonance.

6.2.3 Quantized Radiation Fields

An important application of occupation number representations for bosons is the quantization of electromagnetic fields. Up to now we have considered these fields to be classical and described by a vector potential $\mathbf{A}(\mathbf{r}, t)$. This $\mathbf{A}$ is obtained by solving the wave equation

$$\nabla^2 \mathbf{A} = \frac{1}{c^2} \frac{\partial^2 \mathbf{A}}{\partial t^2} \tag{6.9}$$

subject to certain boundary and gauge conditions.

To quantize the field, it is convenient to expand $\mathbf{A}$ in terms of cavity modes. The cavity boundary conditions limit the allowed frequencies to a discrete set, much like what occurs in a violin string. Under these circumstances it is convenient to expand $\mathbf{A}$ in terms of cavity states as follows:

$$\mathbf{A}(\mathbf{r}, t) = \sqrt{\frac{4\pi}{V}}\, c \sum_{l\sigma} q_{l\sigma}(t) \mathbf{u}_{l\sigma}(\mathbf{r}) \tag{6.10}$$

where $\mathbf{u}_{l\sigma}(\mathbf{r})$ is a spatial cavity state for mode l and polarization σ, $q_{l\sigma}$ is the time-dependent amplitude associated with that state, and $(4\pi/V)^{1/2}c$ is a normalization factor that will prove convenient later.

Equations determining both $\mathbf{u}_{l\sigma}$ and $q_{l\sigma}$ can be obtained from the wave equation by substituting $\mathbf{A}$ therein and separating variables. This gives us

$$\nabla^2 \mathbf{u}_{l\sigma} + \frac{\omega_l^2}{c^2} \mathbf{u}_{l\sigma} = 0 \tag{6.11}$$

$$\frac{d^2 q_{l\sigma}}{dt^2} + \omega_l^2 q_{l\sigma} = 0 \tag{6.12}$$

The types of solutions we are interested in are free-wave solutions where

$$\mathbf{u}_{l\sigma}(\mathbf{r}) = \boldsymbol{\varepsilon}_{l\sigma} e^{i\mathbf{k}_l \cdot \mathbf{r}} \tag{6.13}$$

Here $\boldsymbol{\varepsilon}_{l\sigma}$ is the polarization vector, so that $\mathbf{k}_l \cdot \boldsymbol{\varepsilon}_{l\sigma} = 0$. That $\mathbf{u}_{l\sigma}$ satisfies the wave equation is easily verified provided that we set $k_l^2 = \omega_l^2/c^2$.

Normalization of $\mathbf{u}_{l\sigma}$ is given by

$$\int_V \mathbf{u}_{l\sigma}^*(\mathbf{r}) \cdot \mathbf{u}_{l'\sigma'}(\mathbf{r})\, d^3\mathbf{r} = V \delta_{ll'} \delta_{\sigma\sigma'} \tag{6.14}$$

Note also that $\boldsymbol{\varepsilon}_{l\sigma} e^{-i\mathbf{k}_l \cdot \mathbf{r}}$ is also a solution to Eq. (6.11). In the present treatment we always write $e^{i\mathbf{k}_l \cdot \mathbf{r}}$ and include both positive and negative $\mathbf{k}_l$'s by expanding the sum to consider $+l$ and $-l$ (letting $\mathbf{k}_l = -\mathbf{k}_{-l}$).

Now consider the classical hamiltonian [Eq. (5.12)]:

$$H = \frac{1}{8\pi} \int (|\mathbf{E}|^2 + |\mathbf{B}|^2) \, d^3\mathbf{r} \tag{6.15}$$

Substituting Eqs. (5.1) and (5.2), this becomes

$$H = \frac{1}{8\pi} \int_V \left[\frac{1}{c^2} \left(\frac{\partial \mathbf{A}}{\partial t} \right)^2 + (\nabla \times \mathbf{A})^2 \right] d^3\mathbf{r} \tag{6.16}$$

and upon substituting Eq. (6.10) we get

$$\begin{aligned}
H = \frac{1}{2V} \sum_{l\sigma, l'\sigma'} \dot{q}_{l\sigma} \dot{q}_{l'\sigma'} \int_V \mathbf{u}_{l\sigma}^* \cdot \mathbf{u}_{l'\sigma'} \, d^3\mathbf{r} \\
+ \frac{c^2}{2V} \sum_{l\sigma, l'\sigma'} q_{l\sigma} q_{l'\sigma'} \int (\nabla \times \mathbf{u}_{l\sigma})^* \cdot (\nabla \times \mathbf{u}_{l'\sigma'}) \, d^3\mathbf{r}
\end{aligned} \tag{6.17}$$

Now invoke orthonormality in the first integral. In the second, use the fact that if $\mathbf{u}_{l\sigma} = \boldsymbol{\varepsilon}_{l\sigma} e^{i\mathbf{k}_l \cdot \mathbf{r}}$, then $\nabla \times \mathbf{u}_{l\sigma} = i\mathbf{k}_l \times \mathbf{u}_{l\sigma}$, and

$$(\nabla \times \mathbf{u}_{l\sigma})^* \cdot (\nabla \times \mathbf{u}_{l'\sigma'}) = (\mathbf{k}_l \times \mathbf{u}_{l\sigma})^* \cdot (\mathbf{k}_{l'} \times \mathbf{u}_{l'\sigma'}) \tag{6.18}$$

From the identity

$$(\mathbf{A} \times \mathbf{B}) \cdot (\mathbf{C} \times \mathbf{D}) = (\mathbf{A} \cdot \mathbf{C})(\mathbf{B} \cdot \mathbf{D}) - (\mathbf{A} \cdot \mathbf{D})(\mathbf{B} \cdot \mathbf{C}) \tag{6.19}$$

we can convert Eq. (6.18) into

$$(\mathbf{k}_l \times \mathbf{u}_{l\sigma})^* \cdot (\mathbf{k}_{l'} \times \mathbf{u}_{l'\sigma'}) = (\mathbf{k}_l \cdot \mathbf{k}_{l'})(\mathbf{u}_{l\sigma}^* \cdot \mathbf{u}_{l'\sigma'}) - (\mathbf{k}_l \cdot \mathbf{u}_{l'\sigma'})(\mathbf{k}_{l'} \cdot \mathbf{u}_{l\sigma}^*) \tag{6.20}$$

Now integrate over all space. The second term in (6.20) integrates to zero because $\mathbf{k}_l \cdot \mathbf{u}_{l\sigma} = 0$ from the gauge condition. Using the orthonormality conditions for different $\mathbf{u}_{l\sigma}$'s, we get

$$\begin{aligned}
H &= \frac{1}{2} \sum_{l\sigma} \dot{q}_{l\sigma}^2 + \frac{1}{2} c^2 \sum_{l\sigma} q_{l\sigma}^2 k_l^2 \\
&= \frac{1}{2} \sum_{l\sigma} (\dot{q}_{l\sigma}^2 + \omega_l^2 q_{l\sigma}^2)
\end{aligned} \tag{6.21}$$

This is just the sum of a set of uncoupled harmonic oscillator hamiltonians $H_{l\sigma} = \frac{1}{2} (\dot{q}_{l\sigma}^2 + \omega_l^2 q_{l\sigma}^2)$.

The equivalence between the classical electromagnetic and harmonic oscillator hamiltonians can be used to convert to the corresponding quantum hamiltonian very easily. Simply define $b_{l\sigma}$, $b_{l\sigma}^+$ as used previously [Eq. (6.6a) but without mass factors]:

$$b_{l\sigma} = \frac{1}{\sqrt{2\hbar\omega_l}} (\omega_l q_{l\sigma} + i p_{l\sigma}) \tag{6.22a}$$

$$b_{l\sigma}^+ = \frac{1}{\sqrt{2\hbar\omega_l}} (\omega_l q_{l\sigma} - i p_{l\sigma}) \qquad (6.22b)$$

In this case,

$$q_{l\sigma} = \sqrt{\frac{\hbar}{2\omega_l}} (b_{l\sigma}^+ + b_{l\sigma}) \qquad (6.23a)$$

$$p_{l\sigma} = \dot{q}_{l\sigma} = i \sqrt{\frac{\hbar\omega_l}{2}} (b_{l\sigma}^+ - b_{l\sigma}) \qquad (6.23b)$$

Then postulate that the usual boson commutation relations are obeyed:

$$[b_{l\sigma}, b_{l'\sigma'}^+] = \delta_{ll'}\delta_{\sigma\sigma'} \qquad (6.24)$$

$$[b_{l\sigma}, b_{l'\sigma'}] = [b_{l\sigma}^+, b_{l'\sigma'}^+] = 0 \qquad (6.25)$$

Then the hamiltonian operator for the field is

$$H = \sum_{l\sigma} \hbar\omega_l(b_{l\sigma}^+ b_{l\sigma} + \tfrac{1}{2}) \qquad (6.26)$$

and $\hat{n}_{l\sigma} = b_{l\sigma}^+ b_{l\sigma}$ is the number operator for the number of photons in the $(l\sigma)$th mode. The states of the field are thus

$$\psi = |n_{1\sigma}, n_{2\sigma}, \ldots, n_{l\sigma}, \ldots \rangle \qquad (6.27)$$

with the usual boson constraint $n_{l\sigma} \geq 0$.

6.2.4 Coupling of Radiation to Matter Using Second Quantization

As we learned previously, the coupling of the radiation field to matter can be described by the interaction

$$V(t) = \frac{-e}{mc} \mathbf{A} \cdot \mathbf{p} \qquad (6.28)$$

In this case, $\mathbf{A}(\mathbf{r}, t)$ is given by Eq. (6.10). Substituting the second quantized expression for q [Eq. (6.23a)], we get

$$\mathbf{A}(\mathbf{r}, t) = \sqrt{\frac{4\pi}{V}} c \sum_{l\sigma} \sqrt{\frac{\hbar}{2\omega_l}} (b_{l\sigma}^+(t) + b_{l\sigma}(t))\boldsymbol{\varepsilon}_{l\sigma} e^{i\mathbf{k}_l \cdot \mathbf{r}} \qquad (6.29)$$

Recall (Chapter 4, Problem 4) that the Heisenberg operators $b_{l\sigma}^+(t)$ and $b_{l\sigma}(t)$ for a harmonic oscillator are given by $b_{l\sigma}^+(t) = e^{i\omega_l t}b_{l\sigma}^+$ and $b_{l\sigma}(t) = e^{-i\omega_l t}b_{l\sigma}$. Substituting these into the expression for $\mathbf{A}$, and changing $l \rightarrow -l$ in the $b_{l\sigma}^+(t)$ term (which replaces $\mathbf{k}_l$ by $-\mathbf{k}_l$, but leaves $\boldsymbol{\varepsilon}_{l\sigma}$, ω_l, $b_{l\sigma}^+$ unchanged), we get

$$\mathbf{A}(\mathbf{r},\,t) = \sum_{l\sigma} \sqrt{\frac{2\pi\hbar c^2}{\omega_l V}} \left(b_{l\sigma} e^{i(\mathbf{k}_{l}\cdot\mathbf{r} - \omega_l t)} + b_{l\sigma}^{+} e^{-i(\mathbf{k}_{l}\cdot\mathbf{r} - \omega_l t)} \right) \boldsymbol{\varepsilon}_{l\sigma} \quad (6.30)$$

For a given mode $l\sigma$, this expression is nearly the same as the classical expression we developed [Eqs. (5.8) and (5.14)]. The only difference is that here the field operators $b_{l\sigma}$ and $b_{l\sigma}^{+}$ appear in place of $\sqrt{N}$. Substitution of this expression for $\mathbf{A}(\mathbf{r},\,t)$ into $V(t)$ gives us our matter–field interaction. Note that it operates on both matter and field states. Note also that as we have it, the interaction is expressed in a hybrid representation which is Schrödinger in the matter and Heisenberg in the field. If expressed entirely in the Schrödinger representation, we get (for a given $l\sigma$)

$$V = \frac{-e}{mc} \sqrt{\frac{2\pi\hbar c^2}{\omega_l V}} (b_{l\sigma} e^{i\mathbf{k}_{l}\cdot\mathbf{r}} + b_{l\sigma}^{+} e^{-i\mathbf{k}_{l}\cdot\mathbf{r}})(\boldsymbol{\varepsilon}_{l\sigma} \cdot \mathbf{p}) \quad (6.31)$$

The complete matter–field hamiltonian is given by

$$H = H^{\text{molecule}} + H^{\text{field}} + V \quad (6.32)$$

and the unperturbed states are products of matter and field states.

$$\psi_m = |m\rangle |n_i \cdots n_{l\sigma} \cdots\rangle \quad (6.33a)$$

$$\psi_k = |k\rangle |n_i' \cdots n_{l\sigma}' \cdots\rangle \quad (6.33b)$$

6.2.5 Application of Fermi's Golden Rule

Starting with the general rate expression

$$w_{km} = \frac{2\pi}{\hbar} \rho(E_m) |V_{km}|^2 \quad (6.34)$$

we see that the allowed first-order transitions involve the matrix elements

$$\langle n_i' \cdots n_{l\sigma}' \cdots | \langle k | (b_{l\sigma} e^{i\mathbf{k}_{l}\cdot\mathbf{r}} + b_{l\sigma}^{+} e^{-i\mathbf{k}_{l}\cdot\mathbf{r}})(\boldsymbol{\varepsilon}_{l\sigma} \cdot \mathbf{p}) |m\rangle |n_i \cdots n_{l\sigma} \cdots\rangle \quad (6.35)$$

As far as the field is concerned, the nonzero matrix elements will occur for (1) the $b_{l\sigma}$ term if $n_{l\sigma}' = n_{l\sigma} - 1$ and (2) the $b_{l\sigma}^{+}$ term if $n_{l\sigma}' = n_{l\sigma} + 1$. All occupation numbers in modes other than $l\sigma$ are unchanged. Obviously, the first term corresponds to absorption and the second emission. These are just the same results as were obtained previously from the semiclassical theory.

Let's evaluate the emission rate. Clearly, $\langle n_{l\sigma} + 1 | b_{l\sigma}^{+} | n_{l\sigma} \rangle = \sqrt{n_{l\sigma} + 1}$, so

$$w_{km} = \frac{2\pi}{\hbar} \rho(E_m) \frac{e^2}{m^2} \frac{2\pi\hbar}{\omega_l V} (n_{l\sigma} + 1) |\langle k | e^{-i\mathbf{k}_{l}\cdot\mathbf{r}} (\boldsymbol{\varepsilon}_{l\sigma} \cdot \mathbf{p}) |m\rangle|^2 \quad (6.36)$$

and in the long-wavelength approximation, using $\mathbf{p} \rightarrow im\omega_l\mathbf{r}$, we get

$$w_{km} = 2\pi\rho \, \frac{e^2}{m^2} \, \frac{2\pi}{\omega_l V} \, m^2\omega_l^2(n_{l\sigma} + 1)|\langle k|\boldsymbol{\varepsilon}_{l\sigma} \cdot \mathbf{r}|m\rangle|^2$$

$$= 2\pi\rho \, \frac{2\pi(n_{l\sigma} + 1)}{V} \, \omega_l|\boldsymbol{\varepsilon}_{l\sigma} \cdot \boldsymbol{\mu}_{km}|^2 \tag{6.37}$$

Substituting in Eq. (5.49) for ρ and replacing w_{km} by dw_{km} (see discussion in Section 5.4.4), we get

$$dw_{km} = \frac{(n_{l\sigma} + 1)\omega^3}{2\pi\hbar c^3} \, |\boldsymbol{\varepsilon}_{l\sigma} \cdot \boldsymbol{\mu}_{km}|^2 \, d\Omega \tag{6.38}$$

This is the same as was obtained in the semiclassical theory [Eq. (5.51)] *except* that the factor $n_{l\sigma} + 1$ has appeared naturally rather than in an *ad hoc* fashion. This leads to *spontaneous emission*, which can be thought of as arising from the interaction of matter with the "vacuum fluctuations" of the electromagnetic field.

6.3

Occupation Number Representations for Electrons

6.3.1 Fermion Creation and Destruction Operators

The occupation number representation that we have just discussed applies in situations where the number of excitations in a given oscillator is unlimited; that is, the eigenvalues n of the number operator b^+b can be any nonnegative integer. There is a fundamental symmetry to such excitations. They obey Bose–Einstein statistics, are called bosons, and their creation and annihilation operators obey the commutation relations (6.6c,d). We can rewrite the latter as

$$b_q b_{q'} = b_{q'} b_q \tag{6.39a}$$

or, taking adjoints,

$$b_{q'}^+ b_q^+ = b_q^+ b_{q'}^+ \tag{6.39b}$$

and interpret this as meaning that the order in which the excitations are created or destroyed is irrelevant; this is equivalent to the statement of the Pauli principle for integer-spin systems in the form that the overall wavefunction must be symmetric upon interchange of the coordinates of any two bosons.

This insight is highly suggestive. For particles such as electrons (or positrons or muons) that have half-integer spin, the Pauli principle requires that the wavefunction be antisymmetric (change sign) upon interchange of the coordinates of any two particles. Following (6.39a), this would imply that we can define creation and destruction operators a_k^+ and $a_{k'}$ for electrons in states k and k', and that they must satisfy the antisymmetry requirement

$$a_k a_{k'} = -a_{k'} a_k \qquad (6.40a)$$

or, defining the anticommutator as $[x, y]_+ \equiv xy + yx$,

$$[a_k, a_{k'}]_+ = 0 \qquad (6.40b)$$

By comparison with (6.6c,d) we also write the two other general fermion anticommutation relations as

$$[a_k^+, a_{k'}^+]_+ = 0 \qquad (6.40c)$$

$$[a_k, a_{k'}^+]_+ = \delta_{kk'} \qquad (6.40d)$$

The result (6.40c) follows directly from (6.40b), while (6.40d) follows by analogy with (6.6c). Additional insight into how these operators work comes from considering the situation $k = k'$. Then (6.40c) implies that

$$a_k^+ a_k^+ = 0 \qquad (6.41a)$$

This is in accord with the Pauli principle statement that no two fermions can occupy the same state; the operator product $a_k^+ a_k^+$ results in the creation of two electrons in state k, which is not permitted, so that the effect of such an operator on any function yields zero.

Analogously, (6.40d) implies that

$$a_k a_k^+ |n_k\rangle + a_k^+ a_k |n_k\rangle = |n_k\rangle \qquad (6.41b)$$

where we take the arbitrary function $|n_k\rangle$ as an eigenfunction of the fermion number operator $a_k^+ a_k$. Rewriting, we have

$$a_k a_k^+ |n_k\rangle = (1 - a_k^+ a_k)|n_k\rangle = (1 - n_k)|n_k\rangle \qquad (6.41c)$$

The number n_k can clearly be equal to zero, in which case

$$a_k a_k^+ |0\rangle = |0\rangle \qquad (6.41d)$$

or, from (6.41c),

$$a_k^+ a_k |0\rangle = 0 \qquad (6.41e)$$

We can also take $n_k = 1$, in which case

$$a_k a_k^+ |1\rangle = 0 \qquad (6.41f)$$

or

$$a_k^+ a_k |1\rangle = |1\rangle \tag{6.41g}$$

The interpretation of these results is straightforward: The number n_k of electrons permitted in level k is either zero or 1. If level k is already occupied, trying to create another electron in this level yields zero [from (6.41f)]. On the other hand, if the state is unoccupied, trying to annihilate an electron yields zero [Eq. (6.41e)]; this last property also holds for bosons, as Eq. (1.43) shows. Thus the eigenvalues of $a_k^+ a_k$ are zero and 1: Trying to create a state with $n_k = 2$ produces instead a zero [Eq. (6.41f)]. Thus the operators a_k and a_k^+ truly are fermion destruction and creation operators. They remove and introduce electrons into levels labeled by k, and their product, $a_k^+ a_k$, is a hermitian number operator whose eigenvalues, zero and unity, are the possible numbers of electrons in level k.

It was possible in the case of harmonic oscillators, using Eq. (6.6a), to express any observable operator such as momentum, kinetic energy, dipole moment, or hamiltonian using the set of b_q and $b_{q'}^+$. Similarly, it is possible, for electronic systems, to define any hermitian operator, in a given basis set, in the occupation number representation once its form in terms of the dynamical variables p_k and $x_{k'}$ is known. Even operators such as spin projection, which have no simple classical analog, can be written straightforwardly in terms of the a_k and $a_{k'}^+$. Moreover, as discussed in Section 6.1, expressing the operators in occupation number representation facilitates dynamical calculations, discussion of interactions, model construction and interpretation, and direct use of basis-set methods. In Section 6.3.3 we consider general electronic operators, but first the generalization to multielectron systems will be discussed.

6.3.2 Slater Determinants and Electron Creation Operators

The most straightforward scheme, without use of the occupation number representation, for assuring that electronic wavefunctions are antisymmetric is to write them as Slater determinants or linear combinations of such determinants. In the discussion following Eq. (6.40), it was pointed out that use of the a_k, $a_{k'}^+$ operators automatically assured that antisymmetry is present. This implies a relationship between this occupation number representation and Slater determinants, and in this subsection that relationship will be delineated.

We take $\{\phi_{k\mu}\}$ as a set of mutually orthonormal basis spin orbitals with space index k and spin projection μ ($\mu = \pm\frac{1}{2}$, corresponding to α or β spin). Then a Slater determinant

$$\psi = ((2N)!)^{-1/2} \begin{vmatrix} \phi_{1\alpha}(1) & \phi_{1\beta}(1) & \phi_{2\alpha}(1) & \cdots & \phi_{N\beta}(1) \\ \phi_{1\alpha}(2) & \phi_{1\beta}(2) & \cdots & \cdots & \phi_{N\beta}(2) \\ \vdots & & & & \vdots \\ \vdots & \cdots & \cdots & \cdots & \vdots \\ \phi_{1\alpha}(2N) & \phi_{1\beta}(2N) & \cdots & \cdots & \phi_{N\beta}(2N) \end{vmatrix} \quad (6.42a)$$

represents a closed-shell state in which the N orbitals $\phi_1, \phi_2, \ldots, \phi_N$ are each doubly occupied. The expansion of the determinant in fact includes $(2N)!$ terms. We can also denote this state rather clumsily, as

$$\psi = |1_{1\alpha}1_{1\beta}1_{2\alpha}1_{2\beta}1_{3\alpha} \cdots 1_{N\alpha}1_{N\beta}0_{N+1\alpha}0_{N+1\beta} \cdots\rangle \quad (6.42b)$$

in which the 1's and 0's are the occupation numbers for the basis spin orbitals. The 0's are really not necessary, since only the occupied orbitals are important. The Slater determinant wavefunction is entirely specified once we define the basis orbitals $\phi_{k\mu}$ and their occupation numbers.

The wavefunction ψ of (6.42b) can be written

$$\psi = a_{1\alpha}^+ a_{1\beta}^+ a_{2\alpha}^+ a_{2\beta}^+ \cdots a_{N\alpha}^+ a_{N\beta}^+ |\text{vac}\rangle$$

$$= \prod_{k=1}^{N} a_{k\alpha}^+ a_{k\beta}^+ |\text{vac}\rangle \quad (6.42c)$$

$$= \prod_{k=1}^{N} \prod_{\mu=\alpha,\beta} a_{k\mu}^+ |\text{vac}\rangle \quad (6.42d)$$

where $|\text{vac}\rangle$ denotes a vacuum state, with no electrons present. Just as Pauli antisymmetry is guaranteed from the property of antisymmetry under exchange of the rows of determinant (6.42a), so it is guaranteed by the antisymmetry property (6.40b) of the fermion creation operators appearing in (6.42c). Thus one can make Slater determinants by the product operator of (6.42d).

Consider the operation

$$a_{k\mu}^+ a_{k\mu} | \cdots n_{k\mu} \cdots\rangle = \begin{cases} (-1)^{\nu} a_{k\mu}^+ a_{k\mu} a_{k\mu}^+ | \cdots 0_{k\mu} \cdots\rangle & n_{k\mu} = 1 \\ a_{k\mu}^+ a_{k\mu} | \cdots 0_{k\mu} \cdots\rangle & n_{k\mu} = 0 \end{cases}$$

$$= \begin{cases} (-1)^{\nu} a_{k\mu}^+ | \cdots 0_{k\mu} \cdots\rangle & n_{k\mu} = 1 \\ 0 & n_{k\mu} = 0 \end{cases}$$

$$= \begin{cases} | \cdots n_{k\mu} \cdots\rangle & n_{k\mu} = 1 \\ 0 & n_{k\mu} = 0 \end{cases}$$

$$(6.43)$$

where the ket $|\cdots n_{k\mu}\cdots\rangle$ is a Slater determinant with $n_{k\mu} = 0$ or 1 electrons in the spin orbital $\phi_{k\mu}$. (The factor $\nu = \Sigma_{j>k}\, n_j$ is a phase factor arising from antisymmetry.) Then (6.43) shows that this ket is an eigenstate of the orbital number operator $a_{k\mu}^+ a_{k\mu}$ with eigenvalue 0 or 1, corresponding to $\phi_{k\mu}$ either empty or occupied. Similarly, the ket (6.42d) is an eigenfunction of the total number operator $\Sigma_k \Sigma_\mu\, a_{k\mu}^+ a_{k\mu}$, with eigenvalue $2N$.

Now consider excited configurations. To be specific, we take the Hückel model for ethylene and assume the two $2p\pi$ basis functions to be orthogonal. Then there are two π electrons, and the molecular orbitals are

$$\phi_1 = \frac{1}{\sqrt{2}}\,(u_l + u_r) \qquad \phi_2 = \frac{1}{\sqrt{2}}\,(u_l - u_r) \tag{6.44}$$

where u_l and u_r are $p\pi$ basis functions on the left and right carbons, respectively. The ground-state molecular orbital wavefunction is

$$\psi^{(g)}(^1A_{1g}) = \frac{1}{\sqrt{2}} \begin{vmatrix} \phi_{1\alpha}(1) & \phi_{1\beta}(1) \\ \phi_{1\alpha}(2) & \phi_{1\beta}(2) \end{vmatrix} \tag{6.45a}$$

$$= a_{1\alpha}^+ a_{1\beta}^+ |vac\rangle \tag{6.45b}$$

In the D_{2h} point group, the MOs ϕ_1 and ϕ_2 transform like a_u and b_g, respectively. The singly excited configurations with $m_s = 0$ can be written

$$\psi^{(1)} = a_{2\alpha}^+ a_{1\alpha} \cdot \psi^{(g)}(^1A_{1g}) \qquad \psi^{(2)} = a_{2\beta}^+ a_{1\beta} \cdot \psi^{(g)}(^1A_{1g}) \tag{6.46a}$$

while the doubly excited configuration is just

$$\psi^{(3)} = \psi^{(x)}(^1A_{1g}) = a_{2\alpha}^+ a_{1\alpha} a_{2\beta}^+ a_{1\beta} \psi^{(g)}(^1A_{1g}) \tag{6.46b}$$

If we choose to include interactions among the π electrons [as is done in the PPP model of Section (6.5.4)], the single determinant of (6.45a) is no longer an eigenfunction but, rather, an approximate (the Hartree–Fock) ground-state solution. A more accurate solution (indeed, a complete one within this basis set) is offered by the configuration interaction function

$$\psi^{CI}(^1A_{1g}) = C_g\psi^{(g)}(^1A_{1g}) + C_x\psi^{(x)}(^1A_{1g}) \tag{6.47}$$

with C_g and C_x coefficients to be determined using the variational principle, subject to the normalization condition $C_x^2 + C_g^2 = 1$. Note that the single determinants are eigenfunctions of the orbital occupation numbers

$$a_{1\beta}^+ a_{1\beta}\psi^{(g)}(^1A_{1g}) = \psi^{(g)}(^1A_{1g})$$

$$a_{2\alpha}^+ a_{2\alpha}\psi^{(x)}(^1A_{1g}) = \psi^{(x)}(^1A_{1g}) \tag{6.48a}$$

$$a_{2\alpha}^+ a_{2\alpha}\psi^{(g)}(^1A_{1g}) = 0$$

but that the multideterminant configuration interaction wavefunction is not

$$a_{2\alpha}^{+} a_{2\alpha} \psi^{CI}(^1A_{1g}) = C_x \psi^{(x)}(^1A_{1g}) \qquad (6.48b)$$

We have just seen that while single Slater determinants are eigenfunctions of the spin-orbital number operators, linear combinations of Slater determinants used to form CI wavefunctions lose this eigenfunction property. One can calculate the expectation values giving the population of a given orbital in a CI function. Thus we find, for the ethylene example just discussed,

$$\langle \psi^{(g)}(^1A_{1g}) | a_{k\mu}^{+} a_{k\mu} | \psi^{(g)}(^1A_{1g}) \rangle = \delta_{k1} \qquad (6.49a)$$

$$\langle \psi^{CI}(^1A_{1g}) | a_{k\mu}^{+} a_{k\mu} | \psi^{CI}(^1A_{1g}) \rangle = \delta_{k1} C_g^2 + \delta_{k2} C_x^2 \qquad (6.49b)$$

Thus the excited (antibonding) orbital ϕ_2, which is unoccupied in the molecular orbital ground state $\psi^{(g)}(^1A_{1g})$ is partially full in the CI state, while some electronic occupation has disappeared from the ϕ_1 orbital, which was full in the MO state.

This behavior is characteristic: In many fermion systems, including the many-electron states of atoms, molecules, and solids, one can always define a best independent particle, or Hartree–Fock, state. This single-determinant state has a well-defined Fermi energy ε_F, such that all orbitals with orbital energy below ε_F are full and all those with orbital energy above ε_F are empty. When correlation effects arising from interelectronic repulsions are included, the single-determinant state is generalized to a multideterminant, CI, state. In this case there is still discontinuity in occupation at the Fermi energy, with orbitals with energy levels below ε_F having higher partial occupation than orbitals above ε_F, but occupation in both sets of orbitals (the ones below ε_F are called hole levels in solids, those above ε_F are called particle levels) can now be partial.

Since use of the occupation number representation assures antisymmetry automatically, explicit use of determinant notation or antisymmetrization operators is not necessary when wavefunctions are expressed in terms such as (6.42d), (6.45), and (6.46), while operators are also written in terms of creation and destruction operators. For boson (harmonic oscillator) systems, the operators can be constructed directly using the equivalence of Eq. (6.2). For electrons, the construction involves the idea of basis-set representations.

6.3.3 Manipulation of Fermion Operators; Commutators and Anticommutators

The anticommutation relations of Eq. (6.40) define, formally, the behavior of electron operators $\{a_k, a_{k'}^+\}$ for orthogonal orbitals ϕ_k and $\phi_{k'}$. In general, and especially in connection with use of the Heisenberg representation, it is important to evaluate products and commutators involving the a_k and $a_{k'}^+$. Note that since a_k^+ and a_l, respectively, create or destroy one electron, any operator product $a_i^+ a_j^+ \cdots a_s^+ a_t a_u \cdots a_w$ that does not contain the same number of a^+ operators and a operators will not conserve the number of electrons. Such non-number-conserving operators are of importance in several condensed-phase phenomena, including superconductivity, and in particular atomic and molecular processes such as Auger spectroscopy and photoemission. For the most part, however, the operators with which we will be concerned contain equal numbers of a and a^+ factors, and conserve electron number.

We consider, then, the commutator $[X, Y]$, where X and Y contain M_X, $M_Y = 1, 2, 3, \ldots a$ or a^+ operators, respectively.

A. $M_X = M_Y = 1$:

$$[a_k, a_l] = a_k a_l - a_l a_k = -a_l a_k - a_l a_k \tag{6.50a}$$
$$= -2a_l a_k$$

$$[a_k, a_l^+] = a_k a_l^+ - a_l^+ a_k = (\delta_{kl} - a_l^+ a_k) - a_l^+ a_k \tag{6.50b}$$
$$= \delta_{kl} - 2a_l^+ a_k$$

$$[a_k^+, a_l^+] = -2a_l^+ a_k^+ \tag{6.50c}$$

Result (6.50a) follows from replacing $a_k a_l$ by $-a_l a_k$, according to the fundamental relationship (6.40b). Similarly, (6.40c) and (6.40d) produce the last two results. We note that when $M_X = M_Y = 1$, commutation results in a new operator that is a product.

B. $M_X = 1, M_Y = 2$:

$$[a_k, a_l a_m] = a_k a_l a_m - a_l a_m a_k = a_l a_m a_k - a_l a_m a_k = 0 \tag{6.50d}$$

$$[a_k, a_l^+ a_m] = a_k a_l^+ a_m - a_l^+ a_m a_k = (\delta_{kl} - a_l^+ a_k)a_m - a_l^+ a_m a_k = \delta_{kl} a_m \tag{6.50e}$$

$$[a_k, a_l^+ a_m^+] = a_k a_l^+ a_m^+ - a_l^+ a_m^+ a_k = (\delta_{kl} - a_l^+ a_k)a_m^+ - a_l^+ a_m^+ a_k$$
$$= \delta_{kl} a_m^+ - a_l^+(\delta_{km} - a_m^+ a_k) - a_l^+ a_m^+ a_k = \delta_{kl} a_m^+ - \delta_{km} a_l^+ \tag{6.50f}$$

$$[a_k^+, a_l^+ a_m] = a_k^+ a_l^+ a_m - a_l^+ a_m a_k^+$$
$$= -a_l^+(\delta_{km} - a_m a_k^+) - a_l^+ a_m a_k^+ = -\delta_{km} a_l^+ \qquad (6.50g)$$

C. $M_X = 2$, $M_Y = 2$:

$$[a_k^+ a_l^+, a_m^+ a_n^+] = 0 \qquad (6.50h)$$

$$[a_k^+ a_l, a_m^+ a_n^+] = \delta_{lm} a_k^+ a_n^+ - \delta_{ln} a_k^+ a_m^+ \qquad (6.50i)$$

D. $M_X = 1$, $M_Y = 4$:

$$[a_k^+, a_l^+ a_m^+ a_n a_p] = a_l^+ a_m^+ a_p \delta_{kn} - a_l^+ a_m^+ a_n \delta_{pk} \qquad (6.50j)$$

E. $M_X = 2$, $M_Y = 4$:

$$\begin{aligned}[a_k^+ a_l, a_m^+ a_n^+ a_p a_q] = \ &\delta_{lm} a_k^+ a_n^+ a_p a_q - \delta_{ln} a_k^+ a_m^+ a_p a_q \\ &+ \delta_{kp} a_m^+ a_n^+ a_q a_l - \delta_{kq} a_m^+ a_n^+ a_p a_l \end{aligned} \qquad (6.50k)$$

A convenient rule of thumb for evaluating these commutators in all cases except $M_X = M_Y = 1$ is as follows: Starting with the first a or a^+ to the left of the commutator comma, move the a or a^+ past each term (a or a^+) on the right of the comma. If an a^+ moves past an a^+ or an a past an a, change the sign. If an a_i moves past an a_j^+ or an a_i^+ past an a_j, change sign, and the move contributes a δ_{ij} to the commutator. When the first a or a^+ from the left of the comma has moved past all the operators to the right, start with the next operator to the left. Continue until all operators have been moved past the comma. This seems clumsy at first, but is in fact much faster than actually evaluating the commutators. The reader is urged to try to reproduce (6.50d)–(6.50k) using this rule. These commutators will be of real importance in the study of electronic structure and dynamics.

6.3.4 Arbitrary Electronic Operators in the Occupation Number Representation

The electronic operators a_k and a_k^+ were introduced in terms of a basis function ϕ_k. Operators such as the number operator $a_k^+ a_k$ or the operator $a_k^+ a_l$ that moves an electron from basis function l to basis function k are defined in terms of their effects on basis functions. Indeed, the occupation number formalism permits rewriting of an arbitrary operator acting in momentum and coordinate space in terms of its equivalent behavior in a basis set. In this subsection the rules for writing electronic operators in the occupation number formalism will be stated.

We first consider operators such as momentum or kinetic energy or displacement, which do have classical analogs, that is, that can be written

$$\theta = \theta(p, q) \qquad (6.51)$$

Such operators are generally of two types: One-electron operators can be written

$$\theta = \sum_i \theta_i \tag{6.52a}$$

where the index i numbers the electrons in the system; another class of operators can be written

$$T = \sum_{i<j} T_{ij} \tag{6.52b}$$

where i and j are electron-counting indices. Examples of the type (6.52a), called one-electron operators, include dipole moment, kinetic energy, momentum, and angular momentum. By far the most important operator of type (6.52b) is interelectronic interaction in the potential, such as the Coulomb potential

$$V_{\text{Coul}} = \sum_{i<j} \frac{e^2}{|\mathbf{r}_i - \mathbf{r}_j|} \tag{6.52c}$$

One-electron and two-electron operators are represented quite differently.

Consider first one-electron operators (6.52a). Since all exact wavefunctions of the electronic system can always be written in the CI form as an arbitrary linear combination of single Slater determinants, if we can show that

$$\langle D_1|\theta|D_2\rangle = \langle D_1|\theta'|D_2\rangle \tag{6.53}$$

for arbitrary Slater determinants D_1 and D_2, the operators θ and θ' are equivalent in the set of orbitals defining the Slater determinants.

Consider first the very simple case of one-electron wavefunctions. We then wish to show that

$$\langle 000\ 1_s 00 \cdots |\theta|000 \cdots 1_r 00 \cdots\rangle$$
$$= \langle 000\ 1_s 00 \cdots |\theta'|000 \cdots 1_r 00 \cdots\rangle \tag{6.54}$$

The form for any one-electron operator is

$$\theta = \sum_{kl} \theta_{kl} a_k^+ a_l \tag{6.55}$$

where

$$\theta_{kl} = \langle \phi_k(i)|\theta'(q_i, p_i)|\phi_l(i)\rangle \tag{6.56}$$

and i labels the electron. Substituting (6.55) into the left side of (6.54), we have

$$\sum_{kl} \langle 000\ 1_s 00 \cdots |a_k^+ a_l|000 \cdots 1_r 00 \cdots\rangle \theta_{kl} = \sum_{kl} \delta_{rl}\delta_{sk}\theta_{kl} = \theta_{sr}$$

$$\tag{6.57a}$$

Similarly, the right side of (6.54) becomes

$$\langle 000\ 1_s 00 \cdots |\theta'|000 \cdots 1_r 00 \cdots \rangle = \langle \phi_s(i)|\theta'(q_i, p_i)|\phi_r(i)\rangle = \theta_{sr}$$

(6.57b)

where the last identification follows from (6.56). Comparing (6.57b) with (6.57a) shows that the form (6.55) is correct, at least for matrix elements between one-electron states. In fact, (6.55) is entirely general, though the proof is a bit messy. [See P. A. M. Dirac, *The Principles of Quantum Mechanics,* 4th Ed. (Clarendon Press, Oxford, 1958), p. 250, 251.]

An arbitrary two-particle operator is similarly expressed in the occupation number formalism as

$$T = \sum_{klmn} a_k^+ a_l^+ a_m a_n \langle kn|T'|lm\rangle$$

(6.58a)

where the indices run over spin orbitals. When T' is spin independent, as it is for the most important two-particle operator of Eq. (6.52c), then T becomes

$$T = \sum_{klmn} \sum_{\mu,\nu=\pm 1/2} a_{k\mu}^+ a_{l\nu}^+ a_{m\nu} a_{n\mu} \langle kn|T'|lm\rangle$$

(6.58b)

with the matrix element defined by analogy as

$$\langle kn|T'|lm\rangle = \langle \phi_k(1)\phi_n(1)|T'_{12}|\phi_l(2)\phi_m(2)\rangle$$

(6.58c)

The forms (6.55) and (6.58b) for, respectively, one- and two-particle operators are quite general. Note that they contain no indices that describe the actual enumerated electrons, in contrast to the configuration space forms of (6.52).

The most obvious and important example of an operator with both one and two-electron parts is the Coulomb hamiltonian for the electronic structure of an atom or molecule, in the Born–Oppenheimer limit. The hamiltonian is, then,

$$H = H_1 + H_2$$

(6.59a)

$$H_1 = \sum_i \frac{p_i^2}{2m_e} - \sum_i \sum_\alpha \frac{Z_\alpha e^2}{r_{i\alpha}}$$

(6.59b)

with the ith electron at position $\mathbf{r}_i$ and the αth nucleus having charge Z_α. The two-electron Coulomb repulsion is

$$H_2 = \sum_{i<j} \frac{e^2}{r_{ij}} = \frac{1}{2} \sum_{i\neq j} \frac{e^2}{|\mathbf{r}_i - \mathbf{r}_j|}$$

(6.59c)

Then the overall hamiltonian, in the occupation number representation, becomes

$$H = \sum_{lm} \sum_{\mu} h_{lm} a_{l\mu}^{+} a_{m\mu} + \frac{1}{2} \sum_{\mu\nu} \sum_{klmn} a_{k\mu}^{+} a_{l\nu}^{+} a_{m\nu} a_{n\mu} \langle kn|lm \rangle \quad \text{(6.59d)}$$

Here μ and ν are spin projector labels, taking the values $\pm\frac{1}{2}$, while k, l, m, n are spatial orbital indices that run over the set of orbitals involved in the basis. If, for instance, $klmn$ label atomic orbitals, then (6.59d) is in the atomic orbital representation. Alternatively, $klmn$ could be molecular orbitals or symmetry orbitals or floating basis functions; in all such cases, the matrix elements

$$h_{lm} = \langle \phi_l(1)| \frac{p_1^2}{2m_e} - \sum_{\alpha} \frac{Z_{\alpha} e^2}{r_{\alpha 1}} |\phi_m(1)\rangle \quad \text{(6.60a)}$$

$$\langle kn|lm \rangle = \langle \phi_k(1)\phi_n(1)| \frac{e^2}{r_{12}} |\phi_l(2)\phi_m(2)\rangle \quad \text{(6.60b)}$$

are well defined, and in each case the form (6.59d) is a correct expression of H in the selected set of orbitals.

Since both the one-electron part (first sum) and the two-electron part (second sum) of (6.59d) contain equal numbers of a^{+} and a operators, the total number of electrons is conserved. Note also that both of the terms in (6.59d) conserve the total M_s quantum number, this hamiltonian, like (6.59a–c), is coulombic and contains no spin-orbit couplings.

Exercise: Show explicitly that the number operator for up-spin electrons, $\sum_i a_{i\alpha}^{+} a_{i\alpha}$ commutes with H of (6.59). From the Heisenberg relationship, this means that the number of up-spin electrons is constant.

The interpretation of (6.59d) is both important and straightforward. Unlike arbitrary space-dependent operators such as $p^2/2m$, the Coulomb hamiltonian of (6.59d) acts only within a set of basis states; this set is defined by the space orbital set $\{\phi_l\}$ or, equivalently, by the spin-orbital set $\{\phi_{l\mu}\}$. Once these orbitals are defined, and once values are selected for the matrix elements h_{lm} and $\langle kn|lm \rangle$, the specification of H is fixed. This corresponds to a matrix representation of the full hamiltonian operator in the fixed basis $\{\phi_{l\mu}\}$ and is not necessarily identical to the standard expression for $H(p, q)$, which is not restricted to a given basis. For example, for the C_2H_4 molecule, one can write many basis-set hamiltonians of the form (6.59d). If the $\{\phi_l\}$ set is restricted to the two $2p\pi$ orbitals, and all two-electron integrals are set to zero, the Hückel model is recovered, in the form

$$H = \sum_{\mu} \sum_{i=1,2} \alpha a_{i\mu}^{+} a_{i\mu} + \sum_{\mu} \beta(a_{1\mu}^{+} a_{2\mu} + a_{2\mu}^{+} a_{1\mu}) \quad \text{(6.61)}$$

Another model hamiltonian, a single-zeta or minimum-basis *ab initio* model, chooses $\{\phi_l\}$ to consist of 14 basis functions (1*s* on *H*, 1*s*, 2*s*, 2*p* on *C*), and takes exactly the form (6.59d) with the sums over *k*, *l*, *m*, *n* running from 1 to 14, and the matrix elements evaluated according to (6.60). Note that this model is specified only if the actual forms of the basis functions are given; for example, if Slater-type orbitals are used, the exact form, including numerical values for the screening exponents, must be given for (6.60) to be evaluated. If, as is more common in actual electron structure studies, $\{\phi_l\}$ is described in terms of expansions in gaussian-type basis functions, the contraction coefficients and gaussian exponents must be defined.

The definition and use of different model hamiltonians for electronic structure are discussed in Section 6.5. The form of (6.59c), however, suggests a general interpretation. One-electron operators, the first term of (6.59d), move an electron from ϕ_m to ϕ_l, while two-electron operators, the second term, move two electrons from (ϕ_m, ϕ_n) to (ϕ_k, ϕ_l). Thus the action of operators like *H* can be directly understood, when the occupation number formalism is used. This is particularly helpful in situations in which the hamiltonian does not have a classical analog (i.e., in which classical mechanics does not suggest a form for $H(p, q)$).

A common example of this situation is offered by electron spin. There are classical analogies to the spin of the electron, but its anomalous magnetic moment and the Pauli principle, as well as Dirac's derivation of spin properties from the relativistic quantum mechanics of single-particle systems, all suggest that spin is a quantum phenomenon without a precise classical analog. Nevertheless, the occupation number formalism permits the discussion and computation of spin properties just as it does for dynamical systems with classical analogs. For example, consider a single electron. One can then define $\{a_\mu, a_\nu^+\}$ operators, with $\mu, \nu = \pm\frac{1}{2}$ defining the spin component. Such operators as $a_\mu^+ a_\mu$, which counts the number of spin-μ electrons, are then clearly defined. A particular set of four operators

$$a_\alpha^+ a_\alpha + a_\beta^+ a_\beta = 1$$

$$\frac{\hbar}{2}(a_\alpha^+ a_\beta + a_\beta^+ a_\alpha) \equiv S_x$$

$$\frac{\hbar}{2}(-i)(a_\alpha^+ a_\beta - a_\beta^+ a_\alpha) \equiv S_y \tag{6.62}$$

$$\frac{\hbar}{2}(a_\alpha^+ a_\alpha - a_\beta^+ a_\beta) \equiv S_z$$

can be defined, with (α, β) denoting $m_z = +\frac{1}{2}, -\frac{1}{2}$ as usual. Then the three operators (S_x, S_y, S_z) have the interesting properties

$$S_x^2 = S_y^2 = S_z^2 = \frac{\hbar^2}{4} \ (1) \tag{6.63a}$$

$$[S_x, S_y] = i\hbar S_z \tag{6.63b}$$

$$[S_x, S_y]_+ = 0 \tag{6.63c}$$

Property (6.63b) means that the three S operators correspond to the elements of an angular momentum. In fact, the set (6.62) is equivalent to the Pauli spin matrices

$$\frac{2}{\hbar} S_x = \sigma_x = \begin{pmatrix} 0 & 1 \\ 1 & 0 \end{pmatrix}$$

$$\frac{2}{\hbar} S_y = \sigma_y = \begin{pmatrix} 0 & -i \\ i & 0 \end{pmatrix} \tag{6.64}$$

$$\frac{2}{\hbar} S_z = \sigma_z = \begin{pmatrix} 1 & 0 \\ 0 & -1 \end{pmatrix}$$

which are defined in terms of their actions on the basis vectors $\begin{pmatrix} \alpha \\ \beta \end{pmatrix}$. Once again the verbal interpretation of (6.62) in terms of counting and flipping spins makes the occupation number representation attractive; for example, σ_z describes the net spin polarization.

Exercise: Consider the $^2P_{1/2}, \ ^2P_{3/2}$ terms of potassium, and write the spin-orbit operator $H_{so} \cong \zeta \mathbf{l} \cdot \mathbf{s}$ in terms of $\{a_{k\mu}, a_{k\nu}^+\}$, where $k = (-1, 0, 1)$ are the three m_l orbital angular momentum levels.

Since all operators corresponding to observables must be hermitian, operators involving $\{a, a^+\}$ have hermitian conjugates. Following the results for matrices, we define

$$[\theta_1\theta_2]^+ = \theta_2^+ \theta_1^+ \tag{6.65}$$

for any two operators θ_1 and θ_2. Thus the operators $a_{i\mu}^+ a_{i\mu}$, $(a_{i\mu}^+ + a_{i\mu})$, $i(a_{j\nu}^+ - a_{j\nu})$, and $(a_{i\sigma}^+ a_{l\sigma} + a_{l\sigma}^+ a_{i\sigma})$ are all hermitian, in that $\theta^+ = \theta$. Similarly, the Pauli operators of (6.62) are hermitian. Note that since the operators $a_{i\lambda}^+$ or $a_{j\mu}$ are not hermitian, they are not observables (and the same is true of the boson operators b_q^+ and b_{-q}).

6.4

Fermion Field Operators and Second Quantization

We have introduced the electron creation and destruction operators a_l^+, a_k by analogy with the harmonic oscillator operators b^+ and b; the orbitals, or basis functions ϕ_n, are well-defined spatial structures. An alternative way to introduce these operators involves the theory of quantized fields. In this picture, one defines field operators $\psi(\mathbf{r})$ for all values of the coordinate $\mathbf{r}$; these field operators operate on a Hilbert space, in which any vector is a state of the quantized field. Since the field itself, rather than the energies or actions or angular momenta, is quantized, this method is often referred to as second quantization. For most molecular or solid-state problems of interest, operators are most conveniently used in the occupation number representation, as discussed in Section 6.3. However, the occupation number scheme can be derived, formally, from the quantized field. For this reason, occupation number representation is very often called "second quantization" in the literature, and indeed we will also use this term. Quite apart from this issue of nomenclature, the quantized field description is a very general approach; we will use it here as an alternative method for deriving the occupation number representation.

For simplicity, then, consider a set of spinless fermions, such as electrons (spin can be added subsequently). The field operators are defined by the anticommutation relations

$$[\psi(\mathbf{r}), \psi(\mathbf{r}')]_+ = 0$$
$$[\psi^+(\mathbf{r}), \psi^+(\mathbf{r}')]_+ = 0 \qquad (6.66)$$
$$[\psi(\mathbf{r}), \psi^+(\mathbf{r}')]_+ = \delta(\mathbf{r} - \mathbf{r}')$$

with $\delta(\mathbf{r} - \mathbf{r}')$ the Dirac δ function of Appendix A. The one- and two-electron operators are then conveniently *defined* as

$$\theta = \int d\mathbf{r}\, \psi^+(\mathbf{r})\theta(\mathbf{r})\psi(\mathbf{r})$$
$$T = \frac{1}{2}\int d\mathbf{r} \int d\mathbf{r}'\psi^+(\mathbf{r})\psi^+(\mathbf{r}')T(\mathbf{r}, \mathbf{r}')\psi(\mathbf{r}')\psi(\mathbf{r}) \qquad (6.67a)$$

where $\theta(\mathbf{r})$ and $T(\mathbf{r}, \mathbf{r}')$ are the operators in (p, q) space. For example, the number operator and kinetic energy are just

$$N_{op} = \int d\mathbf{r} \, \psi^+(\mathbf{r})\psi(\mathbf{r})$$

$$KE_{op} = \int d\mathbf{r} \left(\frac{-\hbar^2}{2m} \right) \psi^+(\mathbf{r})\nabla^2\psi(\mathbf{r})$$

(6.67b)

while the Coulomb interaction is

$$V_{coul} = \frac{1}{2} \int d\mathbf{r} \int d\mathbf{r}' \, \psi^+(\mathbf{r})\psi^+(\mathbf{r}') \frac{e^2}{|\mathbf{r} - \mathbf{r}'|} \psi(\mathbf{r}')\psi(\mathbf{r})$$

(6.68)

As in Section 6.3,

$$[N_{op}, KE_{op}] = 0$$

$$[N_{op}, V_{coul}] = 0$$

so that both kinetic and potential energy operators conserve particle number.

The use of $\psi^+(\mathbf{r})$, $\psi(\mathbf{r})$ is precisely equivalent to the use of the often-called "first quantization" form $\theta(p, q)$ for operators; in this sense, second quantization introduces no new physics, no new phenomena; it is merely a convenient, powerful, mnemonic notation.

Now suppose that we choose to expand the field operators $\psi(\mathbf{r})$ in terms of a set of orthonormal single-electron wavefunctions $\phi_j(\mathbf{r})$. We write

$$\psi(\mathbf{r}) = \sum_j \phi_j(\mathbf{r})a_j$$

(6.69a)

$$\psi^+(\mathbf{r}) = \sum_i \phi_i^*(\mathbf{r})a_i^+$$

The operators a_i^+, and a_j must then satisfy

$$[a_i^+, a_j^+]_+ = 0$$

$$[a_i, a_j]_+ = 0$$

(6.69b)

$$[a_i, a_j^+]_+ = \delta_{ij}$$

if the field operator anticommutation relations (6.66) are to hold. Thus we see that the anticommutation relations (6.40) for creation and destruction operators follow from (6.66) for the field operators. Similarly, the forms (6.55) and (6.58) for one- and two-electron operators follow from (6.67).

We have discussed spinless field operators. One can generalize the consideration by writing $\psi_\mu(\mathbf{r})$ to denote a field operator with spin component μ, in which case (6.66) becomes

$$[\psi_\mu(\mathbf{r}), \psi_\nu^+(\mathbf{r}')]_+ = \delta_{\mu\nu}\delta(\mathbf{r} - \mathbf{r}')$$

(6.70)

Two significant points should be raised here that are generally omitted in presentations of second quantization. The first is that the sum in

(6.69) is not necessarily over a complete basis set $\{\phi_j\}$. If the basis is complete, the operators (6.55) and (6.58) are precisely equivalent to the operators (6.67a). In nearly all chemical situations, however (exceptions occur for spin systems), the basis is not complete, and in that sense the hamiltonian of (6.59d) represents a finite basis approximation to the exact Coulomb hamiltonian of (6.59a). In general, larger basis sets will make (6.59d) closer, in some sense, to (6.59a), but only for a complete set is equivalence exact. Thus, in general, any form (6.59d) represents a finite-basis model system for the true hamiltonian. As stressed in Section 6.4.4, these model systems are defined by fixing the basis sets $\{\phi_i\}$ and the matrix elements of (6.59d), but even for the same full hamiltonian (6.59a), different choices of basis set correspond to different models, whose calculated properties will differ. For example, the calculated dipole moment for the HF molecule will differ with different basis sets, and thus different model hamiltonians are used, even though the Born–Oppenheimer electronic hamiltonian

$$ H^{(el)}(p, q) = \frac{-\hbar^2}{2m_e} \sum_i \nabla_i^2 - \sum_i \left(\frac{e^2}{|\mathbf{r}_{iH}|} + \frac{9e^2}{|\mathbf{r}_{iF}|} \right) + \sum_{i<j} \frac{e^2}{|\mathbf{r}_{ij}|} \qquad (6.71) $$

(with $|\mathbf{r}_{iF}|$, $|\mathbf{r}_{iH}|$, and $|\mathbf{r}_{ij}|$, respectively, the distances between electron i and the F nucleus, between electron i and the proton, and between electrons i and j) is unchanged.

The second point is that the basis functions $\{\phi_j\}$ in terms of which the field operators are expanded are not necessarily orthogonal. If there is finite overlap, so that

$$ \int d\mathbf{r} \ \phi_i^*(\mathbf{r})\phi_j(\mathbf{r}) = S_{ij} \qquad (6.72) $$

with the overlap matrix element S_{ij} not necessarily equal to δ_{ij}, it is consistent with (6.66) to set

$$ [a_i, a_j^+]_+ = S_{ij} \qquad (6.73) $$

as a generalization of (6.40d). In many applications of second-quantization, or occupation number formalism, in chemistry one chooses to orthogonalize the set of basis functions $\{\phi_j\}$ so that (6.40) can be used, but if the set is in fact not orthonormal, (6.73) should be used instead.

6.5

Molecular Electronic Structure: Model Hamiltonians and Occupation Number Representations

Once a hamiltonian is constructed for any physical situation, the eigenvalues and eigenfunctions can, in principle, be obtained. A central issue in chemistry is the description of the electronic structure of a molecule; in the usual case in which vibronic interactions can be neglected, this means solving the electronic Schrödinger equation

$$H_{el}(q; Q)\psi_{el}(q; Q) = E_{el}(Q)\psi_{el}(q; Q) \tag{6.74}$$

where q and Q label all the electronic and nuclear coordinates, respectively, and the electronic wavefunction $\psi_{el}(q; Q)$ and energy $E_{el}(Q)$ each depend parametrically on nuclear positions. The electronic hamiltonian $H_{el}(q; Q)$ contains Coulomb interactions, electronic kinetic energy, and whatever other terms (spin interactions, external fields, spin-orbit coupling) may be appropriate. In this section we construct a number of model hamiltonians for molecular electronic structures, and discuss their properties and solutions.

It is important to distinguish between an electronic structure model hamiltonian and the (approximate or exact) solution to that model. For any given physical problem, such as the electron paramagnetic resonance (EPR) spectrum of naphthalide anion or the nonlinear optical response of p-nitroaniline or the dipole moment of CO, a particular model hamiltonian may contain enough of the full electronic structure to describe the desired response property, but the solution to that model may be quite demanding. In the next two subsections we describe the models themselves, and in the following subsections methods of solution will be described.

6.5.1 Model Hamiltonians: Basis Sets and Matrix Elements

The general form for the electronic hamiltonian, with Coulomb interactions, was given in Section 6.3 as

$$H = \sum_{\sigma} \sum_{lm} h_{lm} a_{l\sigma}^+ a_{m\sigma} + \frac{1}{2} \sum_{klmn} \sum_{\sigma\rho} \langle kn|lm \rangle a_{k\sigma}^+ a_{l\rho}^+ a_{m\rho} a_{n\sigma} \tag{6.75}$$

where the indices $klmn$ label the orbital basis functions. The first term arises from the one-electron terms, electronic kinetic energy and nuclear attraction. The second term is due to interelectronic repulsion.

If the matrix elements h_{lm} and $\langle kn|lm \rangle$ are evaluated properly, by performing the appropriate three- and six-dimensional integrals, the resulting hamiltonian is referred to as an *ab initio* model. It corresponds to a matrix representation of the electronic structure problem in the orbital basis set selected. As this basis set becomes larger, the representation (6.75) becomes a closer and closer description of the Coulomb electronic hamiltonian (6.59a). The *ab initio* model for any given molecule requires specification of the nuclear configuration Q for which the electronic structure is to be determined, and of the basis set. To specify the basis, one requires both the number and type of basis functions [to fix the summation range of (6.75)] and the mathematical (analytic or numerical) form of the basis functions (to compute the matrix elements). For example, an *ab initio* model for the ground state of H_2O might involve a minimum-basis description in terms of Slater-type orbital (STO) basis functions, so that k, l, m, n would run from 1 to 7, with ϕ_1 and ϕ_2 the two $1s$ STOs on the two H atoms, and ϕ_3 to ϕ_7 the $1s$, $2s$, $2p_x$, $2p_y$, and $2p_z$ STOs on O. There would then be 49 matrix elements h_{lm} and 2401 elements $\langle kn|lm \rangle$, not all of which are independent.

Other *ab initio* model hamiltonians can be defined for the same molecule at the same geometry. For instance, a split-valence Slater basis for H_2O would include, besides the seven orbitals of the minimal basis, two more s-type functions on the two protons, plus one additional s and three additional p's on the O, giving a total of 13 functions, and leading to 169 h_{lm} integrals and 26,561 $\langle kn|lm \rangle$, not all of which, again, are independent. Alternatively, an STO-3G basis can be defined, in which case there are only seven terms in the sums of Eq. (6.75), but each basis function is itself the sum of three simple primitive gaussian orbitals. The coefficients in this sum are not varied, but are fixed once and for all when the basis set is selected; the notation STO-3G means that a valence Slater-type orbital is approximated by a fixed linear combination of three gaussian functions.

It is useful to consider some of the terms entering into this *ab initio* model. For example, the term $h_{33}a_{3\mu}^+ a_{3\mu}$ describes the energy of electrons with spin μ in basis orbital 3; it is the equivalent of the Hückel α term for this orbital. Similarly, $\langle 33|33 \rangle \, a_{3\alpha}^+ a_{3\beta}^+ a_{3\beta} a_{3\alpha}$ describes the repulsion integral between an up-spin and a down-spin electron, both present in the third basis orbital, while $h_{42}a_{4\alpha}^+ a_{2\alpha}$ is the energy contribution from moving an up-spin electron from basis orbital 2 to 4.

In these *ab initio* models, the number of two-electron integrals $\langle kn|lm \rangle$ is proportional to the fourth power of the number of basis orbitals. It is clear, therefore, that such models become computationally very demanding for molecules containing more than, say, 100 electrons. This fact, coupled with important intuitive concepts about the separation of valence and core electrons, the approximate separability of sigma and pi electrons, and the different importance of Coulomb and exchange interactions, has led to the development of methods for defining other types of simplified model hamiltonians. Such models are nearly always *semiempirical*. When used in the context of electronic structure theories, this term means that in contrast with *ab initio* models, the sums in (6.75) run only over a selected set of (usually valence) orbitals, and the matrix elements h_{lm} and $\langle kn|lm \rangle$ are evaluated not necessarily by performing integrals but, often, by fitting their values to experiment or to other theoretical values. Such semiempirical model hamiltonians are of great utility in chemistry and in the study of solids, especially for describing spectra or bonding or response behavior. In the remainder of this section we discuss such models and the most common method of solving the Schrödinger equation for noninteracting electrons.

6.5.2 Noninteracting Electrons: Hückel, Extended Hückel, and Free-Electron Models

The simplest model descriptions of molecular electronic structure ignore interelectronic repulsions altogether. Such models, called one-electron models, are of real importance in solid-state problems and in chemistry. Often, one speaks of these models (and indeed, of semiempirical models in general) as "neglecting" particular terms in the hamiltonian. It is, perhaps, both more precise and more intuitive to say that these models assume particular choices both for the basis set [i.e., for the terms in the summations of Eq. (6.75)] and for the values of the matrix elements. Different choices define different models. The notion of neglecting certain integrals [and the associated operators in (6.75)] is a deep-seated and traditional one in chemistry, and indeed some of the model hamiltonians to be discussed below have names in which the word "neglect" occurs.

The Hückel model hamiltonian, used for π-electron hydrocarbons and heterocycles, is one of the oldest semiempirical electronic structure models. In this model, the sum over basis functions is limited to one $p-\pi$ basis orbital on each carbon or heterocycle atom. All two-electron integrals $\langle kn|lm \rangle$ are set to zero. Then the hamiltonian can be written

$$H_{\text{huck}} = \sum_{i,j=1}^{N_\pi} \sum_{\sigma=\alpha,\beta} h_{ij} a_{i\sigma}^+ a_{j\sigma} \qquad (6.76)$$

with N_π equal to the number of π-basis orbitals. These orbitals are taken as orthonormal, so that the anticommutation relations (6.40) hold, and the matrix elements h_{ij} are set to

$$h_{ij} = \begin{cases} \alpha & i = j \\ \beta & i, j \text{ neighbors} \\ 0 & \text{else} \end{cases} \tag{6.77}$$

We discuss this hamiltonian, and the relation to molecular orbitals, more extensively in the next subsection.

The idea of using one-electron models actually predates modern quantum mechanics; Sommerfeld used free-electron ideas to discuss properties of metals. The free-electron molecular orbital model (FEMO) is often useful as a guide to the electronic properties of such strongly delocalized species as planar aromatic hydrocarbons or metal clusters. In this model hamiltonian, all interelectronic repulsions are set equal to zero. The basis functions u_j are taken as the eigenstates of a particle-in-a-box hamiltonian, for a box (in one, two, or three dimensions) chosen to enclose the molecule. Then the model hamiltonian becomes

$$H_{\text{FEMO}} = \sum_{\sigma = \alpha, \beta} \sum_{j=1}^{N_\pi} \varepsilon_j a_{j\sigma}^+ a_{j\sigma} \tag{6.78}$$

where $a_{j\sigma}^+$ creates an electron of spin σ in the jth orbital, which is just the jth eigenstate of the relevant particle-in-a-box problem, with eigenvalue ε_j.

At first glance, the FEMO model appears to be even simpler than the Hückel model in that no terms of the sort $a_{i\sigma}^+ a_{j\sigma} (i \neq j)$, called off-diagonal terms, ever occur. The total energy, computed as the expectation value of the hamiltonian, is just

$$E_{\text{FEMO}} = \langle H \rangle = \sum_j \varepsilon_j \sum_\sigma \langle a_{j\sigma}^+ a_{j\sigma} \rangle \tag{6.79a}$$

$$= \sum_j \sum_\sigma \varepsilon_j n_{j\sigma} \tag{6.79b}$$

with $n_{j\sigma} = \langle a_{j\sigma}^+ a_{j\sigma} \rangle$ the number of electrons of spin σ occupying the jth FEMO. Thus, as is expected for noninteracting electrons, the total energy is the sum of the energies for all the electrons. The FEMO model is written in the simple form (6.78), directly in terms of eigenstates, because the basis functions u_j are, in the FEMO case, not localized atomic functions but delocalized solutions to the particle-in-a-box hamiltonian. We shall see in the next subsection that the Hückel hamiltonian can be rewritten in terms of delocalized one-electron eigenfunctions and that, when this is done, the form is the same as (6.78).

The very simple and intuitive form of the Hückel expression can be extended in several ways. One of the simplest is to use different β values for different bond lengths; this is the idea behind the Sandorfy C model.

An extremely useful model hamiltonian is the extended Hückel model, developed and widely applied by Hoffmann. Here one writes exactly the same form as (6.76):

$$H_{\text{EXHUC}} = \sum_{\sigma=\alpha,\beta} \sum_{i,j=1}^{N_B} h_{ij} a_{i\sigma}^+ a_{j\sigma} \qquad (6.80)$$

This hamiltonian extends the Hückel concept in three ways: First, the basis orbital set $\{u_j\}$ includes all N_B valence atomic orbitals on the atoms considered; second, the atoms can be any atom from the periodic chart rather than only carbon or heterocycles, as in the Hückel model; and third, the basis functions $\{u_j\}$ are taken to be Slater-type orbitals, and the overlap integrals $S_{ij} = \langle u_i | u_j \rangle$ are calculated in the extended Hückel model, whereas the basis functions were taken to be orthonormal in the ordinary Hückel hamiltonian. Thus the extended Hückel model for LiH would be

$$H_{\text{EXHUC}} = \sum_{\sigma=\alpha,\beta} \{\alpha_H a_{1\sigma}^+ a_{1\sigma} + \alpha_{\text{Li},s} a_{2\sigma}^+ a_{2\sigma}$$
$$+ \beta_{\text{H,Li}} (a_{1\sigma}^+ a_{2\sigma} + a_{2\sigma}^+ a_{1\sigma})\} \qquad (6.81)$$

where u_1 and u_2 are, respectively, the $1s$ atomic Slater-type orbital (STO) on H and the $2s$ STO on Li. The parameters α and β are chosen semiempirically in terms of ionization potentials, electron affinities, and the $1s$–$2s$ overlap.

The one-electron model hamiltonians that we have written can all be interpreted in terms of energy terms of type $\beta_{ij} a_{i\sigma}^+ a_{j\sigma}$, corresponding to a lowering of the total energy by allowing electron motion between the ith and jth basis functions, and terms like $\alpha a_{i\sigma}^+ a_{i\sigma}$, giving the energy of an electron in the basis orbital u_i. The clearest and simplest form is one like (6.78), in which no off-diagonal terms appear, and the total energy can be interpreted, as in (6.79), as the sum of one-electron energies. The model form (6.76) can be rewritten in diagonal form, and we now discuss how to do so.

6.5.3 Molecular Orbitals for Noninteracting Electrons

In our previous treatment of the Hückel model in Section 2.2, we first wrote a matrix representation of the hamiltonian in an atomic orbital basis and then reexpressed the hamiltonian in a diagonal matrix, represented in terms of molecular orbitals. This same scheme can be used for any one-electron model hamiltonian, and we shall do so now, using the occupation number representation.

To be very explicit, the Hückel hamiltonian for 1,3-butadiene would be written

$$H = \sum_{\sigma=\alpha,\beta} \left\{ \alpha \sum_{i=1}^{4} a_{i\sigma}^{+} a_{i\sigma} + \beta[a_{1\sigma}^{+} a_{2\sigma} + a_{2\sigma}^{+} a_{1\sigma} + a_{2\sigma}^{+} a_{3\sigma} \right. \\ \left. + a_{3\sigma}^{+} a_{2\sigma} + a_{3\sigma}^{+} a_{4\sigma} + a_{4\sigma}^{+} a_{3\sigma}] \right\} \tag{6.82}$$

This hamiltonian is written in the representation of $p-\pi$ basis functions; as such, it is not diagonal (i.e., terms such as $\beta a_{3\sigma}^{+} a_{4\sigma}$ which would appear off the diagonal in a matrix representation of this hamiltonian, do not vanish). We wish, in general, to rewrite the hamiltonian in diagonal form. To do so, we define a new set of orbitals as (with u_j a local $p-\pi$ basis function)

$$\psi_\lambda = \sum_{j=1}^{N_\pi} c_{\lambda j} u_j \tag{6.83}$$

with arbitrary coefficients $c_{\lambda j}$; the ψ_λ are molecular orbitals, and this expansion represents the usual LCAO–MO form. We wish to find the expansion coefficients $c_{\lambda j}$ and the MO energies ε_λ.

To do so, we use the fact that the MOs are eigenstates of the one-electron hamiltonian H, and write

$$H\psi_\lambda = \varepsilon_\lambda \psi_\lambda \tag{6.84}$$

Then we express the MO as a one-electron function

$$\psi_\lambda = a_\lambda^{+}|\text{vac}\rangle \tag{6.85}$$

with $|\text{vac}\rangle$ the vacuum state containing no electrons, and with the spin index μ suppressed for clarity. Using the hamiltonian form of (6.76), we find, from (6.85),

$$\sum_{ij} h_{ij} a_i^{+} a_j a_\lambda^{+}|\text{vac}\rangle = \varepsilon_\lambda a_\lambda^{+}|\text{vac}\rangle \tag{6.86}$$

Now we define the creation operator for the MO ψ_λ by [in analogy to (6.83)]

$$a_\lambda^{+} = \sum_k c_{\lambda k}^{*} a_k^{+} \tag{6.87}$$

with Greek subscripts (λ, ν) labeling MOs and Latin subscripts labeling AOs. For real MOs, the complex conjugate can be ignored. Then (6.84) becomes

$$\sum_k \left(\sum_{ij} h_{ij} a_i^{+} a_j - \varepsilon_\lambda \right) c_{\lambda k} a_k^{+}|\text{vac}\rangle = 0 \tag{6.88}$$

Now we take the matrix element with the occupied one-electron atomic basis function $(a_l^+|\text{vac}))^* = \langle\text{vac}|a_l$:

$$\sum_k \left(\langle\text{vac}| \sum_{ij} a_l a_i^+ a_j a_k^+ |\text{vac}\rangle c_{\lambda k} h_{ij} - \langle\text{vac}|a_l a_k^+|\text{vac}\rangle c_{\lambda k}\varepsilon_\lambda\right) = 0 \quad (6.89)$$

Now using the identity $a_m|\text{vac}\rangle = 0$ and the anticommutation relations (6.40), we have

$$\sum_k \left(\sum_{ij} \delta_{jk}\delta_{il} h_{ij} c_{\lambda k} - \delta_{lk} c_{\lambda k}\varepsilon_\lambda\right) = 0 \quad (6.90)$$

$$\sum_k (h_{lk} - \varepsilon_\lambda\delta_{lk})c_{\lambda k} = 0$$

The result (6.90) is recognized as the secular equation set for determination of MO coefficients $c_{\lambda k}$ and energies ε_λ; these are determined, respectively, as the eigenfunctions and eigenvalues of the AO-represented hamiltonian matrix h_{ij}.

Now we write the hamiltonian in MO representation, following the general rule (6.55) as

$$H = \sum_{\lambda,\nu} \langle\lambda|H|\nu\rangle a_\lambda^+ a_\nu \quad (6.91)$$

which becomes, using (6.83),

$$H = \sum_{\lambda\nu} \sum_{kl} c_{\lambda k} c_{\nu l}\langle k|H|l\rangle a_\lambda^+ a_\nu = \sum_{\lambda\nu} \sum_l \left(\sum_k h_{kl} c_{\lambda k}\right) c_{\nu l} a_\lambda^+ a_\nu \quad (6.92)$$

which, using the secular equation (6.90), is (since $h_{lk} = h_{kl}$)

$$H = \sum_{\lambda\nu} \sum_l \sum_k \delta_{lk}\varepsilon_\lambda\, c_{\lambda k} c_{\nu l} a_\lambda^+ a_\nu = \sum_{\lambda\nu} \varepsilon_\lambda \sum_k c_{\lambda k} c_{\nu k} a_\lambda^+ a_\nu \quad (6.93)$$

Finally, using the orthonormality of the eigenvectors $\sum_k c_{\lambda k} c_{\nu k} = \delta_{\lambda\nu}$, we have

$$H = \sum_{\lambda\nu} \varepsilon_\lambda\delta_{\lambda\nu} a_\lambda^+ a_\nu = \sum_\lambda \varepsilon_\lambda a_\lambda^+ a_\lambda \quad (6.94)$$

or, reinserting the spin sum,

$$H = \sum_{\sigma=\alpha,\beta} \sum_{\lambda=1}^{N_B} \varepsilon_\lambda a_{\lambda\sigma}^+ a_{\lambda\sigma} \quad (6.95)$$

The hamiltonian expressed in the MO basis set ψ_λ is diagonal, and the total energy is again the sum of molecular orbital energies:

$$E_{\text{tot}} = \langle H\rangle = \sum_{\sigma=\alpha,\beta} \sum_{\lambda=1}^{N_B} \langle a_{\lambda\sigma}^+ a_{\lambda\sigma}\rangle\varepsilon_\lambda = \sum_\sigma \sum_\lambda \varepsilon_\lambda n_{\lambda\sigma} \quad (6.96)$$

with $n_{\lambda\sigma} = \langle a_{\lambda\sigma}^+ a_{\lambda\sigma}\rangle$ electrons of spin σ in the λ MO. In the situation where no interelectronic interactions are included, so that the model hamiltonian in the arbitrary initial basis looks like (6.76) or (6.80), construction of the molecular orbitals simply requires finding the diagonal (MO) form (6.95) by solving the secular equation (6.90). Note that the sum in (6.95) runs over N_B functions, since there will be as many MOs as there were AO basis functions. Note also that the forms (6.76) or (6.80) in the AO representation and (6.95) in the MO representation correspond to the same model hamiltonian but that, clearly, the MO model is more easily interpreted.

For complete clarity, we write explicitly that for the Hückel model of butadiene, the hamiltonian in MO representation, corresponding to (6.82) in the AO representation, is

$$H = \sum_{\sigma=\alpha,\beta} \sum_{\lambda=1}^{4} \varepsilon_\lambda a_{\lambda\sigma}^{+} a_{\lambda\sigma} \tag{6.97}$$

with $\varepsilon_1 = \alpha + 1.618\beta$, $\varepsilon_2 = \alpha + 0.618\beta$, $\varepsilon_3 = \alpha - 0.618\beta$, $\varepsilon_4 = \alpha - 1.618\beta$.

One-electron model hamiltonians, like Hückel, FEMO, and extended Hückel, have the great advantage that they can be diagonalized by simple solution of the secular equation (6.90) (or its generalization with the AO overlap S_{lk} replacing the Kronecker delta δ_{lk} for nonorthonormal AOs), and that the resulting MO form (6.95) is simply the sum of independent one-electron energies. These simplifications arise because the original model did not include interelectronic repulsion—it is actually an independent-electron rather than a many-electron model. More general models are necessary for realistic quantitative descriptions of most properties (bonding energy, electronic spectra, ionization energies, polarizability, potential surfaces). Such interacting-electron models include the ab initio models described in Section 6.4.1, and semiempirical many-electron models.

6.6

Treatment of Interacting Electrons

6.6.1 Model Hamiltonians

When electron interactions are included in the electronic structure model hamiltonian, it is necessary to define the basis functions labeled *klmn* in the second term of Eq. (6.75), as well as the matrix elements

$$\langle ij|lk \rangle = \int d^3r_1 \, d^3r_2 \, u_i(1)u_j(1) \frac{e^2}{r_{12}} u_l(2)u_k(2) \tag{6.98a}$$

For *ab initio* studies, once the basis is defined, the matrix elements are also, simply by evaluating the six-dimensional integral in Eq. (6.98a). Semiempirical methods, on the other hand, generally choose values for these integrals in some approximate way, often using experimental comparison to a standard set of molecules to fix the parameters in the model hamiltonian.

Note that we can rewrite Eq. (6.98a) as

$$\langle ij|lk \rangle = \int d^3r_1 \int d^3r_2 [u_i(1)u_j(1)][u_l(2)u_k(2)] \frac{e^2}{r_{12}} \qquad (6.98b)$$

where the brackets are introduced to demonstrate that, formally, the integrand contains components of overlap integrands, $u_i u_j$ for one electron and $u_l u_k$ for the other. With the assumption that

$$u_i(1)u_j(1) = \delta_{ij} u_i^2(1) \qquad (6.99)$$

the second term in (6.75) becomes

$$\frac{1}{2} \sum_{\sigma\rho} \sum_{ijkl} a_{i\sigma}^+ a_{j\rho}^+ a_{k\rho} a_{l\sigma} \langle il|jk \rangle \delta_{il}\delta_{jk}$$

$$= \frac{1}{2} \sum_{\sigma\rho} \sum_{ij} a_{i\sigma}^+ a_{j\rho}^+ a_{j\rho} a_{i\sigma} \langle ii|jj \rangle$$

$$= \frac{1}{2} \sum_{ij} \sum_{\sigma\rho} \langle ii|jj \rangle \, a_{i\sigma}^+ a_{i\sigma} a_{j\rho}^+ a_{j\rho}$$

$$- \frac{1}{2} \sum_{\sigma} \sum_i \langle ii|ii \rangle \, a_{i\sigma}^+ a_{i\sigma}$$

$$(6.100)$$

The forms in (6.100) include only repulsion of an electron density in basis function i by an electron density in basis function j. Note that when the expectation value of this hamiltonian is evaluated with any wavefunction, the result for the electron repulsion energy is independent of the spin character of that wavefunction.

The assumption of Eq. (6.99) is often called *complete neglect of differential overlap* (CNDO). The differential overlap referred to is the basis function products $u_i u_j$ or $u_k u_l$, which are set equal to zero if the two basis functions in the product are not identical. Two very common semiempirical molecular orbital model hamiltonians use the assumption (6.99). If the basis set $\{u_i\}$ includes all valence Slater-type orbitals, the model is called CNDO; it or its variants are often used for calculations of molecular properties such as geometry, charge distribution, or optical properties.

Alternatively, if the basis set $\{u_i\}$ is restricted to one $2p\pi$ basis function on each carbon or heteroatom center in a π-electron system but (6.99) is still assumed, the resulting model is generally called, after its developers, the PPP (Pariser–Parr–Pople) model. It is very useful for such π-electron properties as electron densities, optical and nonlinear optical spectra, and π-bonding descriptions.

If we make the further assumption that

$$u_i(1)u_i(1)u_j(2)u_j(2) = \delta_{ij}u_i^2(1)u_i^2(2) \qquad (6.101)$$

in (6.98b), as well as the Hückel-like assumption that h_{ij} vanishes unless $i = j$ or i is adjacent to j, the resulting electronic structure model

$$H = \sum_\sigma \sum_i h_{ii}a_{i\sigma}^+ a_{i\sigma} + \sum_\sigma \sum_{ij}{}' t_{ij}a_{i\sigma}^+ a_{j\sigma} + \sum_i U_i a_{i\uparrow}^+ a_{i\uparrow} a_{i\downarrow}^+ a_{i\downarrow} \quad (6.102a)$$

is obtained, with $U_i \equiv \langle ii|ii \rangle$. If, finally, the sites are all assumed equivalent, so that $t_{ij} = t$ for all (i, j) that are neighbors, $U_i = U$ for all sites, and $h_{ii} = \varepsilon$ for all sites, we find that

$$H = \varepsilon \sum_i \sum_\sigma a_{i\sigma}^+ a_{i\sigma} + t \sum_{i,j}{}' \sum_\sigma a_{i\sigma}^+ a_{j\sigma} + U \sum_i a_{i\uparrow}^+ a_{i\uparrow} a_{i\downarrow}^+ a_{i\downarrow} \quad (6.102b)$$

The first term is just εN, where N is the total number of electrons; this is a constant, and by choosing $\varepsilon = 0$ as the energy origin, it vanishes. The prime on the second term in (6.102) indicates that the sum is restricted to neighbors. The hamiltonian (6.102b) is the simplest one that includes electron repulsion; such repulsion is limited to that between opposite spin electrons on the same site. This is often referred to as the Hubbard hamiltonian. Although it has not been widely used in molecular electronic structure problems, it is very popular in such solid-state problems as conductivity in molecular metals and magnetism in narrowband (small t/U) materials.

We have already noted that the CNDO assumption (6.99) produces electronic structure model hamiltonians (PPP, CNDO, Hubbard) that do not distinguish among different spin multiplets because no exchange integrals are retained. This can be changed if the assumption

$$u_i(1)u_j(1) = \delta_{c_i,c_j}u_i(1)u_j(1) \qquad (6.103)$$

is used. Here δ_{c_i,c_j} is defined to be zero unless the basis functions u_i and u_j are on the same atomic center. The approximation (6.103), implemented in a valence-only minimum basis of Slater-type orbitals, defines the so-called INDO (intermediate neglect of differential overlap) model hamiltonian, which is simply

$$H = \sum_{\sigma} \sum_{i,j} h_{ij} a^+_{i\sigma} a_{j\sigma} + \frac{1}{2} \sum_{\sigma,\rho} \sum_{i,j} \gamma_{ij} a^+_{i\sigma} a_{i\sigma} a^+_{j\rho} a_{j\rho}$$

$$+ \frac{1}{2} \sum_{\sigma\rho} \sum_{ik}{}' \sum_{jl}{}' \langle il|jk\rangle a^+_{i\sigma} a^+_{j\rho} a_{k\rho} a_{l\sigma} \tag{6.104}$$

The orbital sums run over the valence Slater basis, γ_{ij} is a semiempirically determined Coulomb repulsion integral between electron densities on sites i, j and the sums in the last term are restricted to the condition that u_i and u_l are different basis functions on the same atom, as are u_k and u_j. The CNDO model hamiltonian differs only in that it omits the last term. This term includes exchange integrals of the type, say, $\langle xy|yx\rangle$, and is important if multiplet analysis is to be undertaken, as well as for magnetic problems and excited-state studies.

Unlike the Hückel models, in which the eigenstates and energies can be found trivially by diagonalizing the matrix representation of the hamiltonian, the model hamiltonians with any inclusion of electron repulsion (Hubbard, PPP, CNDO, ab initio) are true many-body hamiltonians, whose eigenfunctions and eigenvalues can be found only approximately for most problems. The most standard and important approximate solution, and the usual starting point for improved approximate solutions, is the self-consistent field (SCF) approximation.

6.6.2 Self-Consistent Field (SCF) Solution

The self-consistent field scheme for electronic structure calculations was first put forward by Hartree and by Fock for atoms. Its generalization to deal with LCAO–MO problems in molecules, by Roothaan and Hall in 1951, really began the era of molecular orbital calculations.

The idea of self-consistent field, or mean field, calculations is a common one in physics. The essential physical idea is to replace the exact dynamics or energetics of a particle or mode by its evolution in a field averaged over all the other particles or modes.

There are many derivations of the Hartree–Fock equations. A particularly simple and elegant one can be given by approximating the time evolution of the fermion operators a and a^+. That is, since we know from the Heisenberg equation that

$$i\hbar \frac{da_{s\mu}}{dt} = [a_{s\mu}, H] \tag{6.105}$$

a physically reasonable approximation is to write

$$i\hbar \frac{da_{s\mu}}{dt} = \sum_{u} f^\mu_{su} a_{u\mu} \tag{6.106}$$

Combining these last two equations, we find that

$$[a_{s\mu}, H] = \sum_u f^\mu_{su} a_{u\mu} \tag{6.107}$$

or, multiplying first from the right then from the left and adding, we get

$$\langle [a^+_{t\mu}, [a_{s\mu}, H]]_+ \rangle = \sum_u f^\mu_{su} \langle [a^+_{t\mu}, a_{u\mu}]_+ \rangle = f^\mu_{st} \tag{6.108}$$

where the brackets indicate an average over the ground state. Here we have assumed that the orbitals $\{\phi_s\}$ form an orthonormal set.

The operator whose elements are f_{rs} is often called the Fock operator. It is a hermitian operator whose eigenvalues are orbital energies. Using the general form (6.75) for the molecular electronic hamiltonian in an orthonormal basis, we can find f_{st} as

$$f^\mu_{st} = \langle [a^+_{t\mu}, [a_{s\mu}, H]]_+ \rangle = -\langle [a^+_{t\mu}, [H, a_{s\mu}]]_+ \rangle \tag{6.109a}$$

$$= -\left\langle \left[a^+_{t\mu}, \left[\sum_{lm} \sum_\sigma h_{lm} a^+_{l\sigma} a_{m\sigma}, a_{s\mu} \right] \right]_+ \right\rangle$$

$$- \frac{1}{2} \left\langle \left[a^+_{t\mu}, \left[\sum_{kl} \sum_{mn} \sum_{\sigma\rho} \langle kn|lm \rangle a^+_{k\sigma} a^+_{l\rho} a_{m\rho} a_{n\sigma}, a_{s\mu} \right] \right]_+ \right\rangle$$

$$= h_{st} + \sum_{lm} \left\{ \sum_\sigma \langle lm|st \rangle \langle a^+_{l\sigma} a_{m\sigma} \rangle - \langle lt|sm \rangle \langle a^+_{l\mu} a_{m\mu} \rangle \right\} \tag{6.109b}$$

Exercise: Derive (6.109b) from (6.109a). To do so, remember that $\langle kn|lm \rangle = \langle nk|lm \rangle = \langle lm|nk \rangle$.

If the state over which one averages in (6.109b) is of closed-shell type, then

$$\langle a^+_{l\alpha} a_{m\alpha} \rangle = \langle a^+_{l\beta} a_{m\beta} \rangle$$

and then (6.109b) becomes

$$f^\mu_{st} = h_{st} + \sum_{lm} (2\langle st|lm \rangle - \langle sm|lt \rangle) \langle a^+_{l\mu} a_{m\mu} \rangle \tag{6.110}$$

The structure of the Fock matrix elements in (6.110) is of interest. The one-electron term is just the one-electron matrix element between orbitals ϕ_s and ϕ_t. In noninteracting electron models such as FEMO or extended Hückel, then, the Fock operator is simply the hamiltonian. For interacting electrons, the operator f^μ_{st} depends on the orbitals through the $\langle a^+_{l\mu} a_{m\mu} \rangle$ average. Thus the orbitals, which are the eigenfunctions of the Fock operator, determine that operator itself. This is the essential self-consistent field (SCF) aspect of the Hartree–Fock SCF method: The effective one-electron hamiltonian f^μ_{rs} depends on its own eigenfunctions. Ordinarily, then, the equations

$$\psi_\lambda^\mu = \sum c_{\lambda t}^\mu \phi_t^\mu \qquad (6.111a)$$

$$\sum_t (f_{st}^\mu - \varepsilon_\lambda S_{st})c_{\lambda t}^\mu = 0 \qquad (6.111b)$$

define molecular orbitals. The eigenvalues of the Fock matrix f_{rs}^μ are the one-electron energy eigenvalues, and the eigenvalues are the molecular orbitals. The matrix defined by

$$\gamma_{ml}^\mu = \langle a_{l\mu}^+ a_{m\mu} \rangle = \sum_\lambda c_{\lambda l}^{*\mu} c_{\lambda m}^\mu n_\lambda^\mu \qquad (6.112)$$

is often called the charge and bond order matrix, since its diagonal and off-diagonal values relate, in simple one-electron theory, to electron density on a site and to bond orders between sites, respectively. This matrix is computed iteratively in the solution to the SCF equations (6.111). From knowledge of the f_{st}^μ, one finds the wavefunctions ψ_λ, which are then used to find γ_{ml}^μ and then [from (6.110)] to redefine f_{st}^μ. The HF equations (6.111) are the generalization of the independent-particle equations (6.90) that take interelectronic repulsion into account in an averaged way.

The one-electron orbital energies, from (6.110), consist of three parts. The first, due to h_{rs}, is just the kinetic energy and nuclear attraction. The terms proportional to $\langle st|lm \rangle$ and to $\langle sm|lt \rangle$ are, respectively, the Coulomb and exchange contributions. Note that, from (6.109b), the Coulomb repulsion felt by one orbital comes from other electrons of both spins, while the exchange is due only to electrons of the same spin. This is because the exchange arises from the Pauli principle, which requires spatial antisymmetry only between electrons of the same spin. The SCF analysis can be used for any model hamiltonian, either ab initio or semiempirical.

6.6.3 Example: SCF Solution for the Two-Center, Two-Orbital Problem

The classic problem for discussion of chemical binding is H_2, using a minimum basis of one $1s$ orbital on each center. To simplify the description, we assume orthogonality of the basis set, so that

$$\langle u_i | u_j \rangle = \delta_{ij} \qquad i, j = 1, 2 \qquad (6.113a)$$

$$[a_i, a_j^+]_+ = \delta_{ij} \qquad i, j = 1, 2 \qquad (6.113b)$$

then in the atomic orbital representation, we have

$$f_{11} = h_{11} + \sum_{lm} \langle a_{l\mu}^+ a_{m\mu} \rangle (2\langle lm|11 \rangle - \langle l1|1m \rangle)$$

$$f_{22} = h_{22} + \sum_{lm} \langle a_{l\mu}^+ a_{m\mu} \rangle (2\langle lm|22 \rangle - \langle l2|2m \rangle) \qquad (6.114)$$

$$f_{12} = f_{21} = h_{12} + \sum_{lm} \langle a_{l\mu}^+ a_{m\mu} \rangle (2\langle lm|12 \rangle - \langle l2|1m \rangle)$$

The molecular orbitals are, from simple symmetry considerations,

$$\phi_\pm = \frac{1}{\sqrt{2}} (\phi_1 \pm \phi_2) \qquad (6.115)$$

and in this representation the ground MO state is

$$\psi_{MO} = a_{+\beta}^+ a_{+\alpha}^+ |vac\rangle$$
$$= \tfrac{1}{2}\{(a_{1\beta}^+ a_{1\alpha}^+ + a_{2\beta}^+ a_{2\alpha}^+) + (a_{1\beta}^+ a_{2\alpha}^+ + a_{2\beta}^+ a_{1\alpha}^+)\}|vac\rangle \quad (6.116)$$

which is an equal admixture of covalent and ionic structures. One can then show by direct substitution that $\langle a_{2\mu}^+ a_{2\mu} \rangle = \langle a_{1\mu}^+ a_{1\mu} \rangle = \tfrac{1}{2}$ and $\langle a_{-\mu}^+ a_{-\mu} \rangle = 0$, $\langle a_{+\mu}^+ a_{+\mu} \rangle = 1$, Then the Fock operator matrix is

$$\mathbf{f} = \begin{pmatrix} h_{++} + \langle ++|++ \rangle & 0 \\ 0 & h_{--} + 2\langle --|++ \rangle - \langle -+|+- \rangle \end{pmatrix} \quad (6.117)$$

where $h_{++} = h_{11} + h_{12}$, $h_{--} = h_{11} - h_{12}$, and the one-electron energy level of the ground state is

$$\varepsilon_+ = h_{11} + h_{12} + \frac{1}{4}\{2\langle 11|11 \rangle + 8\langle 21|11 \rangle + 4\langle 12|12 \rangle + 2\langle 11|22 \rangle\}$$
$$(6.118)$$

(we have used $\langle 11|11 \rangle = \langle 22|22 \rangle$ and $\langle 11|12 \rangle = \langle 22|21 \rangle$). The total energy in this state is

$$E_{tot} = \langle \psi_{MO}|H|\psi_{MO} \rangle = 2h_{++} + \langle ++|++ \rangle \qquad (6.119)$$

BIBLIOGRAPHY

Avery, J., *Creation and Annihilation Operators* (McGraw-Hill, New York, 1978).

Davydov, A. S., *Quantum Mechanics* (Pergamon Press, Oxford, 1976).

Dirac, P. A. M., *The Principles of Quantum Mechanics*, 4th Ed. (Clarendon Press, Oxford, 1958).

Hehre, W. J., L. Radom, P. v. R. Schleyer, and J. A. Pople, *Ab Initio Molecular Orbital Theory* (Wiley, New York, 1986).

Hirst, D. M., *A Computational Approach to Chemistry* (Blackwell Scientific, London, 1990).

Jørgensen, P., and J. Simons, *Second Quantization-Based Methods in Quantum Chemistry* (Academic Press, New York, 1981).

Linderberg, J., and Y. Öhrn, *Propagators in Quantum Chemistry* (Academic Press, London, 1973).

Murrell, J. N., and A. J. Harget, *Semi-empirical Self-Consistent-Field Molecular Orbital Theory of Molecules* (Wiley-Interscience, New York, 1972).

Pople, J. A., and D. L. Beveridge, *Approximate Molecular Orbital Theory* (McGraw-Hill, New York, 1970).

Richards, W. G., and D. L. Cooper, *Ab Initio Molecular Orbital Calculations for Chemists*, 2nd Ed. (Clarendon Press, Oxford, 1983).

Schaefer, H. F., *Quantum Chemistry: The Development of Ab Initio Methods in Molecular Electronic Structure Theory* (Oxford University Press, Oxford, 1984).

Schiff, L., *Quantum Mechanics* (McGraw-Hill, New York, 1968).

Szabo, A., and N. S. Ostlund, *Modern Quantum Chemistry* (McGraw-Hill, New York, 1989).

Wilson, S., *Electron Correlation in Molecules* (Clarendon Press, Oxford, 1984).

PROBLEMS FOR CHAPTER 6

1. Evaluate

(a) $\langle 0 | a_{i\mu} a_{j\nu} a_{k\mu}^+ a_{l\nu}^+ | 0 \rangle$

(b) $\langle 0 | b_{i\mu} b_{j\nu} b_{k\mu}^+ b_{l\nu}^+ | 0 \rangle$

where the a's and b's are the usual fermion and boson operators. The result should not contain operators.

2. Use the occupation number representation for electromagnetic fields to derive an expression for two-photon spontaneous emission. Assume that the interaction is

$$V(t) = \frac{e^2}{2mc^2} \mathbf{A}_{\omega_{l'}} \cdot \mathbf{A}_{\omega_l}$$

where l and l' specify two modes of the field. Ignore other modes. As part of the solution, specify:

(a) The vector potentials $\mathbf{A}_{\omega_l}$ and $\mathbf{A}_{\omega_{l'}}$ and the interaction V in the occupation number representation.

(b) The relevant matrix element V_{km} for emission of the photons at frequencies ω_l and $\omega_{l'}$.

(c) An expression for the differential rate of emission into solid angles $d\Omega$ and $d\Omega'$.

3. Prove the result (6.73) from (6.66), (6.72), and (6.69a). [*Hint:* Start from (6.69a) by multiplying from the left by $\phi_k^*(\mathbf{r})$ and integrating to find a representation for a_k.]

4. Consider the PPP π-electron model for ethylene. In the atomic basis set consisting of a simple p–π orbital on each site, the hamiltonian is

$$H = H_1 + H_2$$

$$H_1 = \beta \sum_\mu (a_{1\mu}^+ a_{2\mu} + a_{2\mu}^+ a_{1\mu})$$

$$H_2 = \sum_\mu \sum_\nu \sum_{i=1}^{2} \sum_{j=1}^{2} \gamma_{ij} a_{i\mu}^+ a_{i\mu} a_{j\nu}^+ a_{j\nu}$$

(a) Consider the operator $n_{k\mu} = a_{k\mu}^+ a_{k\mu}$, where $k = 1, 2$. Show that $n_{k\mu}$ commutes with H_2. If $\beta = 0$ and $\gamma_{11} = \gamma_{22}$, $\gamma_{12} = \gamma_{21}$, show that the ground state of the system has an energy γ_{12} (provided that $\gamma_{12} < \gamma_{11}$).

(b) Now consider $n_{\pm\mu} = a_{\pm\mu}^+ a_{\pm\mu}$, where $a_{\pm\mu} = 2^{-1/2}(a_{1\mu} \pm a_{2\mu})$. Show that $n_{\pm\mu}$ commutes with H_1. If $H_2 = 0$, what is the ground-state energy of the system?

(c) One way to write the Hartree–Fock hamiltonian for ethylene is

$$H^{\mathrm{HF}} = H_1 + \sum_\mu \sum_\nu \sum_{i=1}^{2} \sum_{j=1}^{2} \gamma_{ij}\{n_{i\mu}\langle n_{j\nu}\rangle + n_{j\nu}\langle n_{i\mu}\rangle - \langle n_{i\mu}\rangle\langle n_{j\nu}\rangle\}$$

Assuming that $\langle n_{i\mu}\rangle = \langle n_{j\nu}\rangle = \frac{1}{2}$, show that H^{HF} is diagonal if expressed in terms of $n_{\pm\mu}$, so that the molecular orbitals of the Hückel problem are also molecular orbitals of the HF problem.

5. Consider the problem of mixed valency in a metal complex such as $a_5\mathrm{Ru}$–pyz–Rua_5^{+5}, with a $= \mathrm{NH}_3$ and pyz $=$ pyrazine. Neglecting interelectronic repulsion, we can write a very simple two-site model hamiltonian (the Fröhlich hamiltonian) as

$$H = H_{\mathrm{el}} + H_{\mathrm{nuc}}$$

$$H_{\mathrm{nuc}} = \frac{P_{\mathrm{nuc}}^2}{2M} + \frac{1}{2}kX^2$$

$$H_{\mathrm{el}} = \beta(a_1^+ a_2 + a_2^+ a_1) + g\hbar\omega(a_2^+ a_2 - a_1^+ a_1)X\left(\frac{2M\omega}{\hbar}\right)^{1/2}$$

where X is the nuclear displacement and M, k, β, ω, and g are nuclear reduced mass, force constant, tunneling integral, frequency [$\omega = (k/M)^{1/2}$], and dimensionless coupling strength, respectively. The displacement X actually is a distortion, equal, for instance, to the difference of the Ru–a distances on the left and right Ru's.

The Born–Oppenheimer potential surface is defined by

$$V_{\mathrm{BO}}(X) = V_{\mathrm{nuc}}(X) + \langle\psi|H_{\mathrm{el}}|\psi\rangle$$

for any electron state ψ.

(a) Compute the Fock operator matrix elements for the electrons. The Fock matrix is a 2×2, with

$$f_{ij} = \langle [[a_i, H], a_j^+]_+ \rangle$$

but now the elements depend on X.

(b) Find the two electronic eigenvalues from (a). Add these to V_{nucl} to find the total BO potential curves $V_{BO}(X)$. There will be two of these. Plot the curves for the specific parameters

$$\beta = -0.02 \text{ eV} \qquad g = 1$$

$$\omega = 500 \text{ cm}^{-1} \qquad M = 10{,}000 m_e$$

How do these curves change when $\beta = -2.0$ eV?

(c) Suppose that in the mixed-valent Ru complex we take a_1^+ as creating an electron in the d_{xy} orbital on the left, and a_2^+ analogously on the right. Assume that at time zero, the electronic configuration is Ru^{II}–Ru^{III}, so that

$$\psi_{el}(t = 0) = a_1^+ |0\rangle$$

Using the Franck–Condon idea, calculate the electronic excitation energy for the parameters of part (b). This transition is normally called the intervalence transfer band.

6. Consider the minimum-basis H_2 molecule as discussed in Section 6.6.3.

(a) From simple group theory considerations, show that if one constructs the CI matrix, the $a_{+\beta}^+ a_{+\alpha}^+ |vac\rangle$ state will interact only with the doubly excited singlet $\langle a_{-\beta}^+ a_{-\alpha}^+ |vac\rangle$.

(b) Show that the energy difference between the singlet state

$$\psi_{sing}^* = \frac{1}{\sqrt{2}} (a_{+\alpha}^+ a_{-\beta}^+ + a_{-\alpha}^+ a_{+\beta}^+)|vac\rangle$$

and the triplet state

$$\psi_{trip}^* = \frac{1}{\sqrt{2}} (a_{+\alpha}^+ a_{-\beta}^+ - a_{-\alpha}^+ a_{+\beta}^+)|vac\rangle$$

is $2K$, where

$$K = \langle -+|+-\rangle$$

is the exchange integral in the MO approximation.

7. Consider an open-shell excited beryllium atom Be ($1s^2 \, 2s \, 2p$). Assume that the p electron is present in the p_z orbital.

(a) Write the singlet and triplet states ($m_s = 0$) of this open-shell configuration. Do this first in terms of Slater determinants, then in terms of the occupation number formalism with operators $\{a_{s\alpha}, a_{p\alpha}, a_{s\beta}, a_{p\beta}\}$ and their adjoints.

(b) Evaluate the singlet and triplet energies assuming four different model hamiltonians: (1) extended Hückel, (2) CNDO, (3) INDO, and (4) ab initio, minimum Slater basis. For each model, evaluate the energy difference between singlet and triplet. Comment on the relative appropriateness of these model hamiltonians for magnetic problems.

8. Consider, as a model for the π electrons of propene, a PPP-type hamiltonian, with different one-electron parameters on the two sites. Then

$$H = \sum_{\mu} [a_{1\mu}^{+} a_{1\mu} \bar{\alpha}_1 + a_{2\mu}^{+} a_{2\mu} \bar{\alpha}_2 + \beta a_{1\mu}^{+} a_{2\mu} + \beta a_{2\mu}^{+} a_{1\mu}]$$

$$+ \bar{\gamma}(n_{1\uparrow} n_{1\downarrow} + n_{2\uparrow} n_{2\downarrow}) + \bar{\Gamma}(n_{1\uparrow} + n_{1\downarrow})(n_{2\uparrow} + n_{2\downarrow})$$

(a) Using the fact that for $m_s = 0$, $n_{1\uparrow} + n_{2\uparrow} = n_{1\downarrow} + n_{2\downarrow} = 1$, rewrite the hamiltonian as a Hubbard model, with $\gamma = \gamma_1 = \gamma_2$ as functions of $\bar{\gamma}$ and $\bar{\Gamma}$.

(b) Now choose the origin of energy as the one-electron energy on site 1, thus expressing the entire hamiltonian in terms of the parameters, γ, β, $\alpha = \bar{\alpha}_2 - \bar{\alpha}_1$.

(c) Now assuming that $\beta = -3.0$ eV, $\gamma = 3.0$ eV, $\alpha_2 - \alpha_1 = 1.0$ eV, compute the Hückel orbital energies and total energy (remember that $E = \langle \psi | H | \psi \rangle / \langle \psi | \psi \rangle$).

(d) Now construct the 2×2 Fock operator matrix for α spin (the one for β spin is the same). By actual iteration, find the Hartree–Fock orbital energies. Also find the ground-state (binding) and first-excited-state energy.

(e) Find the Hartree–Fock total energy. Compare to the total energy calculated using the Hückel wavefunction. Does the result make sense in view of the variational principle?

(f) Qualitatively, does the effect of the Hubbard γ term (minimizing polarity) explain the difference between Hückel and Hartree–Fock bonding orbital energies?

(g) Using the localized set of two-electron functions $\{|1\alpha 1\beta\rangle, |2\alpha 2\beta\rangle, |1\alpha 2\beta\rangle, |2\alpha 1\beta\rangle\}$ as a complete basis, evaluate the total energy of the four $m_s = 0$ levels. Solve the quartic numerically. [Use the same parameter values as in part (c).]

7

Quantum Scattering Theory

7.1

Introduction

In quantum scattering theory one is interested in collisions involving atoms, molecules, other species, and even solids. Such collisions can produce many possible results, ranging from elastic scattering to reaction and fragmentation. Scattering theory can also be used to describe many dynamical processes within solids, such as collisions of collective wave motions (such as phonons) with impurities, or the motions of an ejected photoelectron as it escapes from a crystal. While the basic formalism of quantum scattering theory can be found in a variety of physics-oriented textbooks, many chemical and materials applications require special adaptation of the theory. For example, in problems such as gas–surface collisions and tunneling reactions in liquids and solids, angular momentum conservation is not important (or does not apply) and angularly resolved scattering information is not available. As a result, the theory of scattering in one dimension is more useful than in typical particle scattering problems. Thus we will begin our development by considering one-dimensional scattering. Another issue of importance to chemical problems is the short deBroglie wavelength usually associated with molecular motion. This makes it important to consider semiclassical approximations, so we will discuss this later in this chapter. Finally, because most chemical problems involve initial states that are selected from a Boltzmann distribution, it is very important to develop theories where thermally averaged results are obtained directly. This is the subject of Chapter 8.

7.2

One-Dimensional Scattering

7.2.1 Introduction

The problem that we want to solve in this section is defined by the simple one-dimensional hamiltonian

$$H = \frac{p^2}{2m} + V(x) \tag{7.1}$$

where $V(x)$ is a potential such as that pictured in Fig. 7.1. Note that the potential is flat for $x \to \pm\infty$, with $V(x \to -\infty) = 0$ and $V(x \to +\infty) = V_0$.

Examples where this type of potential are relevant include one-dimensional models of electron, proton, or hydrogen atom transfer reactions, tunneling of electrons at interfaces, and the scattering of molecules from other molecules or from surfaces.

The physical question that we wish to answer is: What is the probability P that a particle incident with an energy E from the left at $x \to -\infty$ will end up moving to the right at $x \to +\infty$. The classical solution to this problem is straightforward, namely

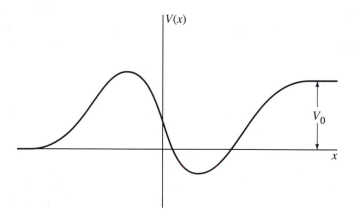

Figure 7.1. Potential associated with the scattering of a particle in one dimension.

1. $P(E) = 0$ if $V(x) \geq E$ for any x.
2. $P(E) = 1$ if $V(x) < E$ for all x.

The quantum solution to this problem is much more difficult, for a number of reasons: First, we need to learn how to define what we mean by a particle moving in a given direction when $V(x)$ is constant; second, we need to determine how much of the particle is moving in any specified direction at any desired location; and third, we need to be able to solve the Schrödinger equation for the potential $V(x)$.

7.2.2 Wavepackets in One Dimension

Consider the case of a free particle for which $V(x) = 0$. In this case the time-dependent Schrödinger equation is

$$i\hbar \frac{\partial \Psi}{\partial t} = -\frac{\hbar^2}{2m} \frac{\partial^2 \Psi}{\partial x^2} \tag{7.2}$$

and if we go through the usual procedure for separating the time and spatial parts of this equation, we can readily show that one possible solution is

$$\Psi_k(x, t) = e^{-iEt/\hbar} e^{ikx} \tag{7.3}$$

where

$$E = \frac{\hbar^2 k^2}{2m} \tag{7.4}$$

is the particle's energy and $\hbar k$ is its linear momentum. Note that both energy and momentum of the particle are exactly specified in this solution. As might be expected from the uncertainty principle, the location of the particle is completely undetermined.

To localize the particle, it is necessary to superimpose wavefunctions Ψ_k with different momenta k. A very general way to do this is to construct a wavepacket, defined through the integral

$$\Psi_{wp}(x, t) = \int_{-\infty}^{\infty} dk \, C(k) \Psi_k(x, t) = \int_{-\infty}^{\infty} dk \, C(k) e^{ikx} e^{-i\hbar k^2 t/2m} \tag{7.5}$$

where $C(k)$ is a function that tells us how much of each momentum $\hbar k$ is contained in the wavepacket. If the particle is to move with roughly a constant velocity, $C(k)$ must be peaked at some k that we take to be k_0. One function that accomplishes this is the gaussian

$$C(k) = \sqrt{\frac{a}{2\pi^{3/2}}} \exp\left[\frac{-a^2(k - k_0)^2}{2}\right] \tag{7.6}$$

where a measures the width of the packet. If we substitute this into Eq. (7.5), the result is

$$\Psi_{wp}(x, t) = \pi^{-1/4}\left[a\left(1 + \frac{i\hbar t}{ma^2}\right)\right]^{-1/2}$$

$$\times \exp\left[-\frac{(x - \hbar k_0 t/m)^2}{2a^2(1 + i\hbar t/ma^2)} + ik_0 x - \frac{i\hbar t}{2ma^2}\right] \tag{7.7}$$

The absolute square of this wavefunction is

$$|\Psi_{wp}|^2 = \pi^{-1/2}a^{-1}\left[1 + \frac{\hbar^2 t^2}{m^2 a^4}\right]^{-1/2}$$

$$\times \exp\left[-\frac{(x - \hbar k_0 t/m)^2}{a^2(1 + \hbar^2 t^2/m^2 a^4)}\right] \tag{7.8}$$

Figure 7.2 shows a plot of $|\Psi_{wp}|^2$ as a function of x, and it should be apparent that this is a gaussian function that peaks at $x = \hbar k_0 t/m$, moving to the right with a momentum $\hbar k_0$. The width of this peak is

$$\Delta = a\left[(\ln 2)\left(1 + \frac{\hbar^2 t^2}{m^2 a^4}\right)\right]^{1/2} \tag{7.9}$$

which starts out at $\Delta = a(\ln 2)^{1/2}$ at $t = 0$ and increases linearly with time for large t. This spreading of the wavepacket reflects the different momentum components present and it is a natural consequence of the uncertainty principle. Note that the wavefunction in Eq. (7.7) still satisfies the Schrödinger equation [Eq. (7.2)].

One can show that the expectation value of the hamiltonian operator for the wavepacket in Eq. (7.7) is

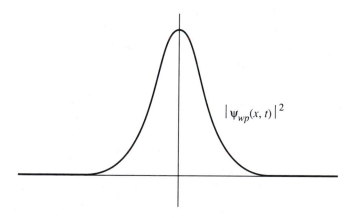

Figure 7.2. $|\Psi_{wp}|^2$ from Eq. (7.8).

$$\langle H \rangle = \frac{\hbar^2 k_0^2}{2m} + \frac{\hbar^2}{4ma^2} \qquad (7.10)$$

The first term is what one would expect to get classically for a particle of momentum $\hbar k_0$, and it is much bigger than the second term provided that $k_0 a \gg 1$. Since the deBroglie wavelength λ is $2\pi/k_0$, this condition is equivalent to the statement that the size of the wavepacket be much larger than the wavelength.

It is also notable that the spreading of the wavepacket can be neglected for times t such that $t \ll ma^2/\hbar$. In this time interval the center of the wavepacket will have moved a distance $(k_0 a)a$. Under the conditions noted above for which $k_0 a \gg 1$, this distance will be many times larger than the width of the packet.

7.2.3 Wavepackets for the Complete Scattering Problem

The generalization of the treatment of the preceding section to the determination of a wavepacket for the hamiltonian in Eq. (7.1) is accomplished by writing the solution as follows:

$$\Psi_{wp}(x, t) = \int_{-\infty}^{\infty} dk \, C(k) \psi_k(x) e^{-iE_k t/\hbar} \qquad (7.11)$$

where ψ_k is the solution of the time-independent Schrödinger equation

$$H\psi_k = E_k \psi_k \qquad (7.12)$$

for an energy E_k. By substituting Eq. (7.11) into the time-dependent Schrödinger equation one can readily show that Ψ_{wp} is a general solution.

However, in contrast to bound-state problems, here we also have to make sure that Ψ_{wp} satisfies the desired boundary conditions initially and finally. Part of this we already know how to handle, since we have already demonstrated in Eqs. (7.3), (7.5), and (7.7) that use of $\psi_k = e^{ikx}$ and a gaussian $C(k)$ gives a gaussian wavepacket that moves with momentum $\hbar k_0$. This is the behavior that we are after initially ($t \to -\infty$) in the limit of $x \to -\infty$.

At the end of the collision ($t \to +\infty$) we expect to see part of the wavepacket moving to the right for $x \to \infty$ (the transmitted part), and part of it moving to the left for $x \to -\infty$ (the reflected part). Both this and the $t \to -\infty$ boundary condition can be satisfied by requiring that

$$\psi_k(x) \underset{x \to -\infty}{=} e^{ikx} + Re^{-ikx} \qquad (7.13a)$$

$$\underset{x \to +\infty}{=} Te^{ikx} \qquad (7.13b)$$

where R and T are unknown coefficients that we will discuss later and $\bar{k} = [2m(E - V_0)/\hbar^2]^{1/2}$ is the wavevector in the flat region of the potential for $x \to \infty$ (the right asymptotic region).

To prove that Eq. (7.13) gives a wavepacket that satisfies the desired boundary conditions, we note that substitution of Eq. (7.13a) into Eq. (7.11) gives us two wavepackets which roughly speaking are given by

$$\Psi_{wp} \underset{x \to -\infty}{\approx} e^{-(x-\hbar k_0 t/m)^2/2a^2} + Re^{-(x+\hbar k_0 t/m)^2/2a^2} \tag{7.14a}$$

In the $t \to -\infty$ limit, only the first term, representing a packet moving to the right, has a peak in the $x \to -\infty$ region (the left asymptotic region). The second term peaks in the right asymptotic region but this is irrelevant, as Eq. (7.14a) does not apply there. Thus in the left asymptotic region the second term is negligible and all we have is a packet moving to the right. For $t \to +\infty$, Eq. (7.14a) still applies, but now it is the second term that peaks in the left asymptotic region, and this packet moves to the left.

Now substitute Eq. (7.13b) into Eq. (7.11). Ignoring various unimportant terms, we get

$$\Psi_{wp} \underset{x \to +\infty}{\approx} Te^{-(x-\hbar\bar{k}_0 t/m)^2/2a^2} \tag{7.14b}$$

This formula represents a packet moving to the right centered at $x = \hbar\bar{k}_0 t/m$. For $t \to -\infty$, this is negligible in the right asymptotic region, so the wavefunction is zero there, while for $t \to +\infty$, this packet is large for $x \to +\infty$, just as wanted.

7.2.4 Fluxes and Probabilities

Now let us use the wavepackets just discussed to extract the physically measurable information about our problem, namely the probabilities of reflection and transmission. As long as the wavepackets do not spread much during the collision, these probabilities are given by the general definition:

$$\text{probability} = \frac{|\text{total flux outgoing for process of interest}|}{|\text{total flux incident}|} \tag{7.15}$$

where the flux is the number of particles per unit time that cross a given point (that cross a given surface in three dimensions), and the total flux is the spatial integral of the instantaneous flux. Classically, the flux is just ρv, where ρ is the density of particles (particles per unit length in one dimension) and v is the velocity of the particles. In quantum mechanics, the flux I is defined as

$$I = \text{Re}[\Psi^* v \Psi] \qquad (7.16)$$

where v is the velocity operator ($v = \dfrac{-i\hbar}{m}\dfrac{\partial}{\partial x}$ in one dimension) and Re implies that only the real part of $\Psi^* v \Psi$ is to be used.

To see how Eq. (7.16) works, substitute Eq. (7.7) into (7.16). Under the condition that wavepacket spreading is small (i.e., $\hbar t/ma^2 \ll 1$), we get

$$I = \frac{\hbar k_0}{m} \pi^{-1/2} a^{-1} \exp\left[\frac{-(x - \hbar k_0 t/m)^2}{a^2}\right] \qquad (7.17)$$

which is just $v_0 |\Psi_{wp}|^2$, where v_0 is the initial velocity ($v_0 = \hbar k_0/m$). In view of Eq. (7.14a), this is just the incident flux. The total flux in this case is $I_{tot}^{inc} = v_0$.

For the reflected wave associated with Eq. (7.13a), the total outgoing flux is $I_{tot}^{out} = |R|^2 v_0$, so the reflection probability P_R is

$$P_R = |R|^2 \qquad (7.18)$$

A similar calculation of the transmission probability gives

$$P_T = \frac{\bar{v}_0}{v_0} |T|^2 \qquad (7.19)$$

where

$$\bar{v}_0 \equiv \frac{\hbar \bar{k}_0}{m}$$

7.2.5 Time-Independent Approach to Scattering

Note from Eqs. (7.18) and (7.19) that all of the physically interesting information about the scattering process involves the coefficients R and T, which are properties of the time-independent wavefunction ψ_k obtained from Eq. (7.12) with the boundary conditions in Eq. (7.13). As a result, we can actually do scattering theory completely in a time-independent picture. This picture can be thought of as related to the time-dependent picture through the superposition of many gaussian wavepackets to form a plane wave. The important point to remember in using time-independent solutions is that the asymptotic solution given by Eq. (7.13) involves waves moving to the left and right that should be treated separately in calculating fluxes since these solutions do not contribute at the same time to the evolution of $\Psi_{wp}(x, t)$ in the $t \to \pm\infty$ limits. As a result, fluxes are evaluated by substituting either the left- or right-moving wavepacket parts of Eq. (7.13) into (7.16). For example, the reflected flux is

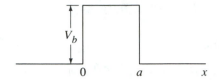

Figure 7.3. Potential $V(x)$ for a square barrier.

$$I^{\text{out}} = \text{Re}\left(R^* e^{ikx} \frac{-i\hbar}{m} \frac{\partial}{\partial x} R e^{-ikx}\right) \qquad (7.20)$$

As an example of the application of time-independent scattering theory, consider transmission through the square barrier pictured in Fig. 7.3. Solutions to the time-independent Schrödinger equation that satisfy the boundary conditions Eqs. (7.13 a, b) are easily written down as follows:

$$\begin{aligned}
\psi &= e^{ikx} + R e^{-ikx} & x < 0 \\
&= A \sin Kx + B \cos Kx & 0 \le x \le a \qquad (7.21)\\
&= T e^{ikx} & x > a
\end{aligned}$$

where

$$K = \left[\frac{2m}{\hbar^2}(E - V_b)\right]^{1/2} \qquad (7.22)$$

Note that K is real for $E > V_b$, and imaginary for $E < V_b$. The coefficients A and B in Eq. (7.21) can be determined by requiring that $\psi(x)$ and $\psi'(x)$ be continuous at $x = 0$ and at $x = a$. In fact, this determines not only A and B but also R and T. The resulting expression for T is

$$T = \frac{e^{-ika}}{\cos Ka - \dfrac{i}{2}\left(\dfrac{k}{K} + \dfrac{K}{k}\right)\sin Ka} \qquad (7.23)$$

From Eq. (7.19) this implies that the transmission probability is

$$P_T = |T|^2 = \frac{1}{\cos^2 Ka + \dfrac{1}{4}\left(\dfrac{k}{K} + \dfrac{K}{k}\right)^2 \sin^2 Ka} \qquad (7.24)$$

For $E < V_b$, it is convenient to reexpress P_T in terms of the real variable $\kappa = iK$, in which case one finds that

$$P_T = \frac{1}{\cosh^2 \kappa a + \dfrac{1}{4}\left(\dfrac{\kappa}{k} - \dfrac{k}{\kappa}\right)^2 \sinh^2 \kappa a} \qquad (7.25)$$

For $\kappa a \gg 1$, this reduces to

$$P_T \approx \frac{4e^{-2\kappa a}}{1 + \frac{1}{4}\left(\frac{\kappa}{k} - \frac{k}{\kappa}\right)^2} \tag{7.26}$$

which shows that the transmission probability decays exponentially with κ at energies well below the barrier top (i.e., the tunneling probability decays exponentially with the square root of $V_b - E$).

For $E \gg V_b$, Eq. (7.24) reduces to $P_T = 1$, which is the same as the classical transmission probability. For energies above but not greatly above V_b, P_T can have either smooth or oscillatory dependence on energy. An example of the latter is presented in Fig. 7.4. The peaks in Fig. 7.4 occur when $\kappa a = n\pi$ ($n = 0, 1, 2, \ldots$), implying that an integral number of wavelengths can be fit on top of the barrier.

7.2.6 Scattering Matrix

It is useful to rewrite the asymptotic part of the wavefunction as

$$\psi_k(x) \underset{x \to -\infty}{=} e^{ikx} + S_{11}e^{-ikx} \tag{7.27a}$$

$$\underset{x \to +\infty}{=} S_{12}\left(\frac{k}{\bar{k}}\right)^{1/2} e^{i\bar{k}x} \tag{7.27b}$$

where the coefficients S_{11} and S_{12} are two elements of a 2×2 matrix known as the scattering (S) matrix. The other two elements are associated with a different scattering solution in which the incident wave at $t \to$

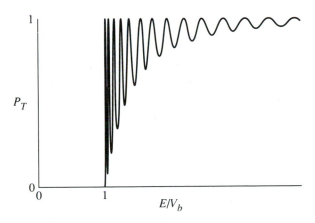

Figure 7.4. Transmission probability versus energy for square barrier where $2mV_b a^2/\hbar^2 = 1000$.

$-\infty$ moves to the left in the $x \to +\infty$ region. The boundary conditions on this solution are

$$\psi_{\bar{k}} \underset{x \to +\infty}{=} e^{-i\bar{k}x} + S_{22}e^{+i\bar{k}x}$$

$$\underset{x \to -\infty}{=} S_{21}\left(\frac{\bar{k}}{k}\right)^{1/2} e^{-ikx} \tag{7.28}$$

The S matrix has a number of important properties, one of which is that it is *unitary*. Mathematically, this means that $\mathbf{S^+S} = 1$, where $\mathbf{S^+}$ is the hermitian conjugate (transpose of complex conjugate) of $\mathbf{S}$. This property comes from the equation of continuity, which says that for any solution Ψ to the time-dependent Schrödinger equation,

$$\frac{\partial|\Psi|^2}{\partial t} + \frac{\partial I}{\partial x} = 0 \tag{7.29}$$

where I is the flux from Eq. (7.16). Equation (7.29) can be proved by substitution of Eq. (7.16) and the time-dependent Schrödinger equation into (7.29).

If $\Psi = \psi_k(x)e^{-iEt/\hbar}$, $|\Psi|^2$ is time independent, so Eq. (7.29) reduces to $\partial I/\partial x = 0$, which implies that I is a constant (i.e., flux is conserved), independent of x. If so, then the evaluation of I at $x \to +\infty$ and at $x \to -\infty$ should give the same result. By directly substituting Eq. (7.27) into (7.16), one finds that

$$I \underset{x \to -\infty}{=} \frac{\hbar k}{m}(1 - |S_{11}|^2)$$

$$\underset{x \to +\infty}{=} \frac{\hbar k}{m}|S_{12}|^2 \tag{7.30}$$

and since these two have to be equal, we find that

$$|S_{11}|^2 + |S_{12}|^2 = 1 \tag{7.31}$$

which indicates that the sum of the reflected and transmitted probabilities has to be unity. This is one of the equations that is implied by unitarity of the S matrix. The other equations can be obtained by using the flux associated with the solution $\psi_{\bar{k}}$ [Eq. (7.28)], and by using the generalized flux

$$I_{k\bar{k}} = \text{Re}(\psi_k^* v \psi_{\bar{k}}) \tag{7.32}$$

which also is a conserved quantity.

Another useful property of the S matrix is that it is *symmetric*. This property follows from conservation of the flux-like expression

$$\tilde{I}_{k\bar{k}} = \text{Re}(\psi_k v \psi_{\bar{k}}) \tag{7.33}$$

which differs from Eq. (7.32) in the absence of a complex conjugate in the wavefunction ψ_k. The symmetry property of S implies that S_{12} in Eq. (7.27) equals S_{21} in Eq. (7.28). Defining the probability matrix $\mathbf{P}$ by the relation

$$P_{ij} = |S_{ij}|^2 \tag{7.34}$$

we see that symmetry of $\mathbf{S}$ implies equal probabilities for the $i \rightarrow j$ and $j \rightarrow i$ transitions. This is a statement of the principle of *microscopic reversibility*, and it arises from the time-reversal symmetry associated with the Schrödinger equation.

7.2.7 Green's Functions for Scattering

Now let's write down the Schrödinger equation [Eq. (7.12)] using Eq. (7.1) for H and assuming that V_0 in Fig. 7.1 is zero. The result can be written

$$\left(\frac{d^2}{dx^2} + k^2\right) \psi_k(x) = \frac{2m}{\hbar^2} V(x)\psi_k(x) \tag{7.35}$$

One way to solve this is to invert the operator on the left-hand side, thereby converting this differential equation into an integral equation. The general result is

$$\psi_k(x) = \phi_k(x) + \frac{2m}{\hbar^2} \int_{-\infty}^{\infty} G_0(x, x')V(x')\psi_k(x') \, dx' \tag{7.36}$$

where G_0 is called the Green's function associated with the operator $d^2/dx^2 + k^2$ and ϕ_k is a solution of the homogeneous equation that is associated with Eq. (7.35), namely

$$\left(\frac{d^2}{dx^2} + k^2\right) \phi_k(x) = 0 \tag{7.37}$$

To determine $G_0(x, x')$, it is customary to use Fourier transforms to solve Eq. (7.35). Letting $F_k(k')$ be the Fourier transform of $\psi_k(x)$, we have

$$\psi_k(x) = \frac{1}{\sqrt{2\pi}} \int_{-\infty}^{\infty} e^{ik'x}F_k(k') \, dk' \tag{7.38a}$$

and its inverse,

$$F_k(k') = \frac{1}{\sqrt{2\pi}} \int_{-\infty}^{\infty} e^{-ik'x}\psi_k(x) \, dx \tag{7.38b}$$

Substituting Eq. (7.38a) into (7.35), we find that

$$\frac{1}{\sqrt{2\pi}} \int_{-\infty}^{\infty} e^{ik'x}(k^2 - k'^2)F_k(k') \, dk' = \frac{1}{\sqrt{2\pi}} \int_{-\infty}^{\infty} e^{ik'x}B(k') \, dk' \quad (7.39)$$

where

$$B(k') = \frac{1}{\sqrt{2\pi}} \int_{-\infty}^{\infty} e^{-ik'x} \frac{2m}{\hbar^2} V(x)\psi_k(x) \, dx \qquad (7.40)$$

Equation (7.39) implies that

$$F_k(k') = \frac{B(k')}{k^2 - k'^2} \qquad (7.41a)$$

so that, from Eq. 7.38a (and ignoring the homogeneous solution),

$$\psi_k(x) = \frac{1}{\sqrt{2\pi}} \int_{-\infty}^{\infty} e^{ik'x} \frac{B(k')}{k^2 - k'^2} \, dk'$$

$$= \frac{2m}{\hbar^2} \int_{-\infty}^{\infty} G_0(x, x')V(x')\psi_k(x') \, dx' \qquad (7.41b)$$

where

$$G_0(x, x') = \frac{1}{2\pi} \int_{-\infty}^{\infty} e^{ik'(x-x')} (k^2 - k'^2)^{-1} \, dk' \qquad (7.42)$$

The evaluation of the integral in Eq. (7.42) needs to be done carefully, as there is a pole at $k' = \pm k$. A standard trick to do it involves replacing k by $k \pm i\varepsilon$, where ε is a small positive constant that will be set to zero in the end. This reduces Eq. (7.42) to

$$G_0(x, x') = \frac{1}{4\pi k} \lim_{\varepsilon \to 0} \int_{-\infty}^{\infty}$$

$$\exp[ik'(x - x')] \left(\frac{1}{k - k' \pm i\varepsilon} + \frac{1}{k + k' \pm i\varepsilon} \right) dk' \quad (7.43)$$

This integral can be done by contour integration. For the $+i\varepsilon$ choice, the contour in Fig. 7.5a is appropriate for $x > x'$, as the circular part has a positive imaginary k' which makes $e^{ik'(x-x')}$ vanish for $|k'| \to \infty$. Similarly, for $x < x'$, we want to use the contour in Fig. 7.5b, as this makes the imaginary part of k' negative along the circular part. In either case, the integral along the real axis equals the full contour integral, and the latter is determined by the residue theorem to be $2\pi i$ times the residue at the pole which is encircled by the contour.

The pole is at $k' = k + i\varepsilon$ for the contour in Fig. 7.5a and at $k' = -k - i\varepsilon$ for Fig. 7.5b. This gives us

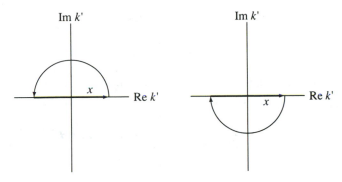

Figure 7.5. Integration contours used to evaluate Eq. (7.43):
(a) for $x > x'$, (b) for $x < x'$.

$$G_0^+(x, x') = \begin{cases} \dfrac{-i}{2k}\, e^{ik(x-x')} & \text{for } x > x' \\[2mm] \dfrac{-i}{2k}\, e^{-ik(x-x')} & \text{for } x < x' \end{cases} \qquad (7.44)$$

which we will call the "plus" wave free-particle Green's function G_0^+. A different Green's function ("minus" wave) is obtained by using $-i\varepsilon$ in the formulas above. It is

$$G_0^-(x, x') = \begin{cases} \left(\dfrac{i}{2k}\right) e^{-ik(x-x')} & \text{for } x > x' \\[2mm] \left(\dfrac{i}{2k}\right) e^{ik(x-x')} & \text{for } x < x' \end{cases} \qquad (7.45)$$

Upon substitution of G_0 into Eq. (7.36) we generate the following integral equation for the solution ψ_k^+ that is associated with G_0^+:

$$\psi_k^+(x) = \phi_k(x) - \int_{-\infty}^{x} \left(\frac{i}{2k}\right) e^{ik(x-x')}\, \frac{2m}{\hbar^2}\, V(x')\psi_k^+(x')\, dx'$$
$$- \int_{x}^{\infty} \left(\frac{i}{2k}\right) e^{-ik(x-x')}\, \frac{2m}{\hbar^2}\, V(x')\psi_k^+(x')\, dx' \qquad (7.46)$$

For $x \to \pm\infty$, it is possible to make ψ_k^+ look like Eq. (7.27) by setting $\phi_k(x) = e^{ikx}$. This shows that the plus Green's function is associated with scattering solutions in which outgoing waves move to the right in the $x \to \infty$ limit. For $x \to -\infty$, Eq. (7.46) becomes

$$\psi_k^+ \underset{x \to -\infty}{=} e^{ikx} - e^{-ikx} \int_{-\infty}^{\infty} \left(\frac{i}{2k}\right) e^{ikx'}\, \frac{2m}{\hbar^2}\, V(x')\psi_k^+(x')\, dx' \qquad (7.47)$$

By comparison with Eq. (7.27a), we see that

$$S_{11} = -\frac{i}{2k} \int_{-\infty}^{\infty} e^{ikx'} \frac{2m}{\hbar^2} V(x')\psi_k^+(x') \, dx' \qquad (7.48)$$

which is an integral that can be used to calculate S_{11} provided that ψ_k^+ is known. One can similarly show that

$$S_{12} = 1 - \frac{i}{2k} \int_{-\infty}^{\infty} e^{-ikx'} \frac{2m}{\hbar^2} V(x')\psi_k^+(x') \, dx' \qquad (7.49)$$

The other S-matrix elements, S_{21} and S_{22}, can be obtained from the G_0^- Green's function.

7.2.8 Born Approximation

If $V(x)$ is "small," ψ_k^+ will not be perturbed much from what it would be if $V(x) = 0$. If so, we can approximate $\psi_k^+ = e^{ikx}$ in Eqs. (7.48) and (7.49) and get

$$S_{11} = -\frac{i}{2k} \int_{-\infty}^{\infty} e^{2ikx'} \frac{2m}{\hbar^2} V(x') \, dx' \qquad (7.50a)$$

$$S_{12} = 1 - \frac{i}{2k} \int_{-\infty}^{\infty} \frac{2m}{\hbar^2} V(x') \, dx' \qquad (7.50b)$$

This is the one-dimensional version of what is usually called the *Born approximation* in scattering theory. Actually, however, this is just Fermi's golden rule [Eq. (4.63)] in a different language! For example, the transition probability obtained from Eq. (7.50a) is

$$P_{11} = \frac{m^2}{\hbar^2 p^2} \left| \int_{-\infty}^{\infty} e^{2ikx'} V(x') \, dx' \right|^2 \qquad (7.51)$$

where $p = \hbar k$ is the momentum. If we now consider Fermi's golden rule, using e^{ikx} as the initial state and e^{-ikx} as the final state, and a density of states $\rho = m/(2\pi\hbar p)$ (as may be easily derived following the method of Section 5.4.2 for a box of unit length), we get a rate

$$w_{11} = \frac{m}{\hbar^2 p} \left| \int_{-\infty}^{\infty} e^{2ikx'} V(x') \, dx' \right|^2 \qquad (7.52)$$

This rate is equal to the probability times the flux per unit length. The latter for a box of unit length is just p/m. Upon substitution of this into Eq. (7.52), we recover Eq. (7.51) for the probability.

A number of improvements to the Born approximation are possible, including *higher-order* Born approximations [obtained by inserting lower-order approximations to ψ_k^+ into Eq. (7.47), then the result into (7.48) and (7.49)], and the *distorted-wave* Born approximation (obtained by replacing the free-particle approximation for ψ_k^+ by the solution to a Schrödinger equation that includes part of the interaction potential). For chemical applications, the distorted-wave Born approximation is the most often used approach, as the approximation of ψ_k^+ by a plane wave is rarely of sufficient accuracy to be even qualitatively useful.

7.3

Semiclassical Theory

7.3.1 WKB Approximation

For many problems in chemistry the masses of the particles are large enough to justify using the $\hbar \to 0$ limit of quantum mechanics in determining energy levels or scattering information. There are actually a number of different ways to develop semiclassical approximations, depending on how time is treated in taking the $\hbar \to 0$ limit. In this section we discuss the most commonly used approach, which is based on time-independent quantum mechanics. For one-dimensional problems this is commonly known as the WKB (Wentzel–Kramers–Brillouin) approximation. At the end of this section, the multidimensional generalization of WKB theory is discussed.

Let's consider the one-dimensional Schrödinger equation of Eq. (7.12), which we write as

$$\frac{d^2\psi}{dx^2} = -\frac{2m}{\hbar^2}[E - V(x)]\psi \tag{7.53}$$

We try a solution of the form $\psi = e^{iS/\hbar}$, where $S(x)$ is a function to be determined, which in general will have real and imaginary parts.* Substituting this into Eq. (7.53), we get

$$S'^2 - i\hbar S'' = 2m(E - V) \tag{7.54}$$

where prime indicates differentiation with respect to x. Now use $S = S_R + iS_I$ in Eq. (7.54), where S_R and S_I are both real. Taking the real and imaginary parts of the resulting equation yields the two coupled equations

$$S_R'^2 - S_I'^2 + \hbar S_I'' = 2m(E - V) \tag{7.55a}$$

---

* This S is entirely unrelated to the S-matrix of the previous section.

$$2S'_R S'_I - \hbar S''_R = 0 \qquad (7.55b)$$

The second equation can be rewritten as

$$\frac{d \ln S'_R}{dx} = \frac{2}{\hbar} S'_I \qquad (7.56)$$

which can be integrated to give

$$S'_R = B \exp \left(\frac{2}{\hbar} S_I \right) \qquad (7.57)$$

where B is a constant.

Substituting Eq. (7.56) and its derivative into Eq. (7.55a) gives us a closed equation for S_R:

$$S'^2_R = 2m(E - V) + \hbar^2 \left[\frac{3}{4} \left(\frac{S''_R}{S'_R} \right)^2 - \frac{1}{2} \left(\frac{S'''_R}{S'_R} \right) \right] \qquad (7.58)$$

Now in the limit $\hbar \to 0$, the term in square brackets (sometimes called the quantum correction potential) can be neglected (provided that $S'_R \neq 0$). This gives us an equation for S'_R that is readily integrated, yielding

$$S_R = \pm \int_{x_0}^{x} [2m(E - V)]^{1/2} \, dx \qquad (7.59)$$

where the limits on this integral will be discussed below. One can then invert Eq. (7.57) to derive an expression for S_I:

$$S_I = \frac{\hbar}{2} \ln \frac{[2m(E - V)]^{1/2}}{B} \qquad (7.60)$$

By substituting Eqs. (7.59) and (7.60) into $\psi = e^{iS/\hbar}$, one finds that the unnormalized wavefunction is

$$\psi(x) = [2m(E - V)]^{-1/4} \exp \left\{ \pm \frac{i}{\hbar} \int_{x_0}^{x} [2m(E - V)]^{1/2} \, dx \right\}$$

$$= p^{-1/2} \exp \left(\pm \frac{i}{\hbar} \int_{x_0}^{x} p \, dx \right) \qquad (7.61)$$

This wavefunction is usually called the WKB or primitive semiclassical solution to the Schrödinger equation. Analogous solutions involving real exponentials can be developed when $E < V$ (i.e., in the classically forbidden regions). Note that for $E > V$, $|\psi|^2 = [2m(E - V)]^{-1/2} = p^{-1}$, which indicates that the probability of a given x is inversely proportional to the momentum (or velocity) of the particle at that x. This is the same as is predicted by classical mechanics. Also note that the argument of the exponential involves the classical action ($\int p\,dx$) evaluated along the classical trajectory associated with the hamiltonian H.

The solution defined by Eq. (7.61) is generally quite accurate in those regions where $S'_R \neq 0$, but it fails badly when $S'_R = 0$. This occurs when $E = V$, corresponding to where there are *turning points* in the classical motion. Indeed, Eq. (7.61) diverges at turning points, so the construction of global WKB solutions for any potential with a turning point cannot be based on just this equation. To get around this problem we take advantage of the fact that the exact solution to the Schrödinger equation can be determined analytically for small intervals in x where the potential can be approximated as a linear function of x. If we develop these exact solutions near each classical turning point and then match these to solutions obtained from Eq. (7.61) on either side, it is possible to connect together WKB solutions on opposite sides of each turning point. This leads to what are known as *connection formulas* that are the key to the WKB method. Figure 7.6 shows the two possible turning points, labeling regions to the "outside" and "inside" of each turning point by I and II, respectively.

The possible connection formulas are as follows:

(a) For a left turning point:

$$\psi_{\mathrm{I}} = \frac{1}{2\sqrt{|p|}} \exp\left(-\int_x^{x_1} |p| \frac{dx}{\hbar} \right) \text{ connects to}$$

$$\psi_{\mathrm{II}} = \frac{1}{\sqrt{p}} \cos\left(\int_{x_1}^x p\,\frac{dx}{\hbar} - \frac{\pi}{4} \right)$$

(7.62a)

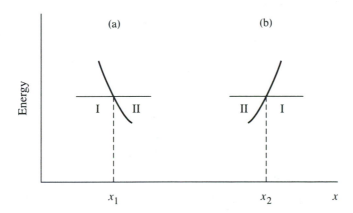

Figure 7.6. Turning points: (a) left; (b) right.

$$\psi_I = \frac{1}{\sqrt{|p|}} \exp\left(+ \int_x^{x_1} |p| \frac{dx}{\hbar} \right) \text{ connects to}$$

(7.62b)

$$\psi_{II} = \frac{-1}{\sqrt{p}} \sin\left(\int_{x_1}^x p \frac{dx}{\hbar} - \frac{\pi}{4} \right)$$

(b) For a right turning point:

$$\psi_I = \frac{1}{2\sqrt{|p|}} \exp\left(- \int_{x_2}^x |p| \frac{dx}{\hbar} \right) \text{ connects to}$$

(7.62c)

$$\psi_{II} = \frac{1}{\sqrt{p}} \cos\left(\int_x^{x_2} p \frac{dx}{\hbar} - \frac{\pi}{4} \right)$$

$$\psi_I = \frac{1}{\sqrt{|p|}} \exp\left(+ \int_{x_2}^x |p| \frac{dx}{\hbar} \right) \text{ connects to}$$

(7.62d)

$$\psi_{II} = \frac{-1}{\sqrt{p}} \sin\left(\int_x^{x_2} p \frac{dx}{\hbar} - \frac{\pi}{4} \right)$$

To see how the connection formulas work, let's consider the penetration of a potential barrier such as that pictured in Fig. 7.7. To satisfy scattering boundary conditions, we want our solution to look like

$$e^{ikx} + S_{11} e^{-ikx} \text{ for } x \to -\infty, \quad \text{and} \quad (k/\bar{k})^{1/2} S_{12} e^{i\bar{k}x} \text{ for } x \to \infty.$$

For $E > V_{\text{max}}$, this can be done by setting

$$\psi = p^{-1/2} \exp\left(\frac{i}{\hbar} \int_{x_0}^x p \, dx \right)$$

(7.63)

where x_0 is a point where $V(x)$ is constant. For $x \to -\infty$, p is a constant, and the wavefunction reduces to $p^{-1/2} \exp(ip(x - x_0)/\hbar)$, which is what

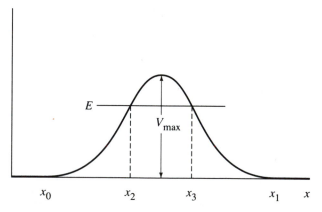

Figure 7.7. Potential barrier with two turning points.

we wanted provided that we take $S_{11} = 0$. For $x \to +\infty$,

$$\int_{x_0}^{x} p \, dx \approx \bar{p}(x - x_1) + \int_{x_0}^{x_1} p \, dx$$

and

$$\psi = \bar{p}^{-1/2} e^{i\bar{p}(x-x_1)/\hbar} \exp\left(i \int_{x_0}^{x_1} p \, \frac{dx}{\hbar}\right)$$

where x_1 is a point on the right side of the maximum such that $V(x)$ is constant for $x \geq x_1$. After normalizing the incoming wave to $e^{ipx/\hbar}$, the outgoing wave equals $(k/\bar{k})^{1/2} S_{12} e^{i\bar{k}x}$ with

$$S_{12} = e^{i(px_0 - \bar{p}x_1)/\hbar} \exp\left(i \int_{x_0}^{x_1} p \, \frac{dx}{\hbar}\right) \tag{7.64}$$

In this case, $|S_{12}|^2 = 1$, indicating that there is unit transmission probability, just as in the classical case, but not in agreement with the exact quantum solution.

For $E < V_{max}$, there are two turning points (at x_2 and x_3). The best way to do this problem is to start with the $x \to +\infty$ part of the solution first, then connect going to the left. It is left for the reader to show that for $x > x_3$,

$$\psi = \frac{1}{\sqrt{p}} \exp\left(i \int_{x_3}^{x} p \, \frac{dx}{\hbar} - \frac{\pi}{4}\right) \tag{7.65a}$$

for $x_2 < x < x_3$,

$$\psi = \frac{1}{2\sqrt{|p|}} \exp\left(- \int_{x}^{x_3} |p| \, \frac{dx}{\hbar}\right) - \frac{i}{\sqrt{|p|}} \exp\left(\int_{x}^{x_3} |p| \, \frac{dx}{\hbar}\right) \tag{7.65b}$$

and for $x < x_2$,

$$\psi = \frac{-2i}{\sqrt{p}} \exp\left(\int_{x_2}^{x_3} |p| \, \frac{dx}{\hbar}\right) \cos\left(\int_{x}^{x_2} p \, \frac{dx}{\hbar} - \frac{\pi}{4}\right) \tag{7.65c}$$

Equation (7.65c) has been derived assuming that $\exp(-\int_{x_2}^{x_3} |p| \, dx/\hbar)$ is small compared to its inverse. After normalizing Eq. (7.65c) so that its incoming part is e^{ikx} for $x \to -\infty$, we find from Eq. (7.65a) that

$$|S_{12}|^2 = \exp\left(- \frac{2}{\hbar} \int_{x_2}^{x_3} |p| \, dx\right) \tag{7.66}$$

Equation (7.66) is an important result of WKB theory. It indicates that the probability of tunneling through a barrier depends exponentially on the imaginary action accumulated between the two turning points.

Note that it is not difficult to turn the potential in Fig. 7.7 upside down and use WKB to determine bound states. In this case one starts with an exponentially decaying solution for $x > x_3$, connects to oscillatory solutions for $x_2 < x < x_3$, and then to exponentially increasing and decreasing solutions for $x < x_2$. Only at certain energies does one get a decaying solution for $x < x_2$, and this occurs when

$$\frac{1}{\pi} \int_{x_2}^{x_3} p\, dx = \left(n + \frac{1}{2}\right)\hbar \qquad (7.67)$$

This is the WKB formula for determining energy levels; n is a nonnegative integer. Note that this formula also depends on the classical action between the turning points. Equation (7.67) works remarkably well for many potentials, yielding the exact eigenvalues for harmonic oscillator, Morse, and Coulomb potentials.

The generalization of WKB to describe multidimensional problems leads in lowest order to S-matrix elements that are of the form [compare with Eq. (7.64)]

$$S_{ij} = \sum_t (\rho_{ij}^t)^{1/2} e^{iS_{ij}^t/\hbar} \qquad (7.68)$$

where the sum is over trajectories t that connect the initial and final state, S_{ij}^t is the classical action along the trajectory path, and ρ_{ij}^t is the classical probability associated with the trajectory t. This theory is sometimes called *classical S-matrix theory* or *Miller–Marcus theory* (see the Bibliography).

Similarly, to determine bound states for multidimensional systems, we use formulas that look like Eq. (7.67):

$$(2\pi)^{-1} \oint_{C_i} \mathbf{p} \cdot d\mathbf{x} = (n_i + \tfrac{1}{2})\hbar \qquad (7.69)$$

where the contours C_i refer to distinct paths in phase space that wrap around the trajectory. This theory is usually called *EBK theory,* after Einstein, Brillouin, and Keller (see the Bibliography).

While classical S-matrix theory and EBK theory are formally well defined and have been applied to a variety of model problems, they are not commonly used in realistic applications. The main reason for this is that the classical mechanics which underlies semiclassical theory very often has complex phase-space structure that is difficult to characterize sufficiently so that Eq. (7.68) or (7.69), or improvements thereto, can be applied. Thus for describing scattering problems, semiclassical theory is most often replaced by classical or quasiclassical theory in which amplitudes are replaced by probabilities (i.e., phase interference is ignored, and states are only coarsely defined).

7.3.2 Semiclassical Wavepackets

Another way to develop semiclassical theory is to approximate the time evolution of wavepackets in such a way that on the average they follow classical trajectories. One such procedure was developed by E. Heller in *J. Chem. Phys.* 62, 1544 (1975) (and many subsequent papers). The basic procedure comes from the fact that wavepackets which are initially gaussian remain gaussian in potentials that are constant, linear, or quadratic functions of the coordinates. In addition, the centers of such wavepackets evolve in time in accord with classical mechanics. We have already seen one example of this with the free-particle wavepacket of Eq. (7.7), but let us now consider the general quadratic hamiltonian

$$H = -\frac{\hbar^2}{2m}\frac{\partial^2}{\partial x^2} + V_0 + V_x(x - x_t) + \tfrac{1}{2} V_{xx}(x - x_t)^2 \qquad (7.70)$$

and the general gaussian wavepacket

$$\Psi(x,t) = \exp\left[\frac{i}{\hbar}\alpha_t(x - x_t)^2 + \frac{i}{\hbar}p_t(x - x_t) + \frac{i}{\hbar}\gamma_t\right] \qquad (7.71)$$

Here x_t and p_t are real time-dependent quantities that specify the average position and momentum of the wavepacket ($p_t = \langle p \rangle$, $x_t = \langle x \rangle$) and α_t and γ_t are complex functions that determine the width, phase, and normalization of the wavepacket.

Inserting Eq. (7.71) into $H\Psi = i\hbar\,\partial\Psi/\partial t$, and using Eq. (7.70) leads to the following relation:

$$[-\dot{\alpha}_t(x - x_t)^2 + (2\alpha_t\dot{x}_t - \dot{p}_t)(x - x_t) - \dot{\gamma}_t + p_t\dot{x}_t]\Psi$$

$$= \left[\left(\frac{2}{m}\alpha_t^2 + \frac{1}{2}V_{xx}\right)(x - x_t)^2 + \left(2\frac{\alpha_t p_t}{m} + V_x\right)(x - x_t)\right.$$

$$\left. + V_0 - \frac{i\hbar\alpha_t}{m} + \frac{p_t^2}{2m}\right]\Psi \qquad (7.72)$$

Comparing coefficients of like powers of $(x - x_t)$ then gives us three equations involving the four unknowns:

$$\dot{\alpha}_t = -\frac{2}{m}\alpha_t^2 - \frac{V_{xx}}{2} \qquad (7.73a)$$

$$2\alpha_t\dot{x}_t - \dot{p}_t = \frac{2\alpha_t p_t}{m} + V_x \qquad (7.73b)$$

$$\dot{\gamma}_t = \frac{i\hbar\alpha_t}{m} + p_t\dot{x}_t - V_0 - \frac{p_t^2}{2m} \qquad (7.73c)$$

To develop an additional equation, we simply make the ansatz that the first term on the left-hand side of Eq. (7.73b) equals the first term on the right-hand side, and similarly with the second term. This immediately gives us Hamilton's equations,

$$\dot{x}_t = \frac{p_t}{m} \qquad (7.74a)$$

$$\dot{p}_t = -V_x \qquad (7.74b)$$

from which it follows that x_t and p_t are related through the classical hamiltonian function

$$H = \frac{p_t^2}{2m} + V_0 = E \qquad (7.75)$$

Equations (7.74) can then be cast in the general form

$$\dot{x}_t = \frac{\partial H}{\partial p_t} \qquad (7.76a)$$

$$-\dot{p}_t = \frac{\partial H}{\partial x_t} \qquad (7.76b)$$

and the remaining two equations in Eq. (7.73) become

$$\dot{\alpha}_t = -\frac{2}{m}\alpha_t^2 - \frac{1}{2}V_{xx} \qquad (7.77a)$$

$$\dot{\gamma}_t = \frac{i\hbar\alpha_t}{m} + p_t\dot{x}_t - E \qquad (7.77b)$$

It is not difficult to show that for a constant potential, Eqs. (7.76) and (7.77) can be solved to give the wavepacket in Eq. (7.7). More generally, one can solve Eqs. (7.76) and (7.77) numerically for *any* potential, even potentials that are not quadratic, but the solution obtained will be exact only for potentials that are constant, linear, or quadratic. The deviation between the exact and gaussian wavepacket solutions for other potentials depends on how close they are to being *locally quadratic*, which means how well the potential can be approximated by a quadratic potential over the width of the wavepacket. Note that although this theory has many classical features, the $\hbar \to 0$ limit has not been used. This circumvents problems with singularities in the wavefunction near classical turning points that cause trouble in WKB theory.

Equation (7.71) is an example of what is commonly called a "thawed" gaussian wavepacket. This means that the wavepacket width is allowed to vary during its time evolution. An even simpler wavepacket is one in which α_t and γ_t are "frozen," meaning that only x_t and p_t are allowed to vary. Such wavepackets do not solve even the quadratic potential Schrödinger equation, but they have a number of advantages from the point of view of practical applications. For bound-state calculations they have the advantage that periodicity of the wavepacket is preserved for arbitrary potentials, whereas the thawed wavepacket is periodic only for a quadratic potential. Several multidimensional methods for determining eigenvalues from wavepackets exist, of which the simplest (conceptually) involves finding periodic orbits that obey conditions analogous to Eq. (7.67) for each vibrational mode.

Both frozen and thawed gaussians can be used to describe scattering problems. In Heller's first paper on the topic, he used thawed gaussians to determine the probabilities of vibrational excitation and deexcitation for $A + BC$ collinear collisions. To do this he wrote the initial wavefunction as a superposition of gaussians in both translational and vibrational coordinates, then propagated each gaussian separately, then projected the superposition onto final vibrational states, and then calculated probabilities as ratios of fluxes. This worked well under a variety of choices of masses and potential parameters. However, note the following incorrect predictions: (1) for one-dimensional barrier penetration problems, at energies above the barrier, $P(E) = 1$; and (2) for energies below the barrier $P(E) = 0$. This is the classical result in both cases, and it occurs because the gaussian wavepackets are (a) forced to follow trajectories, and (b) cannot split in barrier crossing problems into transmitted and reflected waves. These problems can be fixed up, although not without making the theory significantly more complicated.

7.4

Scattering in Three Dimensions

Practically all of the one-dimensional quantum scattering theory of previous sections can be generalized to three dimensions with little or no change, so only the key new features will be discussed here. The biggest change refers to asymptotic boundary conditions. In three dimensions, the initial scattering wavefunction for a single particle should represent a plane wave moving in a direction which we denote with the wavevector $\mathbf{k}$. Scattering then produces outgoing spherical waves as $t \rightarrow \infty$ weighted by an amplitude $f_{\mathbf{k}}(\theta)$, which specifies the scattered intensity as a function of the angle θ between $\mathbf{k}$ and the observation direction. This is depicted schematically in Fig. 7.8. Mathematically, the time-independent boundary condition analogous to Eq. (7.13) is

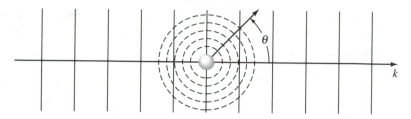

Figure 7.8. Schematic of potential scattering in three dimensions, showing incoming plane wave and outgoing spherical wave.

$$\psi_k(\mathbf{r}) \underset{r \to \infty}{=} e^{i\mathbf{k}\cdot\mathbf{r}} + f_k(\theta) \frac{e^{ikr}}{r} \tag{7.78}$$

Note that for potentials that only depend on the scalar distance r between the colliding particles, the amplitude $f_k(\theta)$ does not depend on the azimuthal angle associated with the direction of observation.

The measurable quantity in a three-dimensional scattering experiment is the differential cross section $d\sigma_k(\theta)/d\Omega$. This is defined as

$$\frac{d\sigma_k(\theta)}{d\Omega} = \frac{|\text{outgoing radial flux}|}{|\text{total incident flux}|} \tag{7.79}$$

where outgoing flux refers to the radial velocity operator $v_r = -i\hbar\, \partial/\partial r$. Substitution of Eq. (7.78) into (7.79) using (7.16) yields

$$\frac{d\sigma_k(\theta)}{d\Omega} = |f_k(\theta)|^2 \tag{7.80}$$

It is usually convenient to expand $f_k(\theta)$ in a basis of Legendre polynomials $P_l(\cos\theta)$:

$$f_k(\theta) = \sum_l a_l^k\, P_l(\cos\theta) \tag{7.81}$$

We call this a *partial wave expansion*. To determine the coefficients a_l^k, one matches asymptotic solutions to the radial Schrödinger equation [Eq. (1.57)] with the corresponding partial wave expansion of Eq. (7.78). It is customary to write the asymptotic radial Schrödinger equation solution as

$$\psi_{lm}(r, \theta, \phi) \underset{r \to \infty}{=} \frac{1}{r} Y_{lm}(\theta, \phi)(e^{-i(kr - l\pi/2)} - S_l e^{i(kr - l\pi/2)}) \tag{7.82}$$

where S_l is the scattering matrix for the lth partial wave and m is the projection quantum number associated with l. Unitarity of the scattering matrix implies that S_l can be written as $\exp(2i\delta_l)$, where δ_l is a real quantity known as the *phase shift*.

The asymptotic partial wave expansion of Eq. (7.78) can be developed using the identity

$$e^{i\mathbf{k}\cdot\mathbf{r}} = e^{ikr\cos\theta} = \sum_{l=0}^{\infty} i^l(2l+1)j_l(kr)P_l(\cos\theta) \tag{7.83}$$

where $j_l(kr)$ is a spherical Bessel function. At large r, the spherical Bessel function reduces to

$$j_l(kr) \underset{r\to\infty}{=} \frac{\sin(kr - l\pi/2)}{kr} \tag{7.84}$$

If Eq. (7.84) is then used to evaluate (7.83) after substitution of the latter into (7.78), and if Eq. (7.81) is also substituted into (7.78) and the result is equated to (7.82), one finds that only $m = 0$ contributes, and that

$$a_l^k = \frac{2l+1}{2ik}(S_l - 1) \tag{7.85}$$

From Eqs. (7.80) and (7.81) one then finds that

$$\frac{d\sigma_k(\theta)}{d\Omega} = \frac{1}{4k^2}\left|\sum_l (2l+1)P_l(\cos\theta)(S_l - 1)\right|^2 \tag{7.86a}$$

$$= \frac{1}{k^2}\left|\sum_l (2l+1)P_l(\cos\theta)e^{i\delta_l}\sin\delta_l\right|^2 \tag{7.86b}$$

The differential cross section may be integrated over scattering angles to define an integral cross section σ as follows:

$$\sigma = 2\pi\int_0^\pi \frac{d\sigma_k(\theta)}{d\Omega}\sin\theta\,d\theta$$

$$= \frac{\pi}{k^2}\sum_l (2l+1)|S_l - 1|^2 \tag{7.87a}$$

$$= \frac{4\pi}{k^2}\sum_l (2l+1)\sin^2\delta_l \tag{7.87b}$$

Equations (7.86b) and (7.87b) are in a form that is convenient to use for potential scattering problems. One needs only determine the phase shift δ_l for each l, then substitute into these equations to determine the cross sections. Note that in the limit of large l, δ_l must vanish so that the infinite sum over partial waves l will converge. For most potentials of interest to chemistry, the calculation of δ_l must be done numerically.

Equation (7.87a) is also useful; this is in a form that enables easy generalization of the potential scattering theory that we have just derived to multistate problems. In particular, if we imagine that we are interested in the collision of two molecules A and B starting out in states n_A and n_B and ending up in states n'_A and n'_B, the asymptotic wavefunction analogous to Eq. (7.78) is

$$
\psi_{n_A n_B \to n'_A n'_B} = \exp(i\mathbf{k}_{n_A n_B} \cdot \mathbf{r})|n_A n_B\rangle
+ r^{-1} \sum_{n'_A n'_B} f_{n_A n_B \to n'_A n'_B}(\theta) \exp(ik_{n'_A n'_B} r)|n'_A n'_B\rangle
\tag{7.88}
$$

where the scattering amplitude f is now labeled by the initial and final state indices. Integral cross sections are then obtained using the following generalization of Eq. (7.87a):

$$
\sigma_{n_A n_B \to n'_A n'_B} = \frac{\pi}{k^2_{n_A n_B}} \sum_J (2J + 1)|S^J_{n_A n_B \to n'_A n'_B} - \delta_{n_A n_B, n'_A n'_B}|^2
\tag{7.89}
$$

where S is the multichannel scattering matrix, δ is the Kronecker delta function, and J is the total angular momentum. The orbital angular momentum l is not a conserved quantity (because of coupling with angular momenta in the molecules A and B), so the sum over l is replaced by a sum over J.

BIBLIOGRAPHY FOR CHAPTER 7

Chemically oriented books on scattering theory include:

Child, M. S., *Molecular Collision Theory* (Academic Press, New York, 1974).

Levine, R. D., *Quantum Mechanics of Molecular Rate Processes* (Oxford University Press, London, 1969).

Levine, R. D., and R. B. Bernstein, *Molecular Reaction Dynamics and Chemical Reactivity* (Oxford University Press, New York, 1987).

Massey, H. S. W., *Atomic and Molecular Collisions* (Taylor & Francis, London, 1979).

Murrell, J. N., and S. D. Bosanac, *Introduction to the Theory of Atomic and Molecular Collisions* (Wiley, New York, 1989).

Nikitin, E. E., *Theory of Elementary Atomic and Molecular Processes in Gases* (Clarendon Press, Oxford, 1974).

Monographs on molecular scattering theory include:

Baer, M., ed., *The Theory of Chemical Reaction Dynamics* (CRC Press, Boca Raton, Fla., 1985).

Bernstein, R. B., ed., *Atom–Molecule Collision Theory: A Guide for the Experimentalist* (Plenum Press, New York, 1979).

Bowman, J. M., ed., *Molecular Collision Dynamics* (Springer-Verlag, Berlin, 1983).

Clary, D. C., ed., *The Theory of Chemical Reaction Dynamics* (D. Reidel, Boston, 1986).

Truhlar, D. G., ed., *Potential Energy Surfaces and Dynamics Calculations* (Plenum Press, New York, 1981).

Miller, W. H., ed., *Dynamics of Molecular Collisions* (Plenum Press, New York, 1976).

Standard physics textbooks treating scattering theory are:

Adhi Kari, S. K. and K. L. Kowolski, *Dynamical Collision Theory and Its Applications* (Academic Press, New York, 1991).

Davydov, A. S., *Quantum Mechanics* (Pergamon Press, Oxford, 1976).

Merzbacher, E., *Quantum Mechanics* (Wiley, New York, 1970).

Messiah, A., *Quantum Mechanics* (North-Holland, Amsterdam, 1965).

Newton, R. G., *Scattering Theory of Waves and Particles* (McGraw-Hill, New York, 1966).

Schiff, L. I., *Quantum Mechanics* (McGraw-Hill, New York, 1968).

Sakurai, J. J., *Modern Quantum Mechanics* (Benjamin-Cummings, Menlo Park, Calif., 1985).

Specific references to semiclassical theories are:

Miller–Marcus references: W. H. Miller, *J. Chem. Phys.* 53, 1949 (1970); R. A. Marcus, *Chem. Phys. Lett.* 7, 525 (1970).

Review articles: W. H. Miller, *Acct. Chem. Res.* 4, 161 (1971); *Adv. Chem. Phys.* 25, 69 (1974); *Ibid.* 30, 77 (1975).

Gaussian wavepackets: E. J. Heller, *J. Chem. Phys.* 62, 1544 (1975).

EBK references: A. Einstein, *Verh. Dtsch. Phys. Ges.* 19, 82 (1917); L. Brillouin, *J. Phys. Paris* 7, 353 (1926); J. B. Keller, *Am. Phys.* 4, 180 (1958).

PROBLEMS FOR CHAPTER 7

1. Calculate the position and momentum uncertainties for the wavepacket in Eq. (7.7). These are defined by

$$\langle \Delta x^2 \rangle = \langle x^2 \rangle - \langle x \rangle^2$$

$$\langle \Delta p^2 \rangle = \langle p^2 \rangle - \langle p \rangle^2$$

Show that $(\langle \Delta p^2 \rangle \langle \Delta x^2 \rangle)^{1/2} = \hbar/2$.

2. Show that the flux I and the wavefunction Ψ satisfy the continuity equation (in one dimension):

$$\frac{\partial I}{\partial x} + \frac{\partial |\Psi|^2}{\partial t} = 0$$

3. Given the barrier potential pictured in Fig. 7.9, solve for the S-matrix elements S_{11}, S_{12}, and the probabilities P_R and P_T. Plot P_T versus E for $0 \le E \le 3V_b$ using $V_0 = 0.1$ eV, $V_b = 0.5$ eV, $L = 2$ Å, $m = 2$ g/mol.

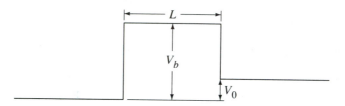

Figure 7.9.

4. For the square barrier problem in Fig. 7.3, show that the exact expression for $P_{11} = 1 - P_{12}$ reduces to the Born result in the limit $V_b \to 0$. This result should hold for arbitrary values of k.

5. Show that WKB theory applied to the harmonic oscillator gives the exact energy formula

$$E_n = \hbar\omega \left(n + \frac{1}{2} \right)$$

6. Use WKB theory to determine the splitting between even and odd pairs of states associated with the symmetric double well potential pictured in Fig. 7.10. Assume that $V(x) = V(-x)$, and that the four turning points pictured satisfy $x_1 = -x_4$, $x_2 = -x_3$. Also assume that the barrier is high and thick so that

$$\int_{x_2}^{x_3} |p| \frac{dx}{\hbar}$$

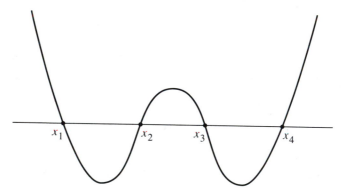

Figure 7.10.

is large, and that the wells are harmonic with frequency ω. The result is

$$\Delta E = \frac{\hbar\omega}{\pi} \exp\left(-\int_{x_2}^{x_3} |p| \frac{dx}{\hbar}\right)$$

7. What is the semiclassical tunneling probability for a parabolic barrier? The potential for this barrier is $V(x) = -\frac{1}{2}kx^2$, where $E = 0$ corresponds to the top of the barrier. You should be able to express the result in terms of energy and the barrier frequency $\omega = (k/m)^{1/2}$.

8

Theories of Reaction Rates

8.1

Introduction

In this chapter we are interested in relating the rate constant for a chemical reaction (either bimolecular or unimolecular) to more fundamental dynamical quantities such as reaction probabilities or reactive fluxes. Much of this will build on what was developed in our discussions of scattering (Chapter 7), but there will be much that is both new and unrelated. In addition to derivations of formally exact expressions for rate constants, we will also develop several approximate theories such as transition-state theory and RRKM theory. We also introduce flux auto-correlation function approaches to evaluating rate constants and the evaluation of autocorrelation functions using path integrals. These topics overlap closely with chapters on correlation function methods in spectroscopy (Chapter 9) and on correlation functions in electron transfer (Chapter 10).

8.2

Rate Constants for Bimolecular Reactions: Cumulative Reaction Probabilities

Consider first a gas-phase bimolecular reaction (A + B → C + D). If we consider that the reagents are approaching each other with a relative velocity v, the total flux of A's moving toward B is just vC_A, where C_A is the concentration of A's (number of A's per unit volume [or per unit length in one dimension]). If σ is the integral cross section for reaction between A and B for a given velocity v (σ is the reaction probability in one dimension), then for every B, the number of reactive collisions per unit time is $\sigma v C_A$. The total number of reactive collisions per unit time per unit volume (or per unit length in one dimension) is then $\sigma v C_A C_B$, where C_B is the concentration of B's. Equating this to the rate constant k times $C_A C_B$ leads us to the conclusion that

$$k = \sigma v \qquad (8.1)$$

This rate constant refers to reactants that all move with a velocity v, whereas the usual situation is such that we have a Boltzmann distribution of velocities. If so, the rate constant is just the average of (8.1) over a Boltzmann distribution P_B:

$$k(T) = \int_0^\infty P_B(v) v \sigma(v) \, dv \qquad (8.2)$$

This expression is still oversimplified, as it ignores the fact that the molecules A and B have internal states, and that the cross section σ depends on these states; σ depends also on the internal states of the products C and D. Letting the indices i and f denote the internal states of the reagents and products, respectively, we find that σ in Eq. (8.2) must be replaced by $\Sigma_f \sigma_{if}$ and the Boltzmann average must now include the internal states. Thus Eq. (8.2) becomes

$$k(T) = \sum_i p_B(i) \int_0^\infty P_B(v_i) v_i \sum_f \sigma_{if}(v_i) \, dv_i \qquad (8.3)$$

where $p_B(i)$ is the internal-state Boltzmann distribution.

Now let's write down explicit expressions for $p_B(i)$, $P_B(v_i)$, and σ_{if}. Denoting the internal energy for a given state i as ε_i, and the relative translational energy as $E_i = \frac{1}{2}\mu v_i^2$, we have (in three dimensions)

$$p_B(i) = \frac{e^{-\varepsilon_i/kT}}{Q_{\text{int}}} \qquad (8.4)$$

and

$$P_B = 4\pi \left(\frac{\mu}{2\pi kT}\right)^{3/2} v_i^2 \exp\left(\frac{-\mu v_i^2}{2kT}\right) \qquad (8.5)$$

where Q_{int} is the internal state partition function.

The cross section σ_{if} is related to the partial wave reactive scattering matrix S_{if}^J through the partial wave sum [i.e., Eq. (7.89) evaluated for $n_A n_B \neq n_{A'} n_{B'}$].

$$\sigma_{if} = \frac{\pi}{k_i^2} \sum_J (2J + 1)|S_{if}^J|^2 \tag{8.6}$$

where $k_i = \mu v_i/\hbar$. Now substitute Eqs. (8.4)–(8.6) into (8.3). Replacing the integral over v_i by one over E_i leads us to the expression

$$k(T) = \frac{(2\pi\hbar)^2}{Q_{\text{int}}(2\pi\mu kT)^{3/2}} \sum_i e^{-\varepsilon_i/kT} \int_0^\infty e^{-E_i/kT} \sum_J (2J + 1) \sum_f |S_{if}^J|^2 \, dE_i \tag{8.7}$$

If we now change the integration variable from E_i to the total energy $E = E_i + \varepsilon_i$, we can rewrite Eq. (8.7) in the remarkably simple expression

$$k(T) = \frac{kT}{h} \frac{1}{Q_{\text{int}}Q_{\text{trans}}} \int_0^\infty e^{-E/kT} P_{\text{cum}}(E) \frac{dE}{kT} \tag{8.8}$$

where Q_{trans} is the translational partition function per unit volume:

$$Q_{\text{trans}} = \left(\frac{2\pi\mu kT}{h^2}\right)^{3/2} \tag{8.9}$$

and P_{cum} is denoted the cumulative reaction probability

$$P_{\text{cum}}(E) = \sum_J (2J + 1) \sum_i \sum_f |S_{if}^J|^2 \tag{8.10}$$

Note that in deriving Eq. (8.8), we have altered the lower integration limit in Eq. (8.7) from zero to $-\varepsilon_i$ by defining S_{if}^J to be zero for $E_i < 0$.

In one physical dimension, Eq. (8.8) still holds, but Q_{trans} is given by its one-dimensional counterpart and the partial wave sum is absent from Eq. (8.10). Note that although the scattering matrix depends on all initial and final states, only the sum over all initial and final states is used in computing rate constants.

8.3

Transition-State Theory

The form of Eq. (8.8) is immediately suggestive of transition-state theory. Indeed, if we assume that the total reaction probability $\sum_f |S_{if}|^2$ is zero for $E < E^\ddagger$ and unity for $E \geq E^\ddagger$, where $E^\ddagger$ is the energy of a critical bottleneck (transition state), then

$$P_{\text{cum}}^\ddagger = \sum_J (2J + 1) \sum_i h(E - E_i^\ddagger) \tag{8.11}$$

where h is a Heaviside (step) function that is unity for positive arguments and zero for negative arguments, and we have added the subscript i to $E_i^\ddagger$

since the bottleneck energies can in general be dependent on internal state.

Equation (8.11) is simply a formula for the number of states energetically accessible at the transition state, and Eq. (8.8) leads to the thermal average of this number. If we imagine that the states of the system form a continuum, then $P^{\ddagger}_{cum}(E)$ can be expressed in terms of a density of states ρ as in

$$P^{\ddagger}_{cum}(E) = \int_0^E \rho^{\ddagger}(\varepsilon) \, d\varepsilon \qquad (8.12a)$$

Substituting this into the integral in Eq. (8.8) and inverting the order of integration, one gets

$$\int_0^\infty e^{-E/kT} \left(\int_0^E \rho^{\ddagger}(\varepsilon) \, d\varepsilon \right) \frac{dE}{kT} = \int_0^\infty \rho^{\ddagger}(\varepsilon) \left(\int_\varepsilon^\infty e^{-E/kt} \frac{dE}{kT} \right) d\varepsilon \quad (8.12b)$$

The inner integral on the right-hand side is just $e^{-\varepsilon/kT}$, so Eq. (8.12b) reduces to the transition state partition function (leaving out relative translation):

$$Q^{\ddagger} = \int_0^\infty e^{-\varepsilon/kT} \rho^{\ddagger}(\varepsilon) \, d\varepsilon \qquad (8.12c)$$

Using this in Eq. (8.8) gives us the usual transition-state theory expression

$$k(T) = \frac{kT}{h} \frac{Q^{\ddagger}}{Q_{int}Q_{trans}} \qquad (8.13)$$

8.4

RRKM Theory

It is now straightforward to derive an expression for the rate of decay in a unimolecular reaction. If we consider the association reaction

$$A + B \underset{k_{-1}}{\overset{k_1}{\rightleftharpoons}} C$$

we note that if A, B, and C are in equilibrium, then detailed balance requires

$$\frac{k_1}{k_{-1}} = K \qquad (8.14)$$

where K is the equilibrium constant. The latter quantity equals the ratio of partition functions per unit volume Q_C/Q_AQ_B. Taking k_1 from Eq. (8.13) and realizing that $Q_{int}Q_{trans} = Q_AQ_B$, we arrive at the following expression for the unimolecular decay rate k_{-1}:

$$k_{-1} = \frac{kT}{h}\frac{Q^{\ddagger}}{Q_C} \qquad (8.15)$$

Writing Q_C as

$$Q_C = \int \rho_C(E)e^{-E/kT}\, dE \qquad (8.16)$$

where ρ_C is the density of states of the molecule C, and using Eqs. (8.8) and (8.12), we can rewrite Eq. (8.15) as

$$k_{-1}(T) = \frac{\int_0^{\infty} e^{-E/kT}P_{cum}^{\ddagger}(E)\, dE}{h\int_0^{\infty} e^{-E/kT}\rho_C(E)\, dE} \qquad (8.17)$$

Equation (8.17) is the canonical ensemble version of a corresponding microcanonical rate constant expression, which is

$$k_{-1}(E) = \frac{P_{cum}^{\ddagger}(E)}{h\rho_C(E)} \qquad (8.18)$$

This is just the standard RRKM theory expression for the unimolecular decay rate.

8.5

Formal Expression for Rates in Terms of Flux Operators

Let's return to Eq. (8.8) and express P_{cum} in terms of fluxes. For simplicity, consider the one-dimensional case where

$$P_{cum} = \sum_{if} |S_{if}|^2 \qquad (8.19)$$

If ψ_i is the complete wavefunction associated with an incoming wave in state i, then $|S_{if}|^2$ is just the outgoing flux in channel f divided by the incident flux v_i. This outgoing flux is obtained by projecting ψ_i onto internal state f, then using the flux formula [Eq. (7.16)]. Denoting the projected wavefunction by $\langle f|\psi_i\rangle_{int} = \psi_{if}$ we can write

$$|S_{if}|^2 = v_i^{-1}\, \text{Re}[\psi_{if}^* v\psi_{if}] \qquad (8.20)$$

Since the velocity operator commutes with the internal coordinates that are implicit in $|f\rangle$, the sum over states f reduces to

$$\sum_f |S_{if}|^2 = v_i^{-1}\, \mathrm{Re}\langle\psi_i \sum_f |f\rangle\langle f|\hat{v}|\psi_i\rangle_{\mathrm{int}} = v_i^{-1}\, \mathrm{Re}\langle\psi_i\hat{v}\psi_i\rangle_{\mathrm{int}} \quad (8.21)$$

Note that the integral in Eq. (8.21) is over the internal state coordinates, but not the translational coordinate, called s, that $\hat{v}$ operates on. It is convenient to integrate over all coordinates, however, and we can do this by adding a Dirac delta function $\delta(s)$ into the integral:

$$\mathrm{Re}\langle\psi_i|\hat{v}|\psi_i\rangle_{\mathrm{int}} = \mathrm{Re}\langle\psi_i|\hat{F}|\psi_i\rangle \quad (8.22)$$

where F is the flux operator:

$$F = \delta(s)\hat{v} = \frac{\delta(s)\hat{p}}{m} \quad (8.23)$$

Thus from Eq. (8.8), for a given J we find (letting $Q = Q_{\mathrm{int}}Q_{\mathrm{trans}}$)

$$k(T) = (hQ)^{-1}\int_0^\infty e^{-E/kT}\sum_i v_i^{-1}\, \mathrm{Re}\langle\psi_i|\hat{F}|\psi_i\rangle\, dE \quad (8.24)$$

Note that although the derivation leading to Eq. (8.24) involves the use of asymptotic coordinates to separate internal and translational variables, the final result is more general, and the variable s can refer to any reaction path coordinate with $\hat{v}$ its conjugate velocity operator.

Now let's change integration variables in Eq. (8.24) from E to p_i, where $E = p_i^2/2m + \varepsilon_i$ and $p_i = mv_i$. In addition, take the $e^{-E/kT}$ factor inside the $\langle\psi_i|\hat{F}|\psi_i\rangle$ integral (replacing E by H). This leads to

$$k(T) = (hQ)^{-1}\int_0^\infty \sum_i \mathrm{Re}\langle\psi_i|e^{-H/kT}\hat{F}|\psi_i\rangle\, dp_i \quad (8.25)$$

Note that only half the momenta p_i produce motion in the right direction for reaction to occur, so the integration limits are taken to be $(0, \infty)$ in Eq. (8.25). (We could just as easily have taken them as positive or negative.)

The factor of h in Eq. (8.25) can be removed at this point by using it to normalize the continuum wavefunctions (i.e., dividing ψ_i and ψ_i^* by $h^{-1/2}$). This leads to the continuum function normalization

$$h^{-1}\langle\psi_i|\psi_{i'}\rangle = \delta(p_i - p_i')$$

which will prove convenient.

Equation (8.25) is almost in the form of a *trace*. For a discrete set of states and any operator Z, $\mathrm{Tr}\, Z$ is simply

$$\mathrm{Tr}\, Z = \sum_n \langle n|Z|n\rangle$$

For continuum states, the trace becomes an integral. For example, for momentum eigenstates, $\psi_{p_i} = h^{-1/2} \exp(ik_i x)$, so one has

$$\text{Tr } Z = \int_{-\infty}^{\infty} \int_{-\infty}^{\infty} h^{-1} \exp(-ik_i x) Z \exp(ik_i x) \, dk_i \, dx \qquad (8.26)$$

Comparison of this with Eq. (8.25) (realizing that the internal states provide the additional sum over i) indicates that Eq. (8.25) differs from a trace only through the lower momentum integration limit. A formal way to get around this "problem" is to define a projection operator $\hat{P}$ that projects out only positive momentum components. Thus if we let $\hat{P}$ satisfy

$$\hat{P}\psi_{p_i} = \begin{cases} \psi_{p_i} & \text{for } p_i > 0 \\ 0 & p_i < 0 \end{cases} \qquad (8.27)$$

Then Eq. (8.25) can be written as

$$k(T) = Q^{-1} \int_{-\infty}^{\infty} \sum_i \text{Re}\langle\psi_i | e^{-H/kT}\hat{F}\hat{P}|\psi_i\rangle \, dp_i = Q^{-1} \text{ Re Tr}(e^{-H/kT}\hat{F}\hat{P})$$

$$= Q^{-1} \text{ Re Tr}(e^{-H/2kT}\hat{F}e^{-H/2kT}\hat{P}) \qquad (8.28)$$

The last form of Eq. (8.28) comes from dividing the $e^{-H/kT}$ into two parts in (8.24) which are placed symmetrically about $\hat{F}$ in Eq. (8.25).

Equation (8.28) is a formally exact expression for the rate constant in terms of operators that was first written down by W. H. Miller in *J. Chem Phys.* **61**, 1823 (1974). It has the form of a Boltzmann average of the operator product $\hat{F}\hat{P}$, and its simple form leads to a number of interesting applications. From a computational point of view, one can imagine using Eq. (8.28) to evaluate rate constants in a direct way simply by evaluating the trace in a basis set. (Note that traces are independent of basis provided that the basis is complete.) Second, one can imagine using Eq. (8.28) to develop a rigorous semiclassical theory of rate constants (through stationary-phase evaluation of the trace). Third, one can use Eq. (8.28) to develop a nonseparable transition-state theory. Miller has pursued all of these possibilities in a number of papers beginning with the 1974 paper.

One other direction of approach using Eq. (8.28) is that presented by W. H. Miller, S. D. Schwartz, and J. W. Tromp, in *J. Chem. Phys.* **79**, 4889 (1983) [and anticipated much earlier by T. Yamamoto (*J. Chem. Phys.* **33**, 281 (1960))] in which $k(T)$ is expressed in terms of flux correlation functions. It is this direction that we now consider. Before we do, however, it is necessary to express P somewhat differently.

8.6

Additional Expressions for Rate Constant

One way to reexpress Eq. (8.27) is in terms of the Heaviside function h [see Eq. (8.11)]. It is straightforward to show that

$$\hat{P} = \sum_i \int_{-\infty}^{\infty} h(p_i)|\psi_{p_i}\rangle\langle\psi_{p_i}|\, dp_i \qquad (8.29)$$

satisfies Eq. (8.27) and thus can be used to represent the projection operator. Now let's play some tricks with Eq. (8.29). ψ_{p_i} is a solution to the time-independent Schrödinger equation, and thus it must be related to the time-dependent solution $\psi_{p_i}(t)$ by

$$\psi_{p_i}(t) = e^{-iEt/\hbar}\psi_{p_i} = e^{-iHt/\hbar}\psi_{p_i} \qquad (8.30)$$

The second two forms of this equation can then be rearranged to write

$$\psi_{p_i} = \lim_{t \to -\infty} e^{-iHt/\hbar}e^{iEt/\hbar}\psi_{p_i} \qquad (8.31)$$

where we have chosen to evaluate this expression at $t \to -\infty$, since in this limit $e^{-iHt/\hbar}\psi_{p_i}$ is identical to $e^{-iHt/\hbar}\phi_{p_i}$, where ϕ_{p_i} is the initial asymptotic momentum eigenfunction. Since ϕ_{p_i} is an eigenfunction of the free-particle operator H_0 with eigenvalue E, we can write $e^{iEt/\hbar}\phi_{p_i} = e^{iH_0t/\hbar}\phi_{p_i}$ and thus

$$\psi_{p_i} = \lim_{t \to -\infty} e^{-iHt/\hbar}e^{iH_0t/\hbar}\phi_{p_i} \qquad (8.32)$$

Relabelling t as $-t$ and inserting the result into Eq. (8.29), we find that

$$\hat{P} = \sum_i \int_{-\infty}^{\infty} \lim_{t \to \infty} e^{+iHt/\hbar}e^{-iH_0t/\hbar}h(p_i)|\phi_{p_i}\rangle\langle\phi_{p_i}|e^{+iH_0t/\hbar}e^{-iHt/\hbar}\, dp_i \quad (8.33)$$

Since ϕ_{p_i} is an eigenfunction of $\hat{P}$, we can write $h(p_i)|\phi_{p_i}\rangle = h(\hat{p})|\phi_{p_i}\rangle$. The sum over i and integral over p_i of $|\phi_{p_i}\rangle\langle\phi_{p_i}|$ is then recognized as the identity, and we find that

$$\hat{P} = \lim_{t \to \infty} e^{iHt/\hbar}h(p)e^{-iHt/\hbar} \qquad (8.34)$$

Substituting this into Eq. (8.28), we find that

$$k(T) = Q^{-1}\lim_{t \to \infty} \text{Re Tr}(e^{-H/2kT}\hat{F}e^{-H/2kT}e^{iHt/\hbar}h(p)e^{-iHt/\hbar}) \quad (8.35)$$

Since the trace is invariant to cyclic permutation of the operators that appear in it and $e^{-H/2kT}$ and $e^{iHt/\hbar}$ commute, we can rewrite Eq. (8.35) as

$$k(T) = Q^{-1} \lim_{t \to \infty} \text{Re } \text{Tr}(\hat{F}e^{iHt/\hbar}\hat{G}e^{-iHt/\hbar}) \qquad (8.36)$$

where

$$\hat{G} = e^{-H/2kT}h(p)e^{-H/2kT} \qquad (8.37)$$

The real part in Eq. (8.36) involves summing the trace plus its complex conjugate. The complex conjugate of a trace is equivalent to the trace of the hermitian conjugate of the operators appearing in the trace in reverse order (as is easily shown). Since $\hat{G}$ is hermitian, this hermitian conjugate can be written as

$$\text{Tr}(\hat{F}e^{iHt/\hbar}\hat{G}e^{-iHt/\hbar})^* = \text{Tr}(e^{iHt/\hbar}\hat{G}e^{-iHt/\hbar}\hat{F}^+) = \text{Tr}(\hat{F}^+e^{iHt/\hbar}\hat{G}e^{-iHt/\hbar}) \qquad (8.38)$$

where F^+ is the hermitian conjugate of F. Summing the original trace with that in Eq. (8.38), the real part reduces to

$$k(T) = Q^{-1} \lim_{t \to \infty} \text{Tr}(\overline{F}e^{iHt/\hbar}\hat{G}e^{-iHt/\hbar}) \qquad (8.39)$$

where, from (8.23),

$$\overline{F} = \frac{1}{2}\left[\delta(s)\frac{\hat{p}}{m} + \frac{\hat{p}}{m}\delta(s)\right] \qquad (8.40)$$

By defining a complex time $t_c = t - i\hbar/2kT$, we can combine Eqs. (8.37) and (8.40) to get

$$k(T) = Q^{-1} \lim_{t \to \infty} \text{Tr}(\overline{F}e^{iHt_c^*/\hbar}h(p)e^{-iHt_c/\hbar}) \qquad (8.41)$$

The advantage of working with the symmetrized flux operator $\overline{F}$ instead of F is that it is hermitian and thus corresponds to the physically measurable flux. Before we can rewrite Eq. (8.41) in terms of flux autocorrelation functions, it is necessary to replace $h(p)$ in Eq. (8.41) by $h(s)$. To do this we need to prove that in the $t \to \infty$ limit, the following two operators are identical:

$$\hat{P}_s = e^{iHt/\hbar}h(s)e^{-iHt/\hbar} \qquad (8.42a)$$

$$\hat{P}_p = e^{iHt/\hbar}h(p)e^{-iHt/\hbar} \qquad (8.42b)$$

Inserting $e^{-iH_0t/\hbar}e^{iH_0t/\hbar}$ before and after $h(s)$ and $h(p)$, we get

$$\hat{P}_s = \Omega^+e^{iH_0t/\hbar}h(s)e^{-iH_0t/\hbar}\Omega \qquad (8.43a)$$

$$\hat{P}_p = \Omega^+e^{iH_0t/\hbar}h(p)e^{-iH_0t/\hbar}\Omega \qquad (8.43b)$$

where $\Omega = e^{iH_0t/\hbar}e^{-iHt/\hbar}$ is an operator called the Møller wave operator. The equivalence of the two P_i's requires that the following are equal (as $t \to \infty$):

$$P_s^0 = e^{iH_0t/\hbar}h(s)e^{-iH_0t/\hbar} \tag{8.44a}$$

$$P_p^0 = e^{iH_0t/\hbar}h(p)e^{-iH_0t/\hbar} \tag{8.44b}$$

Choosing H_0 to be the free-particle hamiltonian ($H_0 = p^2/2m$), we can now explicitly evaluate matrix elements of Eqs. (8.44a) and (8.44b) to show their equivalence. Choosing Eq. (8.44b) and using coordinate eigenfunctions [i.e., $\delta(s' - s) \equiv |s'\rangle$] for the evaluation, we find that

$$\langle s'|\hat{P}_p^0|s\rangle = \langle s'|\exp\left(+\frac{ip^2t}{2m\hbar}\right)h(p)\exp\left(-\frac{ip^2t}{2m\hbar}\right)|s\rangle = \langle s'|h(p)|s\rangle$$

$$= \int_0^\infty \langle s'|p\rangle\langle p|s\rangle\,dp \tag{8.45}$$

Since $\langle s|p\rangle = e^{ips/\hbar}/(2\pi\hbar)^{1/2}$ for translational functions that are normalized as in Eq. (8.26), we find that

$$\langle s'|\hat{P}_p^0|s\rangle = (2\pi\hbar)^{-1}\int_0^\infty e^{ip(s'-s)/\hbar}\,dp \tag{8.46}$$

Now consider matrix elements of Eq. (8.44a). Inserting the identity into this equation, we find that

$$\langle s'|P_s^0|s\rangle = \int_0^\infty \langle s'|e^{iH_0t/\hbar}|s''\rangle\langle s''|e^{-iH_0t/\hbar}|s\rangle\,ds'' \tag{8.47}$$

where $\langle s''|e^{-iH_0t/\hbar}|s\rangle$ is the free-particle propagator. This can be evaluated by inserting complete sets of momentum states, yielding

$$\int dp \int dp'\langle s''|p\rangle\langle p|e^{-iH_0t/\hbar}|p'\rangle\langle p'|s\rangle$$

$$= (2\pi\hbar)^{-1}\int_{-\infty}^\infty dp\,e^{ip(s''-s)/\hbar}e^{-ip^2t/2m\hbar}$$

$$= \left(\frac{m}{2\pi i\hbar t}\right)^{1/2}\exp\left[\frac{im(s'' - s)^2}{2\hbar t}\right] \tag{8.48}$$

Substituting this into (8.47) yields

$$\langle s'|\hat{P}_s^0|s\rangle = \left(\frac{m}{2\pi\hbar t}\right)\int_0^\infty ds''\exp\left\{\frac{im}{2\hbar t}[(s'' - s)^2 - (s' - s'')^2]\right\}$$

$$= \frac{m}{2\pi\hbar t}\exp\left[\frac{im}{2\hbar t}(s^2 - s'^2)\right]\int_0^\infty ds''e^{ims''(s'-s)/\hbar t} \tag{8.49}$$

Now change integration variables from s'' to $p = ms''/t$. This gives

$$\langle s'|\hat{P}_s^0|s\rangle = (2\pi\hbar)^{-1}\exp\left[\frac{im}{2\hbar t}(s^2 - s'^2)\right]\int_0^\infty dp\,e^{ip(s'-s)/\hbar} \tag{8.50}$$

which is equivalent to (8.46) in the $t \to \infty$ limit. Physically, this makes sense since the positive momentum states should in the $t \to \infty$ limit coincide with states that are localized at $s > 0$.

The equivalence of (8.42a) and (8.42b) allows us to write a new expression for $k(T)$ [from Eq. (8.41)]. This is

$$k(T) = Q^{-1} \lim_{t \to \infty} \mathrm{Tr}(\bar{F} e^{iHt^*/\hbar} h(s) e^{-iHt_c/\hbar}) \tag{8.51}$$

8.7

Flux–Flux Autocorrelation Functions

One can show that the trace in Eq. (8.51) vanishes for $t = 0$. (In that limit, the only term that depends on i is $\bar{F}$, but that is purely imaginary, so the trace, which must be real, must vanish.) As a result, one can replace Eq. (8.51) by the following integral:

$$k = Q^{-1} \int_0^\infty dt \, C_f(t) \tag{8.52}$$

where

$$C_f(t) = \frac{d}{dt} \mathrm{Tr}(\bar{F} e^{iHt^*/\hbar} h(s) e^{-iHt_c/\hbar}) \tag{8.53}$$

Evaluation of the time derivative gives

$$C_f(t) = \frac{i}{\hbar} \mathrm{Tr}(\bar{F} e^{iHt^*/\hbar} [H, h(s)] e^{-iHt_c/\hbar}) \tag{8.54}$$

and the commutator in Eq. (8.54) is [using (8.40)]

$$[H, h(s)] = \left[\frac{p^2}{2m}, h(s) \right] = \frac{-i\hbar}{2} \left\{ \frac{p}{m} \delta(s) + \delta(s) \frac{p}{m} \right\} = \frac{\hbar \bar{F}}{i} \tag{8.55}$$

Thus

$$C_f(t) = \mathrm{Tr}(\bar{F} e^{iHt^*/\hbar} \bar{F} e^{-iHt_c/\hbar}) \tag{8.56}$$

which is a kind of flux–flux autocorrelation function. Equations (8.52) and (8.56) indicate that the exact rate constant equals the time integral of this autocorrelation function. This relation bears an important resemblance to expressions that come from linear response theory which relate a variety of dynamical quantities to time integrals of autocorrelation functions (Chapters 9 and 11), and in fact Yamamoto used linear response theory to derive an expression similar to (though not quite the same as) Eqs. (8.52) and (8.56), long before the Miller, Schwarz, Tromp result.

Since traces can be evaluated in any basis, it is attractive to try to evaluate Eq. (8.56) by direct numerical evaluation. Several groups have shown that this works well for problems with small dimensionality such as collinear and three-dimensional $H + H_2$. For problems with large dimensionality, such basis evaluations are impractical. Under these circumstances, evaluation by path integral methods becomes interesting, and it is to this topic that we now direct our attention. First, however, we need to express the flux–flux autocorrelation function in terms of propagator matrix elements.

From Eq. (8.56) it is straightforward to evaluate the trace in terms of coordinate eigenfunctions, inserting identities three times so that all operators become coordinate representation matrix elements. The result is

$$C_f(t) = \int ds \int ds' \int ds'' \int ds''' \, \langle s|\bar{F}|s'\rangle\langle s'|e^{iHt_c^*/\hbar}|s''\rangle\langle s''|\bar{F}|s'''\rangle\langle s'''|e^{-iHt_c/\hbar}|s\rangle$$

(8.57)

The flux operator matrix elements can be evaluated using Eq. (8.40), giving (see Problem 3)

$$\langle s|\bar{F}|s'\rangle = \int ds'' \delta(s'' - s) \left(\frac{-i\hbar}{2m}\right)\left[\delta(s'') \frac{\partial}{\partial s''} + \frac{\partial}{\partial s''} \delta(s'')\right] \delta(s'' - s')$$

$$= \frac{-i\hbar}{2m} [\delta'(s)\delta(s') - \delta(s)\delta'(s')]$$

(8.58)

After substituting this into (8.57) and doing the integrals, we now take the advantage of the following symmetry relations:

$$\langle s|e^{iHt_c^*/\hbar}|s'\rangle = \langle s|e^{-iHt_c/\hbar}|s'\rangle^*$$

(8.59)

(which is easily demonstrated), and

$$\langle s|e^{-iHt_c/\hbar}|s'\rangle = \langle s'|e^{-iHt_c/\hbar}|s\rangle$$

(8.60)

(which follows from time-reversal symmetry). Using these, we arrive at the following expression for $C_f(t)$:

$$C_f(t) = \frac{\hbar^2}{4m^2} \left(\frac{\partial^2}{\partial s \, \partial s'} |\langle s'|e^{-iHt_c/\hbar}|s\rangle|^2 - 4|\frac{\partial}{\partial s'} \langle s'|e^{-iHt_c/\hbar}|s\rangle|^2\right)$$

(8.61)

where the result should be evaluated at $s = s' = 0$. This expression is convenient, as it reduces the evaluation of C_f to the determination of propagator matrix elements and their derivatives at $s = s' = 0$.

Propagators of the form $\langle s'|e^{-iHt/\hbar}|s\rangle$ play a central role in nearly all problems in time-dependent quantum mechanics, and there are a number of different methods for evaluating them, including explicit methods such as were used in Eq. (8.48) for the free particle. For hamiltonians H having a known set of stationary states ϕ_n, one can evaluate the propagator by inserting the identity expressed as $\Sigma_n |n\rangle\langle n|$. This leads to

$$\langle s'|e^{-iHt/\hbar}|s\rangle = \sum_n e^{-iE_n t/\hbar} \phi_n^*(s')\phi_n(s) \qquad (8.62)$$

Still another method for evaluating propagators involves expressing them in terms of path integrals. This approach has the advantage over Eq. (8.62) for problems of high dimension that it allows for evaluation using Monte Carlo integration methods, so it is to this that we now turn our attention.

8.8

Evaluation of Propagator Matrix Elements Using Path Integrals

For the purpose of this discussion we take the matrix element to be that of the propagator between coordinates s_0 and s_N:

$$\langle s_N|e^{-iHt/\hbar}|s_0\rangle \qquad (8.63)$$

where the time t is in general complex.

Following Feynman and Hibbs, we now imagine breaking up $e^{-iHt/\hbar}$ in (8.63) into N pieces, each being $e^{-iHt/\hbar N}$. Now insert $N - 1$ identities in terms of the coordinate basis. This leads to

$$\langle s_N|e^{-iHt/\hbar}|s_0\rangle = \int_{-\infty}^{\infty} ds_1 \int_{-\infty}^{\infty} ds_2 \int_{-\infty}^{\infty} ds_3 \cdots$$

$$\int_{-\infty}^{\infty} ds_{N-1} \langle s_N|e^{-iHt/\hbar N}|s_{N-1}\rangle \qquad (8.64)$$

$$\times \langle s_{N-1}|e^{-iHt/\hbar N}|s_{N-2}\rangle \cdots \langle s_1|e^{-iHt/\hbar N}|s_0\rangle$$

If N is large enough, the exponent in $e^{-iHt/\hbar N}$ is small and can be replaced by $e^{-ip^2 t/2m\hbar N}e^{-iVt/\hbar N}$. The potential term is diagonal when its coordinate space matrix element is taken while the kinetic energy matrix element is the free-particle propagator that can be taken from Eq. (8.48). Thus the typical matrix element in (8.64) is

$$\langle s_{i+1}|e^{-iHt/\hbar N}|s_i\rangle \approx \left(\frac{mN}{2\pi i\hbar t}\right)^{1/2} \exp\left\{i\frac{mN}{2\hbar t}(s_{i+1} - s_i)^2\right\}$$

$$\exp\left\{-\frac{it}{\hbar N} V\left[\frac{(s_{i+1} + s_i)}{2}\right]\right\} \qquad (8.65)$$

Substituting this into Eq. (8.64) leads us to the expression

$$\langle s_N | e^{-iHt/\hbar} | s_0 \rangle = \int_{-\infty}^{\infty} ds_1 \int_{-\infty}^{\infty} ds_2 \int_{-\infty}^{\infty} ds_3 \cdots \int_{-\infty}^{\infty} ds_{N-1} \left(\frac{Nm}{2\pi i\hbar t} \right)^{N/2}$$

$$\times \exp\left\{ i \frac{mN}{2\hbar t} \sum_{i=1}^{N} (s_i - s_{i-1})^2 - \frac{it}{\hbar N} \sum_{i=1}^{N} V\left[\frac{s_i + s_{i-1}}{2} \right] \right\}$$

$$(8.66)$$

At this point it is relatively straightforward to convert Eq. (8.66) into a path integral. If we consider evaluating the integral $\int_0^t V(t') \, dt'$ on a grid in time with N steps equal to t/N in size, the second term in Eq. (8.65) provides a discretized approximation to this integral. Similarly, the first term is approximately the kinetic energy integral $\int_0^t T(t') \, dt'$ since the time derivative $\dot{s}$ at the ith grid point is $(s_i - s_{i-1})/(t/N)$. Thus in the limit of large N, the exponent in Eq. (8.66) becomes

$$\frac{i}{\hbar} \int_0^t (T - V) \, dt' = \frac{i}{\hbar} \int_0^t L(t') \, dt' = \frac{i}{\hbar} S \tag{8.67}$$

where L is the classical lagrangian and S is the action associated with "paths" that connect s_0 initially to s_N at time t. The "paths" in this case involve all possible combinations of $s_1, s_2, \ldots, s_{N-1}$. Equation (8.66) thus represents a sum over all paths of the quantity $\exp((i/\hbar)S)$. Clearly, $\exp((i/\hbar)S)$ is the amplitude associated with each path. Such a sum over paths is often written in the form

$$\langle s_N | e^{-iHt/\hbar} | s_0 \rangle = \int Ds \, \exp\left(\frac{i}{\hbar} S \right) \tag{8.68}$$

where Ds stands for the sum over all paths.

Now let us return to the evaluation of Eq. (8.66). Since the integrals therein involve a large number of dimensions, it is logical (but problematic as we shall see) to attempt its evaluation by Monte Carlo. Since the exponent in (8.66) contains a gaussian function in the coordinates, it is also logical to convert from coordinate variables to new variables that transform the gaussians to constant functions and which change the $(-\infty, \infty)$ limits to $(0,1)$. Such variables are labeled w_i and are defined by the recursion formula

$$s_i = \frac{N - i}{N - i + 1} s_{i-1} + \frac{s_N}{N - i + 1} + \left[\frac{2\pi i\hbar t}{m} \frac{N - i}{N(N - i + 1)} \right]^{1/2} z(w_i) \tag{8.69}$$

where $z(w_i)$ is the inverse of

$$w(z) = \int_{-\infty}^{z} dz' \, e^{-\pi z'^2} \tag{8.70}$$

Substitution of these relations into (8.66) leads to

$$\langle s_N | e^{-iHt/\hbar} | s_0 \rangle = \left(\frac{m}{2\pi i\hbar t} \right)^{1/2} \exp\left[i\,\frac{m}{2\hbar t}\,(s_N - s_0)^2 \right]$$

$$\times \int_0^1 dw_1 \int_0^1 dw_2 \cdots \int_0^1 dw_{N-1} \exp\left[-\frac{it\langle V(\mathbf{w}) \rangle}{\hbar} \right]$$

$$(8.71)$$

with

$$\langle V(\mathbf{w}) \rangle = \frac{1}{N} \sum_{i=1}^{N} V \left[\frac{s_i + s_{i-1}}{2} \right] \qquad (8.72)$$

The factor in front of the integral in this expression is just the free-particle propagator [see Eq. (8.48)] between s_0 and s_N. The integral in (8.71) therefore describes the contribution of the potential to the matrix element, and it is in a form that is easily evaluated by Monte Carlo integration, namely

$$\int_0^1 dw_1 \int_0^1 dw_2 \cdots \int_0^1 dw_{N-1}\, e^{-it\langle V(\mathbf{w})\rangle/\hbar} \approx \frac{1}{M} \sum_{k=1}^{M} e^{-it\langle V(\mathbf{w}_k)\rangle/\hbar} \quad (8.73)$$

In evaluating this integral, one uses a random number generator to generate M sets of w_{ik}'s ($i = 1, \ldots, N - 1$; $k = 1, \ldots, M$). The w_{ik}'s for each k are then used to define s_i's via Eqs. (8.69) and (8.70) and then the s_i's are used to evaluate $\langle V \rangle$ in Eq. (8.72), which is then substituted into Eq. (8.73) and summed over k.

Each set of s_i's can be thought of as defining a path between the initial and final points in the path integral. In the Monte Carlo evaluation, each of these paths has equal probability. Of course, the $e^{-it\langle V\rangle/\hbar}$ in Eq. (8.73) will be largest for paths with the smallest $\langle V \rangle$, so a more efficient way to do the quadrature would involve selecting only the more important paths (i.e., importance sampling). This can be done straightforwardly when the problem being solved looks like an analytically soluble problem for which the optimum paths are known. For purely imaginary times t (i.e., evaluation of Boltzmann operator matrix elements), a number of methods are available for evaluating Eq. (8.73) with reasonable efficiency.

The evaluation of Eq. (8.73) for complex or purely real t provides a much more formidable challenge, since then the $e^{it\langle V\rangle/\hbar}$ contributions are not all real and positive. Brute-force evaluation inevitably leads to imperfect cancellation and poor results. A number of methods have been proposed for improving convergence behavior, but much remains to be done before this approach for calculating propagators will be practical.

BIBLIOGRAPHY

The most closely related references to the topics in this chapter are in the original literature:

Miller, W. H., *J. Chem. Phys.* 61, 1823 (1974).

Miller, W. H., S. D. Schwartz, and J. W. Tromp, *J. Chem. Phys.* 79, 4889 (1983).

Yamamoto, T., *J. Chem. Phys.* 33, 281 (1960).

The standard physicist's text on path integrals is:

Feynman, R. P., and A. P. Hibbs, *Quantum Mechanics and Path Integrals* (McGraw-Hill, New York, 1965).

Modern treatments of gas-phase kinetics include:

Child, M. S., *Molecular Collision Dynamics* (Academic Press, London, 1974).

Gilbert, R. G., and S. C. Smith, *Theory of Unimolecular and Recombination Reactions* (Blackwell Scientific, Oxford, 1990).

Levine, R. D., *Quantum Mechanics of Molecular Rate Processes* (Clarendon Press, Oxford, 1969).

Levine, R. D., and R. B. Bernstein, *Molecular Reaction Dynamics and Chemical Reactivity* (Oxford University Press, New York, 1987).

Murrell, J. N., and S. D. Bosanac, *Introduction to the Theory of Atomic and Molecular Collisions* (Wiley, Chichester, West Sussex, England, 1989).

Smith, I. M. W., *Kinetics and Dynamics of Elementary Gas Reactions* (Butterworth, London, 1980).

Steinfeld, J. I., J. E. Francisco, and W. L. Hase, *Chemical Kinetics and Dynamics* (Prentice-Hall, Englewood Cliffs, N.J., 1989).

Weston, R. S., and H. A. Schwartz, *Chemical Kinetics* (Prentice-Hall, Englewood Cliffs, N.J., 1972).

PROBLEMS FOR CHAPTER 8

1. Given the general formula (8.62) for the propagator,

$$\langle s'|e^{-iHt/\hbar}|s\rangle = \sum_n e^{-iE_n t/\hbar}\phi_n^*(s')\phi_n(s)$$

(a) Show that the partition function $Q = \mathrm{Tr}\, e^{-\beta H}$ is given by the usual formula

$$Q = \sum_n e^{-\beta E_n}$$

where E_n is an eigenvalue of H.

(b) Show how the free-particle propagator (8.48) is derived from (8.62).

2. Feynman and Hibbs show that for simple one-dimensional problems the propagator is

$$\langle s_a | e^{-iHt/\hbar} | s_b \rangle = (\text{function of } t) e^{i/\hbar S(s_a, s_b; t)}$$

where $S(s_a, s_b; t)$ is the action $\int_0^t L(t')\, dt'$ associated with a classical trajectory that moves between points s_a and s_b in time t ($t = t_b - t_a$).

(a) Show that for a harmonic oscillator (where $H = \frac{1}{2}m\dot{s}^2 + \frac{1}{2}m\omega^2 s^2$).

$$S = \frac{m\omega}{2 \sin \omega t} [(s_a^2 + s_b^2) \cos \omega t - 2 s_a s_b]$$

Substituting this into the expression above determines the propagator except for the multiplicative function of t. The latter is

$$\left(\frac{m\omega}{2\pi i\hbar \sin \omega t} \right)^{1/2}$$

(b) Using the propagator from part (a), show that if the wavefunction initially is given by the gaussian function

$$\Psi(x) = \exp\left[-\frac{m\omega}{2\hbar} (s - a)^2 \right]$$

then the wavefunction at time t will still be gaussian. The constant a is arbitrary. Show that the center of the probability density follows the classical equations of motion.

3. Consider the transformation of Eq. (8.57) into (8.61).

(a) As a first step, derive Eq. (8.58), using (8.40). To do this you will need to use several properties of delta functions that are given in Appendix A. Note particularly that

$$\frac{\partial}{\partial s''} \delta(s'' - s') = -\frac{\partial}{\partial s'} \delta(s'' - s')$$

and that $\delta(-s) = \delta(s)$. Two other important clues are (1) that the derivative that comes from the second momentum operator in (8.40) operates on everything to its right, and (2) that the second term in (8.40) ultimately becomes the first term after the last equals sign in (8.58).

(b) Now substitute (8.58) into (8.57) and invoke the relations (8.59) and (8.60). To manipulate this into the form (8.61), you will want to prove the relation

$$\frac{\partial^2}{\partial s\, \partial s'} |\langle s' | e^{-iHt_c/\hbar} | s \rangle|^2 = 2\, \mathrm{Re}\bigg\{ \langle s' | e^{-iHt_c/\hbar} | s \rangle \frac{\partial^2}{\partial s\, \partial s'} \langle s' | e^{-iHt_c/\hbar} | s \rangle^*$$

$$+ \frac{\partial}{\partial s} \langle s' | e^{-iHt_c/\hbar} | s \rangle \frac{\partial}{\partial s'} \langle s' | e^{-iHt_c/\hbar} | s \rangle^* \bigg\}$$

4. Starting with Eq. (8.51), show that

$$k(T) = (\hbar/m)Q^{-1} \lim_{t \to \infty} \int_0^\infty ds'\, \mathrm{Im}\bigg\{ \langle s' | e^{-iHt_c/\hbar} | s \rangle \frac{\partial}{\partial s} \langle s' | e^{-iHt_c/\hbar} | s \rangle^* \bigg\}_{s=0}$$

This expression provides an alternative to Eqs. (8.61) and (8.52) for evaluating rate constants in terms of propagator matrix elements. [*Hint:* The manipulations are similar to those involved in deriving Eqs. (8.57) to (8.61).]

9

Time-Dependent Approach to Spectroscopy: Electronic, Vibrational, and Rotational Spectra

9.1

Introduction

The expression (8.53) for the rate constant in terms of a flux correlation function is one example of the use of time correlation functions to deduce rates of processes. In this chapter we use analogous correlation function expressions to examine electronic spectra, vibration–rotation spectra and rates of molecular motion. In the next chapter we use similar methods to discuss particular types of chemical reactions (nonradiative decay and electron transfer), and in Chapter 11 similar but more general methods are used for subsystem evolution. These time-dependent approaches are often very useful, especially when the short-time behavior of the system largely determines its measured response. As indicated in Chapter 7 with regard to potential scattering problems, the actual time-dependent behavior of real systems can often be very effectively described using time-independent methods. Whether a time-independent or specifically time-dependent approach is best may differ depending on the nature and complexity of the problem. For electronic spectra, whose lineshapes are largely determined by phenomena in the subpicosecond region, a time-dependent approach is especially useful and transparent. Heller pioneered the use of time-dependent wavepackets for calculation of spectra, and we will follow his approach.

9.2

Thermal Averages and Imaginary Time Propagation

In Sections 8.6 to 8.9 we discussed the formal analogy between the density matrix $e^{-\beta H}$, where $\beta = 1/kT$, and the time propagator $e^{-iHt/\hbar}$, which led to the identification of $-i\hbar\beta$ as an imaginary time. In this sense one can view the taking of an average over a thermal ensemble as a propagation in imaginary time. To be more explicit, consider the thermal average of any operator A in a canonical (constant temperature and particle number and volume) ensemble. Then we have

$$\langle A \rangle = \sum_i p_i \langle i|A|i \rangle = \frac{\sum_i e^{-\beta E_i}\langle i|A|i \rangle}{\sum_i e^{-\beta E_i}} = \frac{\sum_i \langle i|e^{-\beta H}A|i \rangle}{\sum_i \langle i|e^{-\beta H}|i \rangle} = \text{Tr}\{\rho A\} \quad (9.1)$$

where $p_i = e^{-\beta E_i}/\sum_i e^{-\beta E_i}$ is the statistical weight, or probability of observing, the state with energy E_i. The density matrix operator ρ is defined from (9.1) by

$$\rho \equiv \frac{e^{-\beta H}}{\text{Tr}\{e^{-\beta H}\}} \quad (9.2)$$

The ket $|i\rangle$ is chosen in (9.1) as an eigenstate of the hamiltonian,

$$H|i\rangle = E_i|i\rangle \quad (9.3)$$

from which clearly follows

$$e^{-\beta H}|i\rangle = e^{-E_i\beta}|i\rangle \quad (9.4)$$

but of course the trace in (9.1)–(9.2) can be evaluated using any other set of states.

In particular, it is convenient to use gaussian wavepackets to provide a basis. Then we can replace the trace by a double integral over p_0 and q_0, the initial momentum and position coordinates at time zero. Thus

$$\langle A \rangle = \text{Tr}\{\rho A\} = c_n h^{-n} \int_{-\infty}^{\infty} d^n q_0 d^n p_0 \langle \chi_{p_0,q_0}|A\rho|\chi_{p_0,q_0}\rangle \quad (9.5)$$

Here c_n is a constant determined by the number of particles in the system. The integral in (9.4) runs over n positions q_0 and momenta p_0, and we denote by the ket $|\chi_{p_0,q_0}\rangle$ a gaussian packet centered at (q_0, p_0) at time zero. Replacing the density operator by its expression (9.2), we then have

$$\langle A \rangle = \frac{\int_{-\infty}^{\infty} d^n q_0 d^n p_0 \langle \chi_{p_0,q_0} | A e^{-\beta H} | \chi_{p_0,q_0} \rangle}{\int_{-\infty}^{\infty} d^n q_0 d^n p_0 \langle \chi_{p_0,q_0} | e^{-\beta H} | \chi_{p_0,q_0} \rangle} \tag{9.6}$$

This expression can be evaluated relatively easily, because the evolution of the wavepacket χ_{p_0,q_0} by the imaginary time propagator $e^{-\beta H}$ is involved. As we stressed in Chapter 7, gaussians are relatively easy to propagate forward in time.

9.3

Electronic Spectra from Time Correlation Functions

Using time-independent, golden rule considerations as outlined in Chapter 5, it is fairly easy to show [combining Eqs. (5.55) and (5.106)] that the molecular absorption cross section $\sigma_{i\nu}(\omega)$ for absorption from an initial vibrational level ν of electronic state i with energy $E_{i\nu}$ and wave function $\chi_{i\nu}$ is

$$\sigma_{i\nu}(\omega) = \frac{4\pi^2 \omega}{3c\hbar} \sum_{\nu'} |\langle \chi_{f\nu'} | \mu | \chi_{i\nu} \rangle|^2 \delta(\omega - \omega_{f\nu',i\nu}) \tag{9.7}$$

Here the final electronic state is labeled f, and $\chi_{f\nu'}$ is the ν' vibrational level of that state, with energy $E_{f\nu'}$; following usual notation,

$$E_{f\nu'} - E_{i\nu} \equiv \hbar \omega_{f\nu',i\nu}$$

The transition dipole operator is denoted by μ. To go to a time-dependent form, we use the Fourier expression (see Appendix A):

$$\delta(\omega) = (2\pi)^{-1} \int_{-\infty}^{\infty} e^{i\omega t} \, dt$$

and the identity

$$e^{-iH_f t/\hbar} = \sum_{\nu'} |\chi_{f\nu'}\rangle e^{-iE_{f\nu'} t/\hbar} \langle \chi_{f\nu'}| \tag{9.8}$$

to rewrite $\sigma_{i\nu}(\omega)$ as

$$\sigma_{i\nu}(\omega) = \frac{2\pi\omega}{3c\hbar} \int_{-\infty}^{\infty} \langle \chi_{i\nu} | \mu^* e^{-iH_f t/\hbar} \mu e^{iH_i t/\hbar} | \chi_{i\nu} \rangle e^{i\omega t} \, dt \tag{9.9}$$

Here H_i and H_f are, respectively, the electronic hamiltonians for initial and final states. Then using the operator

$$A \equiv \mu^* e^{-iH_f t/\hbar} \mu e^{iH_i t/\hbar} \tag{9.10}$$

we define the correlation function

$$C(t) = \text{Tr}\{\rho A\}. \tag{9.11}$$

The total absorption cross section from an equilibrium molecular species is the sum of contributions from all vibrational levels.

$$\sigma(\omega) = \sum_\nu P_{i\nu} \sigma_{i\nu}(\omega) = \frac{\sum_\nu e^{-\beta E_{i\nu}} \sigma_{i\nu}(\omega)}{\sum_\nu e^{-\beta E_{i\nu}}} \tag{9.12a}$$

Then we can express the temperature-dependent cross section as

$$\sigma(\omega) = \frac{2\pi\omega}{3c\hbar} \int_{-\infty}^{\infty} C(t) e^{i\omega t} \, dt \tag{9.12b}$$

Thus the absorption cross section for transition from electronic state i to state f is simply the Fourier transform of the correlation function $C(t)$.

If the transition is taken as occurring within the same electronic state, then redefining μ to be the actual dipole function rather than a transition dipole moment, and replacing H_f by H_i in (9.10), we obtain an expression for the vibrational (infrared) or rotational (microwave) spectrum as

$$\sigma(\omega) = \frac{2\pi\omega}{3c\hbar} \int_{-\infty}^{\infty} e^{i\omega t} \langle \mu^*(t) \mu(0) \rangle \, dt \tag{9.13a}$$

$$= \frac{2\pi\omega}{3c\hbar} \int_{-\infty}^{\infty} \frac{\text{Tr}\{e^{-\beta H} \mu^* e^{-iHt/\hbar} \mu e^{iHt/\hbar}\} e^{i\omega t}}{\text{Tr}\{e^{-\beta H}\}} \, dt \tag{9.13b}$$

[in (9.13a) we have used the Heisenberg representation for the time dependence of μ^*]. We return to this expression in Section 9.5.

9.4

Electronic Spectra: Time Development of the Correlation Functions

The forms (9.12) and (9.13) express spectroscopic properties in terms of time correlation functions of dipole operators; they are expressed in terms of traces (thermal averages) and therefore do not depend on the form chosen for the wavefunction; indeed, often lineshapes can be derived from them in terms of the functional behavior of the time correlations, without introducing wavefunctions at all (see, for example, Section 9.6). For specific systems, however, actual calculation of $\sigma(\omega)$ is carried out starting from forms for the wavefunctions, which are then used to express the trace. In particular, using the gaussian wavepackets χ_{p_0,q_0} (and invariance of the trace to cyclic permutation), we can rewrite the correlation function as

$$C(t) = \frac{\int_{-\infty}^{\infty} d^n q_0 d^n p_0 \langle \chi_{p_0,q_0}|e^{-\beta H_i}e^{iH_i t/\hbar}\mu^* e^{-iH_f t/\hbar}\mu|\chi_{p_0,q_0}\rangle}{\int_{-\infty}^{\infty} d^n q_0 d^n p_0 \langle \chi_{p_0,q_0}|e^{-\beta H_i}|\chi_{p_0,q_0}\rangle} \qquad (9.14)$$

General wavepacket propagation schemes can be used to calculate the matrix elements in (9.14) to any desired precision; for usual spectroscopic situations (with no complexity due to curve crossing or predissociation), frozen gaussian or thawed gaussian approximations, as discussed in Section 7.3.2, are a convenient choice for evaluating these matrix elements.

First, we rewrite $C(t)$ in terms of overlaps of wavepackets. Define time-dependent initial and final wavepackets as

$$|\Phi_i(t + \tau)\rangle \equiv \mu e^{-iH_i t/\hbar} e^{-iH_i \tau/\hbar}|\chi_{p_0,q_0}\rangle \qquad (9.15a)$$

and

$$|\Phi_f(t)\rangle \equiv e^{-iH_f t/\hbar}\mu|\chi_{p_0,q_0}\rangle \qquad (9.15b)$$

with $\tau \equiv -i\beta\hbar$. At time $t = 0$, the wavepackets Φ_i and Φ_f differ only by a statistical factor $e^{-\beta H}$, which is nearly unity for high temperature. As time proceeds, they differ. The initial packet Φ_i propagates the gaussian χ_{p_0,q_0} in time and then includes the dipole operator, whereas in Φ_f this order is reversed. Using the definition of the canonical partition function as

$$Q \equiv \text{Tr}\{e^{-\beta H}\}$$

The correlation function can be expressed as an overlap

$$C(t) = Q^{-1} \int_{-\infty}^{\infty} d^n q_0 d^n p_0 \langle \Phi_i(t + \tau)|\Phi_f(t)\rangle \qquad (9.16)$$

The wavepackets Φ_i and Φ_f evolve in time according to the hamiltonians H_i and H_f, respectively. The lineshape, or frequency dependence of $\sigma(\omega)$, will reflect the temporal evolution of the overlap $\langle \Phi_i|\Phi_f\rangle$. Since electronic spectra are found in the visible or ultraviolet regions with frequency of order 20,000 cm^{-1}, a characteristic timescale for such optical excitation is given by

$$\Delta t \sim \frac{\hbar}{\Delta E} = \frac{1}{2\pi (\Delta \nu)} = \frac{1}{2\pi} \frac{1}{c \ \Delta \bar{\nu}}$$

(9.17)

$$\sim \frac{1}{(20 \times 10^{10} \text{ cm/s})(20 \times 10^3 \text{ cm}^{-1})} = \frac{1}{4} \times 10^{-15} \text{ s}$$

so that the evolution of $C(t)$ on the subpicosecond timescale should be enough to describe the electronic spectrum. Thus we will be concerned with the behavior of (9.16) on the short and intermediate time scales.

The overlap $\langle \Phi_i(t + \tau) | \Phi_f(t) \rangle$, which is close to unity for $t = 0$ and high temperature, will generally start to decay with time, as the initial gaussian χ_{p_0, q_0} evolves according to H_i or H_f. Formally, we can expand the potentials in the Herzberg–Teller form

$$H_f - H_i = T_f - T_i + V_f - V_i = V_f - V_i$$

$$= \sum_{\lambda} \sum_{s=0} \frac{\partial^s (V_f - V_i)}{\partial q_\lambda^s} \bigg|_{q_\lambda = q_\lambda^0} \frac{(q_\lambda - q_\lambda^0)^s}{s!}$$

(9.18)

where the subscript λ labels the nuclear coordinates and s labels the terms in the Taylor series expansion. The $s = 0$ term is just a constant, corresponding to the vertical separation of the potential surfaces at q_0, and contributes a phase factor $\exp\{-i(E_f - E_i)t/\hbar\}$ to the correlation function. If all higher terms in (9.18) are ignored, we find no decay in the magnitude of the overlap. The linear and higher-order terms in (9.18) cause decay in the overlap [integrand of (9.16)]. If these decays are monotonic, we expect to see a single envelope for the electronic spectrum, since the time dependence of $C(t)$ will be a product of the phase factor $\exp\{-it(E_f - E_i)/\hbar\}$ with a monotonically decaying function.

The overlap function $\langle \Phi_i | \Phi_f \rangle$ will often contain maxima due to recurrences following the initial short-time decay. Such recurrences lead to structure in the absorption lineshape. Since the absorption is essentially just a Fourier transform of the overlap function, recurrences occurring on a timescale τ_r will be manifested in a peak in the spectrum at frequency $2\pi/\tau_r$. As temperature increases the thermal average process in (9.13) will include more initial states, each of which should exhibit recurrences in its time correlation function. Averaging over all of these recurrences will tend to broaden and blur the line, in agreement with the usual observation. (This is usually referred to as homogeneous broadening, if medium effects are absent.) Conversely, if the temperature is very low, as is true in jet-cooled experiments, in interstellar space or in matrix isolation experiments, the homogeneous linewidth drops and the features sharpen (though inhomogeneous linewidth, arising from differing local environments, may occur in the matrix experiment).

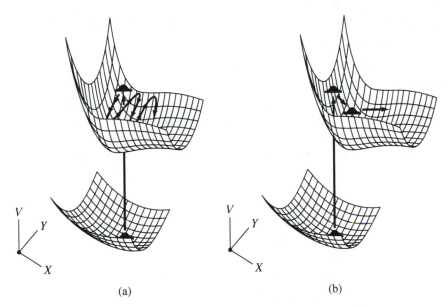

(a)   (b)

Figure 9.1. Schematic of photoabsorption between two Born–
Oppenheimer potential surfaces. On the left, the excited state
Franck–Condon wavepacket $\Phi_f(t)$ vibrates many times before
dissociating, yielding oscillatory structure in the overlap. On the
right, the initial Franck–Condon wavepacket is smoothly accel-
erated to the right, dissociating without any excited-state vibration. [With permission from E. J. Heller, *Acct. Chem. Res.* 14,
368 (1981), Fig. 1.]

Figure 9.1 shows how the predissociative broadening of an absorption spectrum may be understood. In Fig. 9.1a, the initial Franck–Condon packet, once promoted to the upper surface, oscillates repeatedly
before dissociation, while in Fig. 9.1b the upper surface has so moved
with respect to the lower [different first derivative terms in Eq. (9.18)] that
the excited-state packet Φ_f moves directly out to photodissociated products (along the positive x axis). The overlap of $\Phi_f(t)$ with $\Phi_i(t + \tau)$ for
this direct dissociation case will then decay monotonically, and thus the
absorption will be proportional to

$$\int_{-\infty}^{\infty} e^{i\omega t} e^{it(E_i - E_f)/\hbar} D(t)\, dx$$

with $D(t)$ a monotonically decaying overlap. This lineshape will be a
single broad envelope, centered at $\hbar\omega = E_f - E_i$.

The behavior in Fig. 9.1a is more interesting. The shortest time feature, the initial decay of the overlap, occurs on the timescale T_1. By the time–energy uncertainty principle,

$$\Delta t\,\Delta E \sim \hbar$$

it is responsible for the broadest spectral feature, the envelope dashed line in Fig. 9.2. At longer times, the excited-state packet can evolve until it overlaps with its initial position, giving the second maximum in $\langle \Phi_f | \Phi_i \rangle$ at $t = T_2$ in Fig. 9.2, and an associated spacing of $2\pi/T_2$ in the spectrum. Finally, the excited packet can dissociate (the predissociation process,

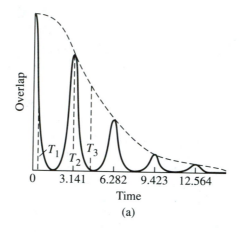

(a)

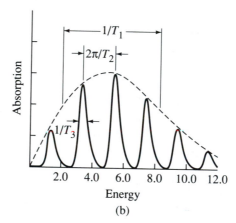

(b)

Figure 9.2. Correlation function $\langle \Phi_i(t) | \Phi_f(t + \tau) \rangle$ for the wavepackets shown in Fig. 9.1 (a) and the resulting absorption spectrum (b). The times T_1, T_2, and T_3 arise physically from initial motion of Φ_f on the excited surface, recurrent overlaps of Φ_f with Φ_i, and eventual photodissociation of the molecule. [With permission from E. J. Heller, *Acct. Chem. Res.* **14**, 368 (1981), Fig. 3.]

corresponding to the wavepacket exiting along the x axis in Fig. 9.1a), giving a remaining intrinsic lifetime of T_3 and subpeak width of $1/T_3$.

We can illustrate the combination of a time-dependent approach to spectroscopy and use of gaussian wavepacket basis sets by considering the cases of harmonic and Morse potentials, following the treatment of J. R. Reimers, K. R. Wilson, and E. J. Heller [*J. Chem. Phys.* **79**, 4749 (1983)].

For harmonic potentials, as was discussed in Section 7.3.2, initially gaussian packets remain gaussian, and their centers evolve according to classical dynamics. The case of harmonic potentials for both initial and final states leads, in the usual time-independent golden rule treatment of Chapter 4, to an absorption cross section simply expressed in terms of harmonic Franck–Condon vibrational overlaps. Using the present, time-dependent formalism, Reimers et al. calculated the cross section $\sigma(\omega)$ in the low-resolution limit by truncating the trajectories after the initial decay. The golden rule results involve energy-conserving δ functions, so that, ordinarily, the spectra occur as a set of δ-function peaks to compare with the low-resolution results obtained from (9.13). Reimers et al. artificially broadened the golden rule results by Fourier transforming into the time domain, truncating the resulting function at the same time that the gaussian wavepacket trajectories were transformed, and Fourier inverting into the frequency domain. The resulting spectra, in a particular high-temperature example, agreed to 1 part in 10^5. This is not surprising, since gaussian packets propagate so simply on harmonic surfaces, but it does indicate that the time-dependent approach is practical and accurate.

A more stringent test is offered by the Morse potential,

$$V_M(q) = D\{1 - \exp[-\alpha(q - q_0)]\}^2 \qquad (9.19)$$

with D the dissociation energy and α a bond stiffness parameter. The eigenfunctions, eigenvalues, and Franck–Condon factors are available in analytic form, so that the golden rule spectrum is available for comparison.

For relatively small differences in the equilibrium position q_0 between ground and excited states, the absorption spectrum will be dominated by transitions to bound, rather than dissociative, vibrational levels of the upper surface; under these conditions one expects the gaussian basis to work well. As Fig. 9.3 shows, the low-resolution spectrum at high temperatures reproduces the golden rule result very well; the thawed gaussian representation works slightly better than the frozen gaussians, with the differences, as expected, becoming smaller as anharmonicity decreases (in the harmonic limit, the gaussians do not thaw).

Medium-resolution results are shown in Fig. 9.4, calculated at the

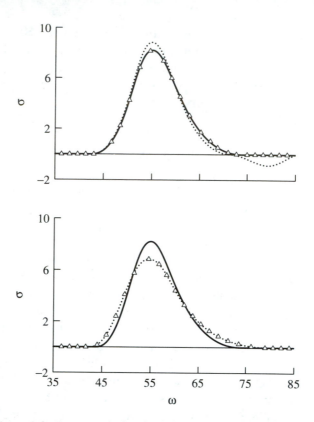

Figure 9.3. Low-resolution band contour for the absorption cross section arising from the electronic absorption for two Morse potentials, with $\mu = 1$. The time-independent (golden rule) result is the solid line, and the dotted line and triangles are, respectively, the frozen and thawed gaussian time-dependent results. Upper and lower panels correspond to different statistical averages over temperature. [With permission from J. R. Reimers, K. R. Wilson, and E. J. Heller, *J. Chem. Phys.* 79, 4749 (1983), Fig. 1.]

same relatively high temperature ($k_BT = [(2\hbar^2 D\alpha^2)/m]^{1/2}$) as Fig. 9.3, but for trajectories long enough for recurrences to be seen. The lineshape is, once again, very accurate, compared to the golden rule result. The thawed gaussian basis has been used in these calculations; it is accurate for low-energy vibrations, but (as discussed in Section 7.3.2) becomes much less accurate as the vibrational energy increases, since the exact wavepackets distort from gaussian behavior, or even fragment, as the local-harmonic approximation to the potential becomes inaccurate. For Morse potentials, this will be serious when the vibrational energy is high enough to reach the part of the surface with $q > q_c$, for critical distance q_c fixed by $(d^2 V_M/dq^2)_{q_c} = 0$.

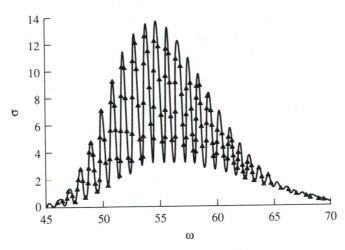

Figure 9.4. Medium resolution spectrum for the same system as Figure 9.3. The triangles are the result of thawed gaussian calculations, and the solid line is from a time-independent calculation. [With permission from J. R. Reimers K. R. Wilson, and E. J. Heller, *J. Chem. Phys.* 79, 4749 (1983), Fig. 3.]

The comparisons that we have discussed here show that, for excitations with relatively small Stokes shifts and moderate anharmonicities, the time-dependent approach of (9.13) is accurate. It has two advantages over the time-independent, golden rule formulation. The first of these involves large systems: When many vibrational degrees of freedom are involved, the time-dependent method is far more efficient than the time-independent one, which involves extensive multiple sums over Franck–Condon factors. The second advantage is in interpretation: The time-dependent method allows facile interpretation of lineshapes in terms of dynamic processes. For an example, we can consider optical spectra with short-pulse and long-pulse excitations.

9.5

Optical Spectra with Narrow-Line or Wide-Line Excitations: Time-Dependent Picture

Time dependence enters naturally into spectroscopy, since the exciting light shone on the molecule will have a pulse shape, defined as a function of time. The shape of this pulse will determine both the nature of the excitation and the absorption spectrum. It is then entirely natural to consider this problem using time-dependent theory. This has been done by M. V. Ramakrishna and R. Coalson [*Chem. Phys.* 120, 327 (1988)], whose formulation we follow.

Consider, then, a light source with arbitrary strength and pulse shape interacting with a molecule. As in Section 9.2, we consider the simple situation in which there are only two electronic states, with no complications from vibronic coupling. In a semiclassical treatment of the coupling of radiation with matter (see Section 5.5.2), a transition dipole coupling $\boldsymbol{\mu} \cdot \mathbf{E}$ will act to effect transitions between the two electronic states; here $\boldsymbol{\mu}$ and $\mathbf{E}$ are the dipole moment operator and electric field vector. Then the two-surface problem leads to the coupled Schrödinger equations for motion on the ground and excited surfaces:

$$i\hbar \frac{\partial \psi_i}{\partial t} = H_i \psi_i(t) + \mu_e E(t) \psi_f(t) \tag{9.20a}$$

$$i\hbar \frac{\partial \psi_f}{\partial t} = H_f \psi_f + \mu_e E(t) \psi_i(t) \tag{9.20b}$$

Here ψ_i and ψ_f are the ground and excited wavefunctions (or, following the notation of the preceding section, initial and final functions), while H_i and H_f are the corresponding hamiltonians. The component of $\boldsymbol{\mu}$ in the direction of $\mathbf{E}$ is denoted μ_e, and

$$\mathbf{E}(t) = A(t)\mathbf{E}^{(0)} \cos \omega t$$

with $A(t)$ the temporal profile of $\mathbf{E}(t)$. We wish to follow the evolution of the system as described by (9.20). To do so, we integrate the equation, obtaining (to lowest order in $\boldsymbol{\mu} \cdot \mathbf{E}$)

$$\psi_i(t) = e^{-iE_i t/\hbar} \psi_i(0) - \frac{i}{\hbar} \int_0^t dt' \, \exp\left\{\frac{-iH_i(t - t')}{\hbar}\right\} \mu_e E(t') \psi_f(t')$$

$$\tag{9.21a}$$

$$\psi_f(t) = \frac{-i}{\hbar} \int_0^t dt' \, \exp\left\{\frac{-iH_f(t - t')}{\hbar}\right\} \mu_e E(t') \psi_i(t') \tag{9.21b}$$

The boundary condition $\psi(t = 0) = \psi_i(0)$ has been used to simplify the second equation.

These coupled equations must, in general, be solved numerically. But they can be simplified in the weak-field limit, where perturbation theory can be used. Then we find that

$$\psi_i(t) = e^{-iE_i t/\hbar} \psi_i(0) - O(\mu E)^2 \tag{9.22a}$$

$$\psi_f(t) = \frac{-i}{\hbar} \int_0^t dt' \, \exp\left\{\frac{-iH_f(t - t')}{\hbar}\right\} E(t')\Phi_i(t') \tag{9.22b}$$

$$\cong \frac{-iE^{(0)}}{\hbar} \int_0^t dt' \, A(t') \cos(\omega t') \, \exp\left\{\frac{-iH_f(t - t')}{\hbar}\right\} e^{-iE_i t'/\hbar}\Phi_i(0) \tag{9.22c}$$

Here $\Phi_i = \mu_e\psi_i$ is the Franck–Condon wavepacket introduced in the preceding section. A further simplification called the *rotating wave approximation* can be made at this point: It consists in retaining only those terms in a sum of complex exponentials whose phases are either smallest (most slowly varying) or closest to resonance; terms with large or substantially different phase factors will have many more oscillations in the integrand and will therefore give a far smaller contribution. Thus we write

$$\cos \omega t' \cong \tfrac{1}{2}e^{-i\omega t'}$$

so that with $E = \hbar\omega + E_i$, the total system energy after optical excitation, (9.22c) becomes

$$\psi_f(t) = \frac{-iE^{(0)}}{2\hbar} \int_0^t dt' \, A(t')e^{-iE_i t'/\hbar}e^{-iH_f(t-t')/\hbar}\Phi_i(0) \tag{9.23}$$

We now consider the limiting case of a very short (δ-function) light source. Then (9.22) becomes

$$\psi_i(t) = e^{-iE_i t/\hbar}\psi_i(0)$$

$$\psi_f(t) = \frac{-iE^{(0)}}{2\hbar}e^{-iH_f t/\hbar}\Phi_i(0) \tag{9.24}$$

Verbally, in this order of perturbation theory and with a δ-function pulse, the initial state propagates freely in time as if there were no interactions. The excited state is produced by promoting the initial wavefunction $\psi_i(0)$ to the excited electronic state with the dipole function μ, and then watching the evolution of the packet, under the propagator $\exp\{-iH_f t/\hbar\}$.

In Section 9.3 we examined excitations between harmonic oscillators and between Morse oscillators. Continuing, we examine here the case of direct photodissociation, in which transfer occurs from a harmonic ground state to an unbound excited level, such as the exponential-repulsive curve of Fig. 9.5. In agreement with the boundary conditions used to derive (9.21), assume that at time zero, the system is in the vibrational ground state of the harmonic well. For simplicity, we take the transition moment operator μ to be a constant ($= 1$) over the vibrational coordinate range between the ground-state turning points; then $\Phi_i = \mu\psi_i$ is just the

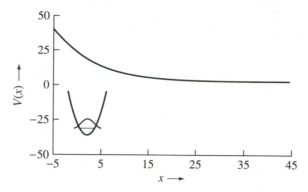

Figure 9.5. Two potential surfaces used for a study of the effect of the excitation line profile on absorption lineshape. The ground state is harmonic, the excited state dissociative (exponential repulsive). [With permission from M. V. Ramakrishna and R. Coalson, *Chem. Phys.* 120, 327 (1988), Fig. 1.]

ground-state harmonic oscillator wavefunction (which is a gaussian). The coupled Schrödinger equations (9.20) can be solved exactly for this problem, with the initial conditions

$$\psi_f(t = 0) = 0,$$

$$\psi_i(t = 0) = \left(\frac{m\omega}{\hbar\pi}\right)^{1/4} \exp\left\{\frac{-(q - q_0)^2}{2q_{zp}^2}\right\}$$

with the zero-point displacement $q_{zp} = (\hbar/m\omega)^{1/2}$; the evolution is carried out straightforwardly using grid techniques. For a gaussian exciting line defined by the profile

$$A_0(t) = \pi \exp\left\{\frac{-(t - t_0)^2}{\Delta_t^2}\right\} \tag{9.25}$$

with $t_0 = 16$ fs, $\Delta_t = 3.9$ fs, the excited-state function $\psi_f(t)$ is shown in Fig. 9.6. At very short times, less than the growth time of A_0 to its maximum (16 fs), the wavefunction ψ_f is small; note also that in agreement with (9.21), the imaginary part of ψ_f starts to grow before the real part. Again in agreement with (9.21), the initial shape of the imaginary part of ψ_f is the same as $\psi_i(0)$—in this case, a simple gaussian. As time continues, the solution of ψ_f [Eq. (9.20b)] contains both a contribution from further excitation and a part due to evolution of the excited-state function ψ_f itself. As time proceeds, both of these contribute to the propagation on the excited-state surface. At 18 fs, the real and imaginary parts of ψ_f are both still nearly gaussian, and the excitation has started to fall. In the period from, say, 10 to 30 fs, the evolution on the excited surface

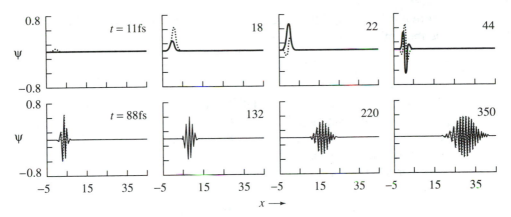

Figure 9.6. Real (solid) and imaginary (dotted) wavefunctions on the upper surface of Fig. 9.5 with a narrow gaussian exciting line. [Reprinted with permission from M. V. Ramakrishna and R. Coalson, *Chem. Phys.* 120, 327 (1988), Fig. 2.]

contains contributions both from the excited-state evolution itself and from further excitation from ψ_i [first and second terms of (9.20b)]. Following the further evolution of the excited state (Fig. 9.6) we see that the function develops nodes and a characteristic wavepacket shape, and that its center moves to the right on the repulsive excited-state surface, as expected from classical mechanics. By 350 fs it has reached its asymptotic shape and is moving smoothly to the right.

The process of excitation does not leave the ground state unchanged; as (9.20a) and (9.20b) show, its evolution contains a contribution due to interaction with ψ_f, although that interaction is relatively small [order $(\mu E)^2$]. In this case, ψ_i develops nodes and imaginary components, and its center starts to move in the ground-state harmonic well.

The gaussian excitation just described is of short duration (full width at half maximum of ~6.5 fs), thus similar to the δ-function shape used to derive (9.24). The other extreme involves a broad band or continuous wave excitation, for which we take $A(t) = 1$, $E^{(0)} = 1$. Then we write for the excited function, from (9.23), for the long time limit

$$\psi_f(t \to \infty) = \lim_{t \to \infty} \left(\frac{-i}{2\hbar} \right) \int_0^\infty dt' \exp\left\{ \frac{-iE_i t'}{\hbar} \right\} e^{-iH_f(t-t')/\hbar} \Phi_i(0)$$

$$= \lim_{t \to \infty} \left(\frac{-i}{2\hbar} \right) e^{-iE_i t/\hbar} \int_0^\infty d\theta \, e^{iE_i \theta/\hbar} \, e^{-iH_f \theta/\hbar} \Phi_i(0)$$

$$= \lim_{t \to \infty} \left(\frac{-i}{2\hbar} \right) e^{-iE_i t/\hbar} |R\rangle \tag{9.26}$$

Here $\theta = t - t'$, and the function $|R\rangle$, called the Raman wavefunction, is defined by

$$|R\rangle \equiv \int_0^\infty d\theta \; e^{iE_i\theta/\hbar} \exp\left\{\frac{-iH_f\theta}{\hbar}\right\} \Phi_i(0) \qquad (9.27)$$

The Raman wavefunction is a half Fourier transform of a wavefunction obtained by propagating $\Phi_i(0)$ on the excited state for time θ. At each time θ, a wavepacket is created on the excited surface with phase factor $e^{iE_i\theta/\hbar}$ and then these packets evolve under the propagator $\exp(-iH_f\theta/\hbar)$; the monochromatic continuous-wave (CW) excitation produces this Raman excited-state wavefunction.

Computationally, the Raman wavefunction is straightforward to calculate: The function $\Phi_i(t)$, the initial-state function multiplied by the dipole operator, is placed on the final-state potential surface and propagated there. By half-Fourier transforming this propagating function, $|R\rangle$ is obtained. This formulation begins by displacing the Franck–Condon wavepacket Φ_i to the excited surface and propagating it according to H_f. This does not mean that in the optical experiment the localized Franck–Condon wavepacket is in fact created on the upper surface. Rather, in accord with (9.27), the light source projects the initial packet upward at various times starting from zero, and these components are added with the proper phase relationships by the half-Fourier integral in (9.27).

9.6

Vibrational Spectra: Correlation Function Approach

Just as the golden rule formula can be converted into a time-dependent expression to calculate electronic spectra, so it can be rewritten into a time-dependent formula for vibration–rotation spectra. Indeed, we will see more generally in Chapters 10 and 11 that response of the system to any external hamiltonian perturbation can be expressed in terms of correlation functions. In this section we deal with vibration–rotation spectra, lineshape generalities, and some issues of symmetry.

We begin by rewriting the result of Eq. (9.13) for the absorption cross section. Using the cyclic invariance of the trace and the fact that the dipole operator μ is real, we can rewrite (9.13), using (9.1), as

$$\sigma(\omega) = \frac{2\pi\omega}{3c\hbar} \int_{-\infty}^\infty dt \langle e^{iHt/\hbar} \mu e^{-iHt/\hbar} \mu \rangle e^{i\omega t} \qquad (9.28a)$$

Then using the Heisenberg representation formula [Eq. (4.42)]

$$\mu(t) = \exp\left(\frac{iHt}{\hbar}\right) \mu \exp\left(\frac{-iHt}{\hbar}\right)$$

we can write

$$\sigma(\omega) = \frac{2\pi\omega}{3c\hbar} \int_{-\infty}^{\infty} dt \; e^{i\omega t} \langle \mu(t)\mu(0) \rangle \qquad (9.28b)$$

Alternatively, one can use (9.13) as it stands, rewrite using $t \to -t$, and obtain

$$\sigma(\omega) = \frac{2\pi\omega}{3c\hbar} \int_{-\infty}^{\infty} dt \; e^{-i\omega t} \langle \mu(0)\mu(t) \rangle \qquad (9.28c)$$

All of these expressions permit calculation of the absorption cross section in terms of the Fourier transform of an autocorrelation function. Just as was true for the optical spectrum, the vibration–rotation spectrum will have structure determined by the behavior of the correlation function, whose falloff will fix the linewidth and the structure of whose recurrences will determine the peak positions.

In general, the dipole operator for a many-particle system can be written

$$\mu = \sum_{i,\text{ particles}} \mu_i \qquad (9.29a)$$

so that the correlation function can be expressed as

$$\langle \mu(t)\mu(0) \rangle = \sum_i \langle \mu_i(t)\mu_i(0) \rangle + \sum_{j \neq i} \langle \mu_i(t)\mu_j(0) \rangle \qquad (9.29b)$$

The first term on the right of (9.29b) is often called the self-correlation, while the second is generally referred to as the cross-correlation term. If the system under study is a dilute one, the cross correlations are generally very small and can be ignored. We will do so here, but for some problems, such as diffusion in concentrated materials, these cross-correlations can contribute importantly.

The Fourier transform of (9.28) can be inverted. Thus we obtain

$$\frac{1}{2\pi} \int_{-\infty}^{\infty} e^{-i\omega t} \sigma(\omega) \, d\omega = \frac{2\pi\omega}{3c\hbar} \langle \mu(t)\mu(0) \rangle \qquad (9.30)$$

Therefore, Fourier analysis of the absorption lineshape, in the infrared or Raman or microwave region, can be used to examine the dynamics of the molecule in terms of dipole autocorrelation. Extensive studies of this model have been done; Fig. 9.7 shows the dipole autocorrelation function for CO in various solvents, as deduced from the infrared lineshape. Note

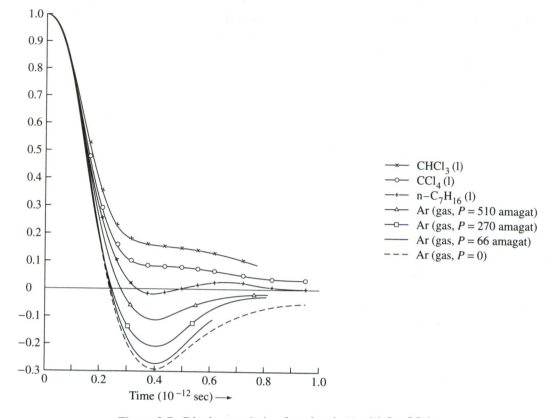

Figure 9.7. Dipole correlation function $\langle \mu(t)\mu(0) \rangle$ for CO in a number of solvents, as deduced from Fourier inversion of the observed vibrational line. As the solvent becomes more complicated, vibrational and rotational relaxation is more efficient, and the correlations decay more rapidly. [With permission from R. G. Gordon, *Adv. Magn. Reson.* 3, 1 (1968), Fig. 1.]

that for liquid argon, the curve shows damped harmonic behavior, whereas in chloroform relaxation is essentially complete in 1 ps, and no oscillations are observed.

Before analyzing relaxation processes and their effects on spectra, we consider some fundamental symmetry properties of correlation functions. Following (9.11), define the correlation function

$$C(t) = \langle \mu(t)\mu(0) \rangle = \langle e^{iHt/\hbar} \mu e^{-iHt/\hbar} \mu \rangle$$

It is then clear, since μ is a real operator, that

$$C^*(-t) = C(t) \tag{9.31}$$

It then also follows that

$$\text{Re } C(t) = \text{Re } C(-t) \tag{9.32a}$$

$$\text{Im } C(t) = -\text{Im } C(-t) \tag{9.32b}$$

so that the real part of the correlation function is even in time. Then the frequency-dependent absorption can be written

$$\frac{3c\hbar\sigma(\omega)}{2\pi\omega} = \int_{-\infty}^{0} e^{i\omega t} C(t)\, dt + \int_{0}^{\infty} e^{i\omega t} C(t)\, dt$$

$$= \int_{0}^{\infty} C(-t)e^{-i\omega t}\, dt + \cdots$$

$$= \int_{0}^{\infty} C^*(t)(e^{i\omega t})^*\, dt + \cdots$$

$$= \int_{0}^{\infty} dt\{C^*(t)(e^{i\omega t})^* + C(t)e^{i\omega t}\} \tag{9.33a}$$

$$= 2 \text{ Re} \int_{0}^{\infty} C(t)e^{i\omega t}\, dt \tag{9.33b}$$

The last result can be rewritten as

$$\sigma(\omega) = \text{Re } \frac{4\pi\omega}{3c\hbar} \int_{0}^{\infty} e^{i\omega t}\langle \boldsymbol{\mu}(t)\boldsymbol{\mu}(0)\rangle\, dt \tag{9.34}$$

This rewriting, with $\sigma(\omega)$ expressed in the form of a half-Fourier transform, or an imaginary Laplace transform, of the time correlation function is related to the more general form of linear response theory (Chapter 11).

The symmetry properties (9.32) and (9.31) follow from the Heisenberg representation, and therefore are a fully quantum result. The analogous classical result is even simpler, and follows from time-reversal invariance of the Newton equations: Classically,

$$C(t) = C(-t) \tag{9.32c}$$

It then follows classically from (9.32c) and quantum mechanically from (9.32a) that the correlation function $C(t)$ can be of gaussian type,

$$C(t) = C(0)e^{-t^2/\tau_g^2} \tag{9.35}$$

at short time, but that the simple damped form

$$C(t) = C(0)e^{-t/\tau_l} \tag{9.36}$$

is formally inappropriate. The loss of time-reversal invariance in classical statistical mechanics normally arises from so-called coarse graining (averaging over intermediate time and particle-number scales) or from introducing a dissipative (velocity-dependent) force. We will see that precisely these sorts of treatments can in fact produce loss of time reversibility in the form of an exponential decay of $C(t)$.

The times τ_g in (9.35) and τ_l in (9.36) are ordinarily called relaxation times; the subscripts g and l stand for gaussian or lorentzian, respectively. The lineshapes corresponding to these decays are obtained directly, from substitution of (9.35) into (9.34), as

$$\sigma_g(\omega) = \frac{4\pi C(0)}{3c\hbar} \sqrt{\pi}\, \tau_g \omega \, \exp\!\left(\frac{-\omega^2 \tau_g^2}{4}\right) \tag{9.37}$$

$$\sigma_l(\omega) = \frac{4\pi C(0)}{3c\hbar} \frac{\tau_l \omega}{1 + \tau_l^2 \omega^2} \tag{9.38}$$

The lineshape function $I(\omega)$ is related to the absorption cross section per unit wavelength, and is defined by

$$I(\omega) \equiv 2\,\mathrm{Re} \int_0^\infty dt\, C(t) e^{-i\omega t} \tag{9.39}$$

so that

$$I(\omega) = \frac{1}{\omega} \frac{3c\hbar}{2\pi} \sigma(\omega) \tag{9.40}$$

Thus an exponential relaxation (9.36) leads to a symmetric lorentzian lineshape

$$I_l(\omega) = \frac{2\tau_l C(0)}{1 + \omega^2 \tau_l^2} \tag{9.41}$$

The correlation time τ_c is ordinarily defined as

$$\tau_c = \int_0^\infty \frac{C(t)}{C(0)}\, dt \tag{9.42}$$

and we see that for exponential decay, $\tau_c = \tau_l$, while for gaussian decay, (9.35), $\tau_c = \sqrt{\pi}\, \tau_g$.

The transfers back and forth between frequency and time space demonstrated in Eqs. (9.28), (9.30), and (9.41) form the basis for pulse techniques in various spectroscopies, including both NMR and optical, as well as Fourier-transform spectroscopies. We discuss the behavior of the spectra in several representative situations.

Consider first the situation of simple harmonic behavior of the dipole correlation function, so that

$$C(t) = C(0) \cos \omega_0 t \tag{9.43}$$

with ω_0 the unperturbed oscillator frequency. Then the lineshape, from (9.39), is

$$I(\omega) = 2\pi\delta(\omega - \omega_0) \qquad (9.44)$$

In real situations, such lines clearly are not observed. Quite apart from factors including the shape of the exciting line, the intrinsic evolution of the dipole in interacting situations will never be a simple single harmonic.

Consider next, then, a situation in which a single dipole is evolving in a brownian situation—that is, in a solution in which many individual collisions lead to loss of memory on a characteristic relaxation timescale τ. Within a purely classical description, the behavior of the dipole can be described by a *Langevin-type equation*

$$\ddot{\boldsymbol{\mu}}(t) = -q\boldsymbol{\nabla}\{U_{\text{tot}}(x)\} + q\mathbf{R}(t) - \gamma\dot{\boldsymbol{\mu}} \qquad (9.45)$$

This is a modified form of the Newton equations. The dipole operator is written

$$\boldsymbol{\mu} = q\mathbf{x} \qquad (9.46)$$

for a charge q and displacement $\mathbf{x}$. Formally, (9.46) corresponds to neglect of electrical anharmonicity, so that the first term on the right of (9.45) is simply the force, derived from a potential $U(\mathbf{x})$. The second and third terms on the right side of (9.45) arise from averaging over multiple collisions with the solvent, and the realm of validity of (9.45) begins, therefore, at times large compared to a single solvent/dipole collision time. The $\mathbf{R}(t)$ term is a random force term that describes the random impacts felt by the dipole due to collisions with the solvent. The last term in (9.45) is a viscous decay term, like that felt by a damped oscillator or a pendulum swinging in a fluid. Formally, it is a velocity-dependent potential and therefore leads to nonconservation of energy. Neglecting memory terms in the solvent itself, one can write $\mathbf{R}(t)$ as a white noise

$$\langle\mathbf{R}(t)\rangle = 0 \qquad (9.47)$$

$$\langle\mathbf{R}(t)\mathbf{R}(0)\rangle = \delta(t)2\pi kT\gamma \qquad (9.48)$$

The result (9.48) is often referred to as the *fluctuation-dissipation theorem:* It relates the size of the random force both to the temperature and to the size of γ, the damping strength. Since the damping and the random force both come from the same physical behavior (the multiple collisions with the solvent), it is reasonable that their strengths be related; (9.48) gives the relationship.

As special cases of Langevin behavior, first consider a free dipole (no restoring force), $U(\mathbf{x}) = 0$. Then we have

$$\ddot{\boldsymbol{\mu}}(t) = -\gamma\dot{\boldsymbol{\mu}}(t) + q\mathbf{R}(t) \qquad (9.49)$$

Multiply from the right by $\boldsymbol{\mu}(0)$ and thermally average, to obtain

$$\langle\ddot{\boldsymbol{\mu}}(t)\boldsymbol{\mu}(0)\rangle = -\gamma\langle\dot{\boldsymbol{\mu}}(t)\boldsymbol{\mu}(0)\rangle + q\langle\mathbf{R}(t)\boldsymbol{\mu}(0)\rangle \qquad (9.50)$$

The last term of this equation vanishes, because there is no correlation between the random force felt by a dipole and the initial position of that dipole. Thus the average of the products becomes the product of independent averages, which vanishes in accord with (9.47). Thus we can write

$$\frac{d^2}{dt^2} \langle \boldsymbol{\mu}(t)\boldsymbol{\mu}(0) \rangle = -\gamma \frac{d}{dt} \langle \boldsymbol{\mu}(t)\boldsymbol{\mu}(0) \rangle \tag{9.51}$$

This equation, rewritten

$$\frac{d^2}{dt^2} C(t) = -\gamma \frac{d}{dt} C(t) \tag{9.52}$$

has a solution given by the simple exponential

$$C(t) = C(0)e^{-\gamma t} \tag{9.53}$$

Then solving for $I(\omega)$ from (9.39), we find that

$$I(\omega) = \frac{2C(0)\gamma}{\omega^2 + \gamma^2} \tag{9.54}$$

This is a lorentzian line, falling off smoothly from a maximum at $\omega = 0$. The full width at half maximum is 2γ.

9.7

Rotational, Raman, and Magnetic Resonance Spectra

Pure rotational spectra, observed in the far-infrared and microwave spectral regions, are of great importance both historically and from an interpretive viewpoint. It is straightforward to rewrite the absorption cross section for pure rotation as

$$\frac{3\hbar c}{2\pi\omega[1 - \exp(-\hbar\omega/kT)]} \sigma(\omega) = \int_{-\infty}^{\infty} dt \, e^{-i\omega t} \cdot [\langle \boldsymbol{\mu}_1(0)\boldsymbol{\mu}_1(t) \rangle$$

$$+ \sum_{i \neq 1} \langle \boldsymbol{\mu}_1(0) \cdot \boldsymbol{\mu}_i(t) \rangle] \tag{9.55}$$

where $\boldsymbol{\mu}_i$ is now the permanent dipole moment of the ith molecule; once again, in dilute solution the second term on the right of (9.55) is negligible compared to the first. For pure rotational spectra, unlike vibrational spectra, the issue of electrical anharmonicity does not arise: The rotational dipole moment $\boldsymbol{\mu}$ can always be written $\boldsymbol{\mu} = \mu_0 \mathbf{u}$, with μ_0 the magnitude of the dipole moment and $\mathbf{u}$ a unit vector along $\boldsymbol{\mu}$. Then in the correlation functions in (9.55), $\boldsymbol{\mu}$ can be replaced by $\mathbf{u}$ if σ is replaced by $\sigma\mu_0^2$.

Let us consider the nature of rotational correlation functions in dilute solution. We will distinguish short-time motion, in which there have not yet been many collisions and the dipole motion can still be followed using mechanics, from the long-time behavior, in which the dipole has suffered many collisions and all phase information has been lost.

Consider first, then, a two-dimensional rotor, with essentially free rotation at short times, but eventually reaching a thermal distribution in the original thermal velocities. Then, defining $f(\omega_r)\,d\omega_r$ as the probability of observing an angular velocity between $(\omega_r,\,\omega_r + d\omega_r)$, we have

$$f(\omega_r) = \left(\frac{I}{2\pi kT}\right)^{1/2} e^{-I\omega_r^2/2kT} \tag{9.56}$$

with I here the moment of inertia. For free rotation we have

$$\langle \mathbf{u}(t)\mathbf{u}(0)\rangle = \int_{-\infty}^{\infty} d\omega_r\, f(\omega_r)\cos\omega_r t \tag{9.57}$$

where the cosine is just the oscillatory projection of $\mathbf{u}(t)$, the unit vector along the dipole at time t, onto $\mathbf{u}(0)$. The integral in (9.57) can be evaluated analytically, leading to

$$\langle \mathbf{u}(t)\mathbf{u}(0)\rangle = e^{-kTt^2/2I} \tag{9.58}$$

So the short-time behavior is gaussian in time, essentially because of the gaussian dependence of the Boltzmann weight on frequency ω_r.

The long-time behavior is a bit more elaborate to calculate. For long times, the behavior of the rigid dipole is expected to be stochastic, or random: The dipole has suffered many collisions with the solvent, or bath. Thus its correlation function will be fixed not by free rotation but by diffusion motion. Considerations in terms of rotational diffusion are appropriate here, but since there is no underlying potential, qualitatively the same behavior is seen using simple translational diffusion. Then the rotor behaves like a brownian particle and is subject to a Langevin-like equation of the general form (9.42), and it follows, using the same arguments as those that gave (9.47), that

$$\langle \mathbf{u}(t)\mathbf{u}(0)\rangle = e^{-\gamma t} \tag{9.59}$$

for long times.

Thus it is clear that the long-time behavior of the correlation function is exponential (9.59), and the short-time behavior is gaussian (9.58). Since the absorption lineshape is just the Fourier transform of the correlation function, it might be expected to be gaussian at high frequencies, and lorentzian at low frequencies. The time t_0 at which the short-time and long-time behaviors correspond is found by equating Eqs. (9.58) and (9.59):

$$-\gamma t_0 = \frac{-kTt_0^2}{2I}$$

or

$$t_0 = \frac{2I\gamma}{kT} \tag{9.60}$$

Thus for larger moments of inertia, the initial relaxation dominates for longer times, while higher temperature causes the stochastic regime to dominate more quickly. The dominant physical phenomena are also different in these two time regimes: For short times, loss of correlation in momentum dominates the decay, as the initial rotational momentum is relaxed. At longer times, loss of space correlations, as described by brownian dynamics, is dominant.

Often, the time dependence of the correlation functions is represented using a moment expansion. Thus one expands the correlation function in a Taylor series, obtaining

$$\langle \mathbf{u}(t)\mathbf{u}(0)\rangle = \sum_{n=0}^{\infty} \frac{t^n}{n!} \left[\frac{d^n}{dt^n} \langle \mathbf{u}(t)\mathbf{u}(0)\rangle \right]_{t=0} \tag{9.61a}$$

$$= \sum_{n} \frac{t^n}{n!} \left\langle \frac{d^n\mathbf{u}(t)}{dt^n} \mathbf{u}(0)\right\rangle \Big|_{t=0} \tag{9.61b}$$

To evaluate the time derivatives, we use the Heisenberg equation

$$\frac{d\mathbf{u}}{dt} = \frac{i}{\hbar} [H, \mathbf{u}] \tag{9.62}$$

and the expansion (9.61) becomes

$$\langle \mathbf{u}(t)\mathbf{u}(0)\rangle = \sum_{n=0} \left(\frac{it}{\hbar}\right)^n \frac{1}{n!} \langle [H, [H, \ldots, [H, \mathbf{u}(0)] \ldots]]\mathbf{u}(0)\rangle \tag{9.63}$$

where there are n commutators in the nth term. To rewrite the expression, we note that the inverse Fourier formula applied to (9.39) gives

$$\langle \mathbf{u}(t)\mathbf{u}(0)\rangle = \frac{1}{2\pi\mu_0^2} \int_{-\infty}^{\infty} e^{i\omega t}I(\omega)\, d\omega \tag{9.64}$$

Rewriting the exponential in Taylor series, we have

$$\langle \mathbf{u}(t)\mathbf{u}(0)\rangle = \frac{1}{2\pi\mu_0^2} \sum_{n} \frac{(it)^n}{n!} \int_{-\infty}^{\infty} \omega^n I(\omega)\, d\omega \tag{9.65}$$

Comparing the right sides of (9.65) and (9.63), we find that

$$\frac{1}{2\pi\mu_0^2} \int_{-\infty}^{\infty} \omega^n I(\omega) \, d\omega = \hbar^{-n}\langle[H, [H, \ldots, [H, \mathbf{u}(0)] \ldots]]\mathbf{u}(0)\rangle \quad (9.66a)$$

$$\equiv M^{(n)} \quad (9.66b)$$

where $M^{(n)}$ is called the nth frequency moment of the normalized lineshape function $\hat{I}(\omega)$, to be defined in Eq. (9.76). Thus we can write

$$\langle\mathbf{u}(t)\mathbf{u}(0)\rangle = \sum_{n=0}^{\infty} \frac{(it)^n}{n!} M^{(n)} \quad (9.67)$$

Therefore, analysis of the moments $M^{(n)}$ of the observed normalized lineshape can be used to discuss motion of the dipoles in solution. For the two-dimensional rotation case just discussed, we have (at short times)

$$\langle\mathbf{u}(t)\mathbf{u}(0)\rangle = \exp\left(\frac{-kTt^2}{2I}\right) \cong 1 - \frac{kTt^2}{2I} + \cdots O(t^4)$$

So that $M^{(1)}$ vanishes and $M^{(2)}$ is kT/I. Again, we note that the second moment depends only on the temperature and the moment of inertia, because it is determined by kinetic energy loss (i.e., by loss of correlations in momentum space). The higher moments in (9.67) reflect longer-time motion, including both the forces due to molecular potentials and the stochastic (diffusive) behavior in solution.

The general approach of using correlation functions to give spectra, and vice versa, by Fourier transforming from frequency to time and back works not only for infrared and optical spectroscopy, but for any other situation in which transitions are observed that can be described using the golden rule, since then the process used to derive (9.13) from (9.7) can be repeated. In particular, both Raman spectra and magnetic resonance lines can be analyzed in this fashion. For light scattering, for instance, the partial cross section is given using the golden rule by [see Eq. (5.111)]

$$\frac{d^2\sigma}{d\Omega \, d\omega} = \left(\frac{2\pi}{\lambda_s}\right)^4 \sum_{i,f} |\langle i|\boldsymbol{\varepsilon}_i \cdot \boldsymbol{\alpha} \cdot \boldsymbol{\varepsilon}_f|f\rangle|^2 P_i \delta(\omega + \omega_i - \omega_f) \quad (9.68)$$

with P_i the initial Boltzmann probability, λ_s the scattered wavelength, i and f the initial and final states, Ω and ω the solid angle and frequency, respectively. The polarization vectors $\boldsymbol{\varepsilon}_i$ and $\boldsymbol{\varepsilon}_f$ describe initial and final states, and α is the polarizability tensor. The correlation function form of (9.68) is

$$\left(\frac{\lambda_s}{2\pi}\right)^4 \frac{d^2\sigma}{d\Omega \, d\omega} = \int_{-\infty}^{\infty} \langle[\boldsymbol{\varepsilon}_i \cdot \boldsymbol{\alpha} \cdot \boldsymbol{\varepsilon}_f](t)[\boldsymbol{\varepsilon}_i \cdot \boldsymbol{\alpha} \cdot \boldsymbol{\varepsilon}_f](0)\rangle e^{i\omega t} \, dt \quad (9.69)$$

One then ordinarily distinguishes the polarized from the nonpolarized Raman components, by writing

$$\alpha = (\tfrac{1}{3}\mathrm{tr}\ \alpha)\cdot\mathbf{1} + [\alpha - \tfrac{1}{3}(\mathrm{tr}\ \alpha)\cdot\mathbf{1}] \tag{9.70a}$$

and defining

$$\bar{\alpha} = (\tfrac{1}{3}\mathrm{tr}\ \alpha) \tag{9.70b}$$

$$\beta = \alpha - \mathbf{1}\cdot\bar{\alpha} \tag{9.70c}$$

Then the polarized or isotropic Raman spectrum is given by

$$\left(\frac{\lambda_s}{2\pi}\right)^4 \frac{d^2\sigma_{\mathrm{pol}}}{d\Omega\ d\omega} = \int e^{i\omega t}\langle\bar{\alpha}(t)\bar{\alpha}(0)\rangle\ dt \tag{9.71a}$$

and the depolarized spectrum by

$$\left(\frac{\lambda_s}{2\pi}\right)^4 \frac{d^2\sigma_{\mathrm{depol}}}{d\Omega\ d\omega} = \int e^{i\omega t}\langle\beta(t)\beta(0)\rangle\ dt \tag{9.71b}$$

The polarizabilities α may be approximated very well as a sum of molecular polarizabilities; this ignores collision-induced polarizabilities, which are expected to be negligible in dilute solution. Then the isotropic spectrum will include both low-frequency contributions due to quasi-elastic scattering, which are centered at the frequency of the incident radiation and are simply the Rayleigh–Brillouin scattering peaks, and higher-frequency vibrational Raman peaks. The Raman contribution may be well approximated by

$$I_{\mathrm{pol}}(\omega) = \int_{-\infty}^{\infty} dt\ e^{i\omega t} \sum_{i,j=1}^{N} \frac{\partial\bar{\alpha}}{\partial q_i} \frac{\partial\bar{\alpha}}{\partial q_j} \langle q_i(t)q_j(0)\rangle \tag{9.72}$$

with q_i the vibrational coordinate on the molecule of interest and the sum running over modes. Clearly, there will be a term for each vibrational mode of the molecule for which $\partial\bar{\alpha}/\partial q_i$ has a nonvanishing matrix element between ground and excited states.

The Raman lineshape, then, will reflect the dynamics of $\langle q_i(t)q_j(0)\rangle$. Since for simple harmonic oscillations $q_i(t)$ contains the phase terms $e^{i\omega_0 t}$ and $e^{-i\omega_0 t}$, the Raman line will have peaks at frequency shifts of $\pm\omega_0$ from the position of the exciting line. Linewidths can reflect either vibrational relaxation, leading to damping of the oscillations, or dephasing, in which phase correlation is lost, but no energy relaxation occurs.

Finally, infrared lineshapes are determined from

$$I(\omega) = \int_{-\infty}^{\infty} dt\ e^{i\omega t}\langle\boldsymbol{\mu}(t)\boldsymbol{\mu}(0)\rangle \tag{9.73}$$

Again expanding the dipole moment and ignoring electrical anharmonicity, we find that

$$I(\omega) = \int_{-\infty}^{\infty} dt\, e^{i\omega t} \sum_{ij} \left\langle \frac{\partial \boldsymbol{\mu}_i}{\partial q_i}\bigg|_{q0} \frac{\partial \boldsymbol{\mu}_j}{\partial q_j}\bigg|_{q0} \cdot q_i(t)q_j(0) \right\rangle \qquad (9.74a)$$

$$= \int_{-\infty}^{\infty} dt\, e^{i\omega t} Q_0^2 \sum_{i,j} \langle \mathbf{u}_i(t)\mathbf{u}_j(0)q_i(t)q_j(0)\rangle \qquad (9.74b)$$

where we have taken a linear approximation to the dipole moment.

$$\boldsymbol{\mu}_i = \boldsymbol{\mu}_i^0 + Q_0 \mathbf{u}_i q_i \qquad (9.75)$$

with $\mathbf{u}_i$ a unit vector and Q_0 an arbitrary charge. We see from (9.74b) that the infrared spectrum depends on both rotational correlations (from the $\mathbf{u}$) and vibrational correlations (from the q). Note that the polarized Raman line and the isotropic infrared line both contain vibrational relaxation contributions, but while the infrared lineshape is affected by rotational correlations, the isotropic Raman lineshape is not. If, therefore, these two lineshapes are similar, the rotational contribution is probably negligible.

9.8

Motional Narrowing and Stochastic Motion

The study of lineshapes can tell a good deal [by Fourier inversions, Eq. (9.39)] about the dynamics of molecules. One especially important case occurs in the use of magnetic resonance to measure such molecular processes as exchange rates. As we have seen in Section 9.7, molecular lineshapes can be fairly well described in terms of correlation functions involving a second moment and a long-range tail. The second moment is defined by

$$M^{(2)} = \frac{\int d\omega\, I(\omega)\, \omega^2}{\int d\omega\, I(\omega)} = \int \hat{I}(\omega)\omega^2\, d\omega \qquad (9.76a)$$

where we have defined the normalized lineshape by

$$\hat{I}(\omega) = \frac{I(\omega)}{\int_{-\infty}^{\infty} I(\omega)\, d\omega} \qquad (9.76b)$$

Defining a new variable by $\omega' = \omega + \omega_0$, we can rewrite (9.76a) as

$$M^{(2)} = \int (\omega' - \omega_0)^2 \hat{I}(\omega' - \omega_0)\, d\omega' \qquad (9.76c)$$

This says simply that moments can be measured from the line center at ω_0.

Suppose now that the frequency ω' contains two contributions,

$$\omega' = \omega_0 + \omega_1(t)$$

where ω_0 is an intrinsic frequency due to the molecular hamiltonian, while $\omega_1(t)$ is a modulating frequency due to some other process, such as interaction with other molecules or a solvent. Then if $P(\omega_1)$ is the relative probability of observing the modulating frequency ω_1, we define the root mean square frequency width as Δ, where

$$\Delta^2 = \int_{-\infty}^{\infty} d\omega_1 \, \omega_1^2 P(\omega_1) \tag{9.77}$$

This width measures the magnitude of the frequency modulation. Similarly, a time correlation expression

$$\frac{1}{\Delta^2} \int_0^{\infty} dt \langle \omega_1(t)\omega_1(0) \rangle = \tau_c \tag{9.78}$$

defines τ_c, the correlation time of the modulation; for $t \gg \tau_c$, there is no average correlation between $\omega_1(t)$ and $\omega_1(0)$.

There are then two important limiting situations. For slow modulation, $\Delta\tau_c \gg 1$, the response of the system to the modulation is coherent and dynamic—the system follows the modulation. The line is broad, reflecting the large value of Δ. In the opposite limit of fast modulation, $\Delta\tau_c \ll 1$, on average the modulation remains constant only for a short time, and the system cannot respond dynamically. Thus the line will be sharp, reflecting the relatively small value of Δ. For this situation, we

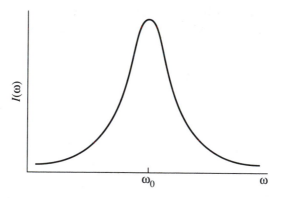

Figure 9.8. Lorentzian lineshape function, observed for a brownian oscillator moving in solution.

observe a lorentzian line of width γ (Fig. 9.8). This situation of line narrowing as modulation rates increase is complementary to the lifetime broadening of spectral lines discussed in Section 9.4: There, the width was determined essentially by the time–energy uncertainty relation. Here, it is fixed by whether the oscillator can respond to the modulation. A modulation ω_1 can influence the system dynamically only if the system oscillates through most of a complete period before the modulation changes significantly. Thus for slow modulation of a line at frequency ω_0, we require that $\tau_c \gtrsim 2\pi/\omega_0$. If this does not hold, we have fast modulation, and the line is not broadened (Fig. 9.8). Important applications of these results are found in magnetic resonance: In NMR, one observes motional narrowing, and in EPR, exchange narrowing.

BIBLIOGRAPHY FOR CHAPTER 9

Specific papers used in this chapter:

> Gordon, R. G., *Adv. Magn. Reson.* 3, 1 (1968).
> Heller, E. J., *Acct. Chem. Res.* 14, 368 (1981).
> Ramakrishna, M. V., and R. Coalson, *Chem. Phys.* 120, 327 (1988).
> Reimers, J. R., K. R. Wilson, and E. J. Heller, *J. Chem. Phys.* 79, 4749 (1983).

Textbooks that cover parts of the material in this chapter:

> Berne, B., and R. Pecora, *Dynamic Light Scattering* (Wiley, New York, 1976).
> Flygare, W. H., *Molecular Structure and Dynamics* (Prentice Hall, Englewood Cliffs, N.J., 1978).
> MacQuarrie, D. A., *Statistical Mechanics* (Harper & Row, New York, 1975).
> Slichter, C. P., *Principles of Magnetic Resonance* (Springer-Verlag, New York, 1990).
> Steinfeld, J., *Molecules and Radiation* (MIT Press, Cambridge, Mass., 1985).
> Waugh, J. S., ed., *Advances in Magnetic Resonance,* Vol. 3 (Academic Press, New York, 1968).

PROBLEMS FOR CHAPTER 9

1. We can consider the lineshape for a damped harmonic oscillator, and some other results, using the correlation function not of displacements, as in Section

9.6, but rather of *velocities*. Consider, then, a harmonic oscillator without electrical anharmonicity, so that

$$\mu(q) = \mu(q_0) + (q - q_0) \frac{\partial \mu}{\partial q}\bigg|_{q_0}$$

with q_0 the equilibrium displacement. One can evaluate the lineshape from the formula

$$I(\omega) = 2\ \text{Re} \int_0^\infty \langle \mu(t)\mu(0)\rangle e^{-i\omega t}\ dt \qquad (9.39)$$

If the oscillator is simply damped by interaction with its environment, the dipole will follow the Langevin equation

$$m\ddot{q} = -\gamma m\dot{q} - m\omega_0^2(q - q_0) + R$$

Now define $x = q - q_0$ as the displacement away from equilibrium. Then defining $\partial\mu/\partial q|_0 = Q_0$, a standard charge; we can then write the velocity $\dot{x}$ in terms of $\dot{\mu}$ and Q_0.

(a) Show that

$$I(\omega) = \frac{2\gamma \langle x_0^2\rangle \omega_0^2 Q_0^2}{(\omega^2 - \omega_0^2)^2 + \gamma^2\omega^2}$$

(b) Evaluate the integral

$$I'(\omega) = 2\ \text{Re} \int_0^\infty dt\ \langle \dot{\mu}(t)\dot{\mu}(0)\rangle e^{-i\omega t}$$

Write it in terms of $\langle \dot{\mu}(0)\dot{\mu}(0)\rangle$. (*Hint:* This is very easy to do using Laplace transforms.)

(c) For harmonic oscillators in a canonical assemble, the equipartition theorem can be written

$$\left\langle \frac{1}{2} m\omega_0^2 x^2\right\rangle = \left\langle \frac{1}{2} mv^2\right\rangle = \frac{1}{2} m\langle \dot{x}^2\rangle = \frac{kT}{2}$$

Use this, with your results for $I'(\omega)$ and for $I(\omega)$, to relate $I'(\omega)$ to $I(\omega)$. The result that you have found for this relationship for the special case of harmonic oscillators turns out to be true in general.

(d) Compute the frequency ω_{max} at which the lineshape maximizes. Is this the value that you expected for a damped oscillator? Comment on the relationship of lineshape position maximum to linewidth.

2. Consider a simple free brownian particle. For the Langevin equation

$$m\ddot{x} = -\gamma\dot{x} + R(x)$$

compute

$$I'(\omega) = 2\ \text{Re} \int_0^\infty e^{-i\omega t}\langle \dot{x}(t)\dot{x}(0)\rangle\ dt$$

Find the value of $I'(0)$. Note the units—what other variable for brownian motion has these units? Can you relate the width of the brownian line $\hat{I}(\omega)$ to this other variable?

3. For molecules and solids, the normal coordinate, or phonon, picture of vibrational spectra is determined by expanding the vibrational potential about the equilibrium geometry in a Taylor series and truncating after the second term. The resulting quadratic hamiltonian is reexpressed in new modes that are linear combinations of the local stretch coordinates fixed by $\xi_\lambda = \Sigma f_{\lambda c} x_c$, where x_c and ξ_λ are, respectively, the displacement along local coordinate and normal coordinate. The vibrational hamiltonians in the two different mode representations can be written

$$H_{\text{nor}} = \sum_\lambda \left(b_\lambda^+ b_\lambda + \frac{1}{2} \right) \hbar\omega_\lambda = \sum_\lambda \left(\frac{\pi_\lambda^2}{2} + \frac{1}{2} \omega_\lambda^2 \xi_\lambda^2 \right)$$

$$H_{\text{loc}} = \sum_c \frac{p_c^2}{2m_c} + \frac{1}{2} \sum_{c,d} k_{cd} x_c x_d$$

with π_λ the momentum conjugate to ξ_λ.

Now suppose that the dipole moment operator is dominated by one local stretch mode, and that electrical anharmonicity is negligible, so that $\mu - \mu_0 \cong Q_0 x_a$, with μ_0 and Q_0 constants, and x_a denoting the active stretch.

(a) Show that the lineshape function defined by (9.39) will, in the normal-mode picture, have a series of lorentzian lines at frequencies $\omega = \omega_\lambda$; find the widths and relative intensities of these peaks.

(b) The real hamiltonian contains anharmonicities and can be written to lowest anharmonic order as

$$H = \sum_\lambda \left(\frac{\pi_\lambda^2}{2} + \frac{1}{2} \omega_\lambda^2 \xi_\lambda^2 \right) + \sum \kappa_{\lambda\mu\nu} \xi_\lambda \xi_\mu \xi_\nu$$

with $\kappa_{\lambda\mu\nu}$ cubic anharmonicity constants. By expanding the time propagator in (9.39), show that at very short times the lineshape is insensitive to the anharmonicity.

4. Correlation function expressions can be used to deduce rates for processes induced by perturbations not arising from an external field; examples include diffusion and rates of chemical reactions (Chapters 8 and 10). Consider transitions between internal eigenstates of a large molecule induced by isolated binary collisons with a monoatomic gas. [See R. D. Levine, *Quantum Mechanics of Molecular Rate Processes* (Oxford University Press, London, 1969) p. 300.] Define the internal states i and f with energy difference $\hbar\omega$ and the kinetic energies of the atom as $e_i = \hbar^2 k_i^2/2\mu$, $e_f = \hbar^2 k_f^2/2\mu$. Then start from the golden rule in the form (summing over final states, averaging over initial states) of a markovian approximation to the transition rate:

$$w_{i\to f} = \frac{2\pi}{\hbar} \sum_{k_i} \sum_{k_f} p(k_i) |\langle k_i, i | V | k_f, f \rangle|^2 \delta(E_f - E_i)$$

with p_i a probability factor for initial state $|k_i, i\rangle$, which is just a Boltzmann factor for thermal equilibrium. Then using energy conservation in the form

$$E_f - E_i = \frac{\hbar}{2\mu} (k_f^2 - k_i^2) + \hbar\omega$$

where $\hbar\omega$ is the energy difference between the two molecular eigenstates, show that

$$w_{i \to f} = \hbar^{-2} \int_{-\infty}^{\infty} \langle V_{if}(t)V_{fi}\rangle \exp(i\omega t) \, dt$$

Interpret this result in terms of a resonance picture—what Fourier component of the correlation function is effective at inducing transitions?

Now write an analogous expression for $w_{f \to i}$. Take the ratio: Does this ratio give the correct (equilibrium constant) ratio for a thermal reaction? Why or why not?

5. In Eq. (9.39) the frequency factor is the complex conjugate of that in (9.34)—does that matter? Prove your answer, and interpret the result spectroscopically.

10

Correlation Functions and Dynamical Processes: Nonadiabatic Intramolecular Electron Transfer

10.1

Introduction

In Chapter 9 we examined the formulation of response properties of molecular systems to applied fields in terms of time-dependent correlation functions. In particular, we began with golden rule results in the general form $w_{if} = 2\pi/\hbar |\langle i|V|f\rangle|^2 \delta(E_i - E_f)$, and introducing the Fourier representation of the δ function, rewrote the rate w_{if} as a Fourier transform of a time correlation function. This general approach will work whenever a rate can be expressed in golden rule form. Usually, the golden rule will be valid under three conditions: The perturbation V must be in some sense small, so that higher-order terms may be neglected; the final state must lie in a fairly dense manifold of states, so that the δ function and the rate expression are meaningful (no recurrences are important); and finally, the perturbation can be written in the form

$$H' = V(x)F_{ex}(t) \tag{10.1}$$

with $V(x)$ an operator of the system and $F_{ex}(t)$ an external field applied to the system. Thus the correlation-function formalism is straightforward for applications to spectroscopy, light scattering, and magnetic phenomena, including magnetic resonance, x-ray and neutron scattering, and so on. In extended systems, the formalism is applied straightforwardly to such processes as conductivity or dielectric phenomena.

There are, however, phenomena in which the chemical system evolves in time due not to an applied external mechanical field, but to a statistical inhomogeneity. For extended systems, diffusion and thermal conduction, which are fluxes due, respectively, to gradients in concentration and temperature, are flow processes that do not arise from external applied mechanical fields (although both can be written in terms of correlation functions). For chemists, the most important rate phenomenon involves the rate of chemical reactions. Here the system evolves, in a general sense, in response to forces arising from the potential energy surface. For a single potential energy surface, we saw in Chapter 8 that it is possible to write the rate constant as a time integral of a flux autocorrelation function [Eq. (8.53)]. In this chapter we consider a slightly more complex case in which there are two potential energy surfaces, and the chemical reaction involves crossing from one surface, corresponding to the electronic configuration of the reactants, to the other surface, with the product electronic configuration. Several rate phenomena in chemistry, including intersystem crossing or internal conversion processes in photoexcited states, some isomerization reactions, and the dynamical Jahn–Teller effect, can be described in the same way. The most important such reactions are electron transfer reactions, and consequently, we will study a particular type of electron transfer reaction to illustrate the use of correlation-function methods to calculate rate phenomena. In this analysis we encounter several useful quantum-mechanical techniques and tricks, including canonical transformations, operator manipulations, use of steepest-descent methods, and interpretation of the Franck–Condon principle.

10.2

Electron Transfer: Some Generalities

We will use, in this chapter, the simplest possible quantum-mechanical model for electron transfer reactions. We assume that the donor and acceptor species have been preassembled into a fixed relative geometry that is favorable for electron transfer (ET), and that initially the system is prepared with the excess electron on the donor; in real intramolecular ET reactions it requires work to assemble this donor–acceptor geometry, and the rate constant includes collision rates, work terms, and the effects of averaging the rates over all accessible donor–acceptor geometries. The precursor state of the donor–acceptor pair is not a stationary state, but will evolve in time due to (weak) mixing with the successor state to which the electron is transferred. We deal with ET reactions between molecules in solution; then the high density of states, due to vibration of the molecules and to solvent modes, in addition to the weak donor–acceptor coupling and initial state preparation, justify the use of the golden rule [Eq. (10.27)] or, equivalently, of a correlation function expression [Eq. (10.29)] to calculate the rate constant k_r.

One usually calls the regime in which the electronic donor–acceptor mixing is really weak that of nonadiabatic electron transfer; it is this situation that we shall study. If the electronic mixing is strong (the splitting $2J$ of Fig. 10.2 substantially greater than k_BT), these considerations must be generalized, and the prefactor in Eq. (10.27) is no longer proportional to the square of the electronic donor–acceptor interaction J^2, but rather to an effective collision number; this regime is normally called adiabatic ET. Finally, if the situation is one in which the solvent modes relax only slowly, the prefactor is essentially a solvent relaxation rate; this regime is often distinguished as that of solvent dynamics control.

We use a Born–Oppenheimer-like picture in which the electronic wavefunctions and the electronic energies depend parametrically on the nuclear coordinates, and we will use Herzberg–Teller expansions of the hamiltonians about arbitrary nuclear positions. The nonadiabatic mixing terms that arise from the action of the nuclear momentum operator on the electronic wavefunctions are not present in our picture; instead, the dynamics of crossing from reactant to product potential curves is described in terms of simple matrix elements of an electronic mixing term. This picture is sometimes called a *quasi-adiabatic representation*. This treatment is often used to compute resistivity in metals (a situation very like electron transfer, in which electronic motion is modulated by vibrations). Under certain assumptions, the picture that we use can be shown to be equivalent to use of nuclear momentum coupling.

Starting with the simple physical model, straightforward quantum mechanics and some very simple equilibrium statistical mechanics permit computation of the full temperature-dependent intramolecular ET rate constant $k_r(T)$. Interestingly, both the activated rate at high temperatures and the essentially temperature-independent rate at $T \to 0$ follow from the golden rule, with no necessity for any physical assumptions, such as the existence of a transition state.

10.3

Molecular Crystal Model

A very important but simple one-electron model for electron transfer called the molecular crystal model was originally developed by Holstein to describe so-called polaron motion in narrowband conductors. It provides a convenient framework in which to discuss electron transfer rates. For concreteness, suppose that we consider electron transfer from H_2 to H_2^+ and freeze the two diatomics both rotationally and translationally. Figure 10.1 shows that the relevant nuclear coordinates are then the displacements x_1 and x_2 of the two diatomics. The Holstein model ignores both anharmonic corrections in the nuclear hamiltonian and electron–electron repulsions in the electronic hamiltonian, and describes electron transfer in terms of displaced harmonic oscillators and motions of single electrons.

$$
\begin{array}{cc}
\text{H} & \text{H}^+ \\
\bigg|\, x_1 & \bigg|\, x_2 \\
\text{H} & \text{H}
\end{array}
$$

Figure 10.1. Coordinates x_1 and x_2 used to describe electron transfer between two diatomic molecules.

The molecular crystal model hamiltonian for the two-site problem that we consider here may be written

$$H = H_v + H_e + H_{ev} \tag{10.2}$$

where the three terms are, respectively, the vibrational, electronic, and interaction energies. The vibrational term can be written in the harmonic approximation as

$$H_v = \frac{-\hbar^2}{2} \sum_{i=1}^{2} \frac{1}{m_i} \frac{\partial^2}{\partial x_i^2} + \frac{1}{2} \sum_i m_i \omega_i^2 x_i^2 + m_1^{1/2} m_2^{1/2} f x_1 x_2 \tag{10.3}$$

with x_i the displacement of the ith bond, whose frequency is ω_i and reduced mass is m_i. The first two terms comprise a harmonic hamiltonian; the last, off-diagonal quadratic term can be made to vanish by defining normal coordinates.

The electronic hamiltonian is expressed in terms of sites, best thought of in our model as the highest-occupied molecular orbitals of the two diatomics. Then the Hückel-type electronic hamiltonian can be written

$$H_{ev} + H_e = \sum_{i=1}^{2} \varepsilon_i(x)a_i^+ a_i - J(x)[a_1^+ a_2 + a_2^+ a_1] \qquad (10.4)$$

with $\varepsilon_i(x)$ the molecular orbital energy and $-J(x)$, the tunneling integral, analogous to the Hückel β. We will neglect the overlap, so that

$$[a_i, a_j^+]_+ = \delta_{ij} \qquad (10.5)$$

The electron–vibration hamiltonian contains several interactions. The easiest derivation entails a formal rewriting of (10.4) with both $J(x)$ and $\varepsilon_i(x)$ taken as functions of the displacements x_i. These can then be expanded in Taylor series:

$$\varepsilon_i(x) = \varepsilon_i + \sum_j \frac{\partial \varepsilon_i}{\partial x_j}\Big|_{x=0} x_j + \cdots \qquad (10.6a)$$

$$J(x) = J + \sum_j \frac{\partial J}{\partial x_j}\Big|_{x=0} x_j + \cdots \qquad (10.6b)$$

so that the hamiltonian becomes

$$H_e = \sum_{i=1}^{2} \varepsilon_i a_i^+ a_i - J(a_1^+ a_2 + a_2^+ a_1) \qquad (10.7)$$

and the linearly coupled electron–vibration hamiltonian is

$$H_{ev} = \sum_{i,j=1}^{2} a_i^+ a_i \frac{\partial \varepsilon_i}{\partial x_j}\Big|_{x=0} x_j + (a_1^+ a_2 + a_2^+ a_1)\left(\frac{\partial J}{\partial x_1}\Big|_{x=0} x_1 + \frac{\partial J}{\partial x_2}\Big|_{x=0} x_2\right) \qquad (10.8)$$

The term $\partial \varepsilon_i / \partial x_j$ for $i \neq j$ is clearly negligible (the MO energy on the left H_2 changes negligibly when the bond length on the right H_2 varies). The second term is, in general, nonvanishing. Within the usual molecular crystal model it is neglected, being considered small compared to the second term of (10.7). Such terms are important in particular cases, such as bandwidth modulation in molecular metals. They can also be important in some electron-transfer situations and will be returned to later. When they are neglected, H_{ev} becomes

$$H_{ev} = 2^{-1/2} \sum_i A_i a_i^+ a_i x_i \qquad (10.9)$$

with the coupling constant $A_i = 2^{1/2} \partial \varepsilon_i / \partial x_i|_{x=0}$.

For the two-site molecular crystal model, the hamiltonian can be simplified by working in relative vibrational coordinates. First we define mass-weighted vibrational coordinates and coupling constants

$$m_i x_i^2 = m \hat{x}_i^2 \qquad (10.10a)$$

$$A_i x_i = A \hat{x}_i \qquad (10.10b)$$

with m some convenient reference mass, such as the average reduced mass for the diatomics. Then H_v becomes

$$H_v = \frac{-\hbar^2}{2m} \left(\frac{\partial^2}{\partial \hat{x}_1^2} + \frac{\partial^2}{\partial \hat{x}_2^2} \right) + \frac{1}{2} m(\omega_1^2 \hat{x}_1^2 + \omega_1^2 \hat{x}_2^2 + 2f\hat{x}_1\hat{x}_2); \qquad (10.11)$$

We have assumed the two oscillators have the same frequency: $\omega_2 = \omega_1$. Analogously, in (10.9) we will, for simplicity, assume that $A_1 = A_2 = A$.

We now choose relative vibrational coordinates

$$X = \sqrt{\frac{1}{2}} (\hat{x}_1 + \hat{x}_2) \qquad (10.12a)$$

$$x = \sqrt{\frac{1}{2}} (\hat{x}_1 - \hat{x}_2) \qquad (10.12b)$$

Then H can be written $H = H_{rel} + H_{cg}$, with

$$H_{rel} = \frac{-\hbar^2}{2m} \frac{\partial^2}{\partial x^2} + \frac{m}{2} \omega_-^2 x^2 - \frac{Ax(a_2^+ a_2 - a_1^+ a_1)}{2}$$
$$- J(a_2^+ a_1 + a_1^+ a_2) + a_1^+ a_1 \varepsilon_1 + a_2^+ a_2 \varepsilon_2 \qquad (10.13)$$

describing relative motion within the two sites, and the center-of-gravity hamiltonian

$$H_{cg} = \frac{-\hbar^2}{2m} \frac{\partial^2}{\partial X^2} + \frac{m\omega_+^2 X^2}{2} + \frac{AX}{2} \qquad (10.14)$$

with

$$\omega_\pm^2 = \omega_1^2 \pm f \qquad (10.15)$$

Note that $n_i = a_i^+ a_i$ is the number operator. The center-of-gravity hamiltonian contains only X; it is therefore entirely decoupled from H_{rel}, is simply an additive term that does not effect the electron dynamics, and can be dropped.

The two-site molecular crystal model for equal frequencies thus is written

$$H_{rel} = \frac{-\hbar^2}{2m} \frac{\partial^2}{\partial x^2} + \frac{m}{2} \omega_-^2 x^2 + \frac{1}{2} Ax(a_1^+ a_1 - a_2^+ a_2) + \varepsilon_1 a_1^+ a_1$$
$$+ \varepsilon_2 a_2^+ a_2 - J(a_1^+ a_2 + a_2^+ a_1) \qquad (10.16)$$

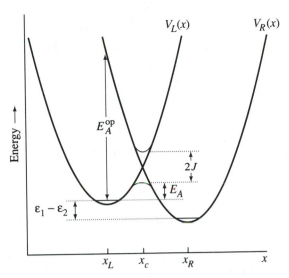

Figure 10.2. Potential energy curves for the model hamiltonian of Eq. (10.17).

We note that, physically, this model is simply a two-level electronic (fermion) system coupling linearly to a vibrational (boson) mode. Such models are generally called spin-boson systems, and will recur in Chapter 11. (For the actual model of Fig. 10.1, $\varepsilon_1 = \varepsilon_2$, but we keep the more general notation.)

If the electronic tunneling term J is set to zero, there are two electronic eigenvalues in (10.16) (remember that $a_1^+ a_1 + a_2^+ a_2 = 1$):

$$H_{\text{rel}} + J(a_1^+ a_2 + a_2^+ a_1) = \frac{-\hbar^2}{2m} \frac{\partial^2}{\partial x^2} + a_1^+ a_1 \left[\frac{m\omega_-^2}{2} x^2 + \frac{1}{2} Ax + \varepsilon_1 \right]$$

$$+ a_2^+ a_2 \left[\frac{m\omega_-^2}{2} x^2 - \frac{1}{2} Ax + \varepsilon_2 \right] \quad (10.17)$$

These potentials are simply parabolas, displaced in energy by $\varepsilon_2 - \varepsilon_1$ and displaced along the x coordinate by $A/m\omega_-^2$. The role of the J-dependent electronic overlap term is to cause transitions between these two wells, which are sketched in Fig. 10.2.

Before analyzing the dynamics in this situation, some comments on the potential energy curves of Fig. (10.2) are useful. The curves cross when $V_L(x_c) = V_R(x_c)$, where we define

$$k \equiv m\omega_-^2 \qquad (10.18a)$$

$$\Delta \equiv \varepsilon_2 - \varepsilon_1 \qquad (10.18b)$$

$$V_L = \tfrac{1}{2} kx^2 + \tfrac{1}{2} Ax + \varepsilon_1 \qquad (10.18c)$$

$$V_R = \tfrac{1}{2} kx^2 - \tfrac{1}{2} Ax + \varepsilon_2 \qquad (10.18d)$$

Thus x_c, the crossing point, is just $x_c = \Delta/A$, and the potential there is

$$V(x_c) = \frac{1}{2} kx_c^2 + \frac{1}{2} Ax_c + \varepsilon_1$$

$$= \frac{1}{2} k \frac{\Delta^2}{A^2} + \frac{1}{2} \Delta + \varepsilon_1$$

The potential at $x_L = -A/2k$, the minimum of V_L, is

$$V_L(x_L) = \frac{1}{2} k \frac{A^2}{4k^2} - \frac{A}{2} \frac{A}{2k} + \varepsilon_1 = -\frac{A^2}{8k} + \varepsilon_1$$

Subtracting, we obtain the potential barrier E_A for crossing from the left as

$$E_A = V(x_c) - V_L(x_L) = \frac{k\Delta^2}{2A^2} + \frac{A^2}{8k} + \frac{\Delta}{2} = \frac{k}{2A^2} \left[\Delta^2 + \frac{A^4}{4k^2} + \frac{A^2\Delta}{k} \right]$$

$$= \frac{1}{4(A^2/2k)} \left[\Delta + \frac{A^2}{2k} \right]^2 \qquad (10.19)$$

It is customary to define an energy λ, called the reorganization energy, as the energy separation between V_R at x_L and V_R at x_R:

$$\lambda = V_R(x_L) - V_R(x_R)$$

$$= \frac{A^2}{8k} + \frac{A^2}{4k} + \varepsilon_2 - \left[\frac{1}{8} \frac{A^2}{k} - \frac{A^2}{4k} + \varepsilon_2 \right] \qquad (10.20a)$$

$$= \frac{A^2}{2k} \qquad (10.20b)$$

Rewriting A in terms of λ in (10.19), we find that

$$E_A = \frac{1}{4\lambda} (\lambda + \Delta)^2 \qquad (10.21)$$

This is the potential barrier for activation from the left minimum to the barrier top.

The vertical energy at the minimum of V_L is often called the optical activation energy E_A^{op} and is just

$$E_A^{op} = V_R(x_L) - V_L(x_L)$$

$$= \frac{3A^2}{8k} + \varepsilon_2 - \left[\frac{-A^2}{8k} + \varepsilon_1 \right] = \frac{A^2}{2k} + \varepsilon_2 - \varepsilon_1 \qquad (10.22)$$

$$= \lambda + \Delta$$

The relevant quantities are sketched in Fig. 10.2. Note that we have taken $\Delta < 0$, the usual situation for exoergic electron transfer.

The two potential curves V_L and V_R describe the situation with J of (10.17) set equal to zero. When the J term is included, the hamiltonian can be written

$$H_{rel} = \frac{-\hbar^2}{2m} \frac{\partial^2}{\partial x^2} + V_{tot}(x)$$

$$V_{tot}(x) = a_1^+ a_1 \left[\frac{1}{2} kx^2 + \frac{1}{2} Ax + \varepsilon_1 \right]$$

$$+ a_2^+ a_2 \left[\frac{1}{2} kx^2 - \frac{1}{2} Ax + \varepsilon_2 \right] - J(a_1^+ a_2 + a_2^+ a_1) \quad (10.23)$$

The total potential curves can then be found as the eigenvalues of the matrix

$$m_V = \begin{bmatrix} \frac{1}{2} kx^2 + \frac{1}{2} Ax + \varepsilon_1 & -J \\[2mm] -J & \frac{1}{2} kx^2 - \frac{Ax}{2} + \varepsilon_2 \end{bmatrix} \quad (10.24)$$

which is just the representation of V_{tot} in the electronic-state basis $(a_1^+|vac\rangle, a_2^+|vac\rangle)$. The eigenvalues are

$$V_{tot}^{\pm}(x) = \frac{1}{2} kx^2 + \frac{\varepsilon_1 + \varepsilon_2}{2} \pm \frac{1}{2} \sqrt{(\Delta - Ax)^2 + 4J^2} \quad (10.25)$$

These curves are sketched in Fig. 10.2. The distance between them, which we can call $2\beta(x)$, is then

$$2\beta(x) = V^+(x) - V^-(x) = \sqrt{(\Delta - Ax)^2 + 4J^2} \quad (10.26a)$$

and

$$2\beta(x_c) = 2J \quad (10.26b)$$

Thus the splitting of these two curves at their point of closest approach is $2J$.

The molecular crystal hamiltonian of (10.17) describes electron transfer in a two-site system. It is also clear from Fig. 10.2 that it can be used to describe optical excitation, simply by choosing $-\Delta$ to be a large energy, in the optical range. Accordingly, analysis of the behavior of the system evolving according to (10.17) should be able to comprehend both electron transfer and optical excitation. In Sections 10.5 and 10.6 we will do such an analysis, introducing a number of techniques for dealing with electron–

vibration (or, more generally, linear fermion–boson) coupling. In Section 10.7 we analyze the electron transfer problem itself in an alternative fashion, based on rate theory. Before considering the dynamics, however, we will use semiclassical theory to differentiate several regimes of the rate process.

10.4

Rate Processes with Vibronic Coupling: Canonical Transformations and Franck–Condon Factors

The rate constant k_r for electron transfer, when the molecular crystal model is applicable, can be written using the golden rule of Eq. (4.66) as

$$k_r = \frac{2\pi}{\hbar} |\langle 2|J(a_1^+ a_2 + a_2^+ a_1)|1\rangle|^2 \delta(E_2 - E_1) \qquad (10.27)$$

with the unmixed electronic states and levels from (10.17),

$$[H_{\text{rel}} + J(a_1^+ a_2 + a_2^+ a_1)]|i\rangle = E_i|i\rangle \qquad (10.28)$$

These two equations can also be used to discuss rates in a number of other situations in which vibronic (electron–vibration) coupling can modeled in terms of two electronic states, with weak nonadiabatic coupling (J is small compared to characteristic vibronic energies). Evaluation of k_r is a fairly complicated issue, however, since the reorganization energy $\lambda = A^2/2m\omega_-^2$ is large compared to J, and therefore the vibronic coupling cannot be treated as a weak perturbation. Since ω_- is the only relevant frequency, we will write $\omega_- = \omega$.

Before evaluating k_r, the qualitative features might be examined. The Franck–Condon principle states generally that transitions between electronic states occur vertically; that is, the nuclear coordinates do not change during the process of electronic transition. The intersecting parabolas of Figure 10.2, normally called diabatic curves, are very close to the true adiabatic energy potentials $V^{\pm}(x)$ of (10.25) except near the crossing point x_c. The vibronic coupling [A-dependent term in (10.23)] simply describes a change in bond length when the electron transfer occurs between sites 1 and 2 of the molecular crystal model. Thus transitions between the two electronic states correspond to changing the value of the equilibrium displacement from the minimum of the left parabola of Fig. 10.2 to the minimum of the right parabola. The Franck–Condon principle suggests that the rate will be controlled by overlaps of the eigenstates of the left parabola with those of the right. Such overlaps will enter into the rate expression, and to evaluate them will involve defining operators that shift from left to right minima in Fig. 10.2.

Formally, the rate constant (10.27) can be rewritten as a correlation function using a Fourier representation, just as was done for the spectroscopic problem in Section 9.2. We then obtain, formally

$$k_r(\omega)_{ap} = \frac{2}{\hbar^2} \operatorname{Re} \int_0^\infty J^2 e^{i\omega_{ap}t} \langle I|(a_1^+ a_2 + a_2^+ a_1)(t)(a_1^+ a_2 + a_2^+ a_1)|I\rangle \, dt$$

(10.29)

where $|I\rangle$ is the initial state for ET, and ω_{ap} the applied field. The frequency dependence of the applied field is artificial, so the result

$$k_r = \frac{2}{\hbar^2} \operatorname{Re} \int_0^\infty J^2 \langle I|(a_1^+ a_2 + a_2^+ a_1)(t)(a_1^+ a_2 + a_2^+ a_1)|I\rangle \, dt \quad (10.30)$$

is the desired expression for the rate constant. The equilibrium displacements appropriate for the electron in electronic state 1 differ from those in electronic state 2. From the form of (10.18c,d), what should be done is to complete the square, writing for instance,

$$V_L(x) = \tfrac{1}{2} k x^2 + \tfrac{1}{2} A x + \varepsilon_1 \tag{10.31a}$$

$$= \tfrac{1}{2} k(x - x_L^0)^2 + [\varepsilon_1 - \tfrac{1}{2} k(x_L^0)^2] \tag{10.31b}$$

with

$$x_L^0 = \frac{-A}{2k} \tag{10.31c}$$

But this operation must be performed quantum mechanically and is relevant only when the electron is in state 1; an analogous transformation with $x_R^0 = -x_L^0$ holds for the electron in state 2.

To perform this readjustment of the vibrational origins, we formally write a transformed hamiltonian

$$\tilde{H}_{rel} = e^{is} H_{rel} e^{-is} \tag{10.32}$$

with the operator s to be determined. The transformation (10.32), called a canonical transformation, will not affect the eigenvalues of H_{rel}; the operator s is chosen to optimize, in some sense, the form of $\tilde{H}_{rel}$. Transformations of this type are widely used in quantum mechanics and are often called van Vleck transformations.

In the case of vibronic coupling, we wish to choose s to effect the change of equilibrium displacement. Formal transformation theory can be used to show that the correct form is $s = x_0 p/\hbar$, since the momentum operator p generates translational motion along the x coordinate. It is more instructive, however, to obtain the form of s directly from the nature of the transformation. We wish $\tilde{H}_{rel}$ to be written as suggested in (10.31b); that is, it should not include the linear x coupling directly. This suggests

$$\tilde{H}_{rel} = \sum \tilde{\varepsilon}_i \tilde{a}_i^+ \tilde{a}_i + \frac{1}{2} k \sum_i (x - x_0^{(i)})^2 \tilde{a}_i^+ \tilde{a}_i - J(\tilde{a}_1^+ \tilde{a}_2 + \tilde{a}_2^+ \tilde{a}_1)$$

$$- \sum_i (\tilde{a}_i^+ \tilde{a}_i) \frac{\hbar^2}{2m} \frac{\partial^2}{\partial x^2} \quad (10.33)$$

with

$$e^{is} a e^{-is} \equiv \tilde{a} \quad (10.34)$$

and $\tilde{\varepsilon}_i$ a new site energy, that should include the $-\frac{1}{2}k(x_0^{(i)})^2$ contribution. To perform the requisite operator algebra, it is convenient to reexpress the vibrational motions in occupation number terms as [Eqs. (6.23) and (6.26)]

$$\frac{-\hbar^2}{2m} \frac{\partial^2}{\partial x^2} + \frac{1}{2} kx^2 = \left(b^+ b + \frac{1}{2}\right) \hbar\omega$$

$$x = (b^+ + b) \sqrt{\frac{\hbar}{2m\omega}}$$

$$p = i(b^+ - b) \sqrt{\frac{m\hbar\omega}{2}}$$

Then the x-dependent part of the hamiltonian is

$$H_{rel} + J(a_1^+ a_2 + a_2^+ a_1) = a_1^+ a_1 \left[\left(b^+ b + \frac{1}{2}\right) \hbar\omega + \varepsilon_1 \right.$$

$$+ \frac{1}{2} (b^+ + b) \sqrt{\frac{\hbar}{2m\omega}} A \right] + a_2^+ a_2 \left[\left(b^+ b + \frac{1}{2}\right) \hbar\omega \right.$$

$$\left. + \varepsilon_2 - \frac{1}{2} (b^+ + b) \sqrt{\frac{\hbar}{2m\omega}} A \right] \quad (10.35)$$

Following the recipe suggested, classically, by (10.31b), we wish to write this as

$$= a_1^+ a_1 \left[(b^+ - \phi_1^*)(b - \phi_1)\hbar\omega + \frac{1}{2} \hbar\omega + \frac{1}{2} A \sqrt{\frac{\hbar}{2m\omega}} (b^+ + b - \phi_1 - \phi_1^*) \right.$$

$$\left. + \varepsilon_1 + d\varepsilon_1 \right] + a_2^+ a_2 \left[(b^+ - \phi_2^*)(b - \phi_2)\hbar\omega \right.$$

$$\left. + \frac{1}{2} \hbar\omega - \frac{1}{2} A \sqrt{\frac{\hbar}{2m\omega}} (b^+ + b - \phi_2 - \phi_2^*) + \varepsilon_2 + d\varepsilon_2 \right] \quad (10.36)$$

with $d\varepsilon_i$ the correction to the site energy caused by electron–vibration coupling, and ϕ_i a constant proportional to the shift in origin. The ϕ_i should be so chosen that the terms proportional to b^+ or b in (10.36) should vanish; this is the operator equivalent of completing the square.

Defining, then,

$$\bar{b} = e^{is}be^{-is} \tag{10.37a}$$

we see that the choice

$$\bar{b} = b - \sum_i \phi_i a_i^+ a_i \tag{10.37b}$$

$$\phi_1 = \frac{A}{\hbar\omega}\sqrt{\frac{\hbar}{8m\omega}}, \qquad \phi_2 = \frac{-A}{\hbar\omega}\sqrt{\frac{\hbar}{8m\omega}} \tag{10.37c}$$

will remove the linear term in (b, b^+). [Remember that $(a_i^+ a_i)^2 = (a_i^+ a_i)$.] Then the right side of (10.35), after canonical transformation, is

$$a_1^+ a_1 \left[\left(b^+ b + \frac{1}{2} \right) \hbar\omega + \varepsilon_1 + \frac{1}{2}(b^+ + b)A\sqrt{\frac{\hbar}{2m\omega}} \right]$$

$$+ a_2^+ a_2 \left[\left(b^+ b + \frac{1}{2} \right) \hbar\omega + \varepsilon_2 - \frac{1}{2}(b^+ + b)A\sqrt{\frac{\hbar}{2m\omega}} \right]$$

Substituting the values from (10.37b,c) here, we obtain (remember that $a_1^+ a_1 + a_2^+ a_2 = 1$, so $a_1^+ a_1 a_2^+ a_2 = 0$)

$$\bar{H}_{\text{rel}} + J(\bar{a}_1^+ \bar{a}_2 + \bar{a}_2^+ \bar{a}_1) = \bar{a}_1^+ \bar{a}_1 \left[\left(b^+ b + \frac{1}{2} \right) \hbar\omega + \varepsilon_1 - \frac{A^2}{8k} \right]$$

$$+ \bar{a}_2^+ \bar{a}_2 \left[\left(b^+ b + \frac{1}{2} \right) \hbar\omega + \varepsilon_2 - \frac{A^2}{8k} \right] \tag{10.38}$$

In this form we have, indeed, removed the linear displacement coupling in x.

To obtain the operator s, we expand (10.37a) to first order and compare to (10.37b), obtaining

$$(1 + is)b(1 - is) = b + i[s, b] = b - \sum_i \phi_i a_i^+ a_i$$

or

$$i[s, b] = -\sum_i \phi_i a_i^+ a_i \tag{10.39}$$

Using the fundamental boson commutation relation $[b, b^+] = 1$, a solution to (10.39) is

$$s = \frac{1}{i} \sum_i \phi_i(b^+ - b)a_i^+ a_i \tag{10.40}$$

This turns out to be the correct choice, precisely the one that is obtained from more formal considerations. The identity (10.37b) is, in this case, valid to all orders in s, though from (10.39) we obtained it using only the first-order term. In other applications, often the transformation cancels the coupling terms only to low orders in s; in that case, the canonical transformation method is called van Vleck perturbation theory.

To complete the expression of $\tilde{H}_{rel}$ we require $\tilde{a}_i$, calculated according to (10.34). To do so, we note that s can be written as

$$s = \sum_j s_j$$

$$s_j = \frac{1}{i} \phi_j(b^+ - b)a_j^+ a_j \qquad (10.41a)$$

Thus

$$\tilde{a}_i = \exp\left(i \sum_j s_j\right) a_i \exp\left(-i \sum_j s_j\right) \qquad (10.41b)$$

$$= e^{is_i} a_i e^{-is_i} \qquad (10.41c)$$

$$= a_i e^{-is_i} \qquad (10.41d)$$

$$= a_i e^{-\phi_i(b^+ - b)} \qquad (10.41e)$$

The equations are justified as follows: (c) follows from (b) because $[a_i, a_j^+ a_j] = 0$ if $i \neq j$; (d) follows from (c) by expanding the first exponential, since every term in the expansion, except the 1, contains $a_i^+ a_i$, and $a_i^+ a_i a_i = 0$; (e) follows from (d) by expansion, since each term beyond the 1 contains an $a_i^+ a_i$, and $a_i a_i^+ a_i = a_i$. By hermitian conjugation,

$$\tilde{a}_i^+ = a_i^+ e^{\phi_i(b^+ - b)} \qquad (10.41f)$$

$$\tilde{a}_i^+ \tilde{a}_i = a_i^+ a_i \qquad (10.41g)$$

Using (10.41e,f,g) with (10.38), we find that

$$\tilde{H}_{rel} = \sum_{i=1}^{2} a_i^+ a_i \{(b^+ b + \tfrac{1}{2})\hbar\omega + \tilde{\varepsilon}_i\} - J[a_1^+ B_1^+ B_2 a_2 + a_2^+ B_2^+ B_1 a_1] \qquad (10.42)$$

The shifted energies are

$$\tilde{\varepsilon}_i = \varepsilon_i - \frac{A^2}{8k} \qquad (10.43)$$

and the vibrational shift operators are

$$B_i = \exp\{-\phi_i(b^+ - b)\} \qquad (10.44)$$

Now we can evaluate the rate constant using (10.30). Formally,

$$\text{Tr}_{\text{vib}}\{e^{-\beta H}e^{iHt/\hbar}(a_1^+ a_2 + a_2^+ a_1)e^{-iHt/\hbar}(a_1^+ a_2 + a_2^+ a_1)\}$$

$$= \text{Tr}_{\text{vib}}\{e^{-\beta \tilde{H}}e^{i\tilde{H}t/\hbar}(\tilde{a}_1^+ \tilde{a}_2 + \tilde{a}_2^+ \tilde{a}_1)e^{-i\tilde{H}t/\hbar}(\tilde{a}_1^+ \tilde{a}_2 + \tilde{a}_2^+ \tilde{a}_1)\} \quad (10.45)$$

So that the trace, or average, is unchanged by the canonical transformation. Since we are treating the smallness parameter J only to second order for this nonadiabatic ET case, the evolution and statistical operators in (10.45) may be evaluated using only the J-independent part of (10.42). Then it follows that

$$\tilde{a}_i(t) = \tilde{a}_i e^{-i\tilde{\varepsilon}_i t/\hbar} \quad (10.46a)$$

$$b(t) = b e^{-i\omega t} \quad (10.46b)$$

$$B_i(t) = \exp\{-\phi_i[b^+(t) - b(t)]\} = \exp\{-\phi_i[b^+ e^{i\omega t} - b e^{-i\omega t}]\} \quad (10.46c)$$

The rate constant, from (10.30), can be written

$$k_r = \frac{2}{\hbar^2} \text{Re} \int_0^\infty dt \, J^2[\langle I|a_1^+ a_2 e^{i\omega_{12}t}a_2^+ a_1 B_1^+(t)B_2(t)B_2^+ B_1|I\rangle$$

$$+\langle I|a_2^+ a_1 e^{-i\omega_{12}t}a_1^+ a_2 B_2^+(t)B_1(t)B_1^+ B_2|I\rangle]$$

where $\hbar\omega_{12} = \tilde{\varepsilon}_1 - \tilde{\varepsilon}_2 = \varepsilon_1 - \varepsilon_2 = -\Delta$. The averages over electronic and vibrational operators can be evaluated separately, since $\tilde{H}_{\text{rel}}$, to zero order in J, is just the sum of x-dependent and electronic parts. The electronic averages $\langle I|a_2^+ a_1 a_1^+ a_2|I\rangle$ and $\langle I|a_1^+ a_2 a_2^+ a_1|I\rangle$ can be evaluated since the initial state $|I\rangle$ is clearly defined. The simplest choice is just $|I\rangle = |l\rangle$, the electron being localized on the left with energy $\tilde{\varepsilon}_1$. Then (since we have assumed the electronic sites orthogonal) $\langle I|a_2^+ a_1 a_1^+ a_2|I\rangle = 0$, $\langle I|a_1^+ a_2 a_2^+ a_1|I\rangle = 1$, and we find that

$$k_r = \frac{2}{\hbar^2} J^2 \text{Re} \int_0^\infty dt \, e^{i\omega_{12}t}\langle B_1^+(t)B_2(t)B_2^+ B_1\rangle \quad (10.47)$$

The rate thus will depend on the overlap factors (averages over B's): The B operators themselves describe a shift of the vibrational origin accompanying local presence of an electron.

Evaluation of the averages in (10.47) is facilitated using two results of formal operator algebra. These are

$$e^A e^B = e^{A+B+(1/2)[A,B]} \quad (10.48)$$

if $[A, B]$ is a constant, and

$$\langle e^{L(b,b^+)}\rangle = \exp\{\tfrac{1}{2}\langle L^2\rangle\} \quad (10.49)$$

for $L(b, b^+)$ any linear form in (b, b^+). Using these, we have, after some messy algebra,

$$\langle B_1^+(t)B_2(t)B_2^+ B_1\rangle = \langle B_2^+(t)B_1(t)B_1^+ B_2\rangle$$

$$= \exp\{-(2n + 1)S\} \exp\{Sn e^{i\omega t} + S(n + 1)e^{-i\omega t}\} \quad (10.50)$$

where we have defined the reorganization energy, in units of $\hbar\omega$, by

$$S = \frac{A^2}{2\hbar\omega k} = \frac{\lambda}{\hbar\omega} \tag{10.51}$$

and

$$n = \frac{1}{e^{\hbar\omega/k_B T} - 1} \tag{10.52}$$

is the Bose–Einstein thermal occupation of vibrational quanta.

The expression for the rate constant now reads

$$k_r = \frac{2J^2}{\hbar^2} \, \text{Re} \int_0^\infty dt \, e^{i\omega_{12}t} e^{-S(2n+1)} \exp\{Sne^{i\omega t} + S(n+1)e^{-i\omega t}\} \tag{10.53}$$

To gain some understanding of this result, we examine the case of low temperature, so that $n \to 0$. Then we find

$$k_r(T \to 0) = \frac{2J^2}{\hbar^2} e^{-S} \, \text{Re} \int_0^\infty dt \, e^{i\omega_{12}t} e^{Se^{-i\omega t}}$$

or, expanding in Taylor series,

$$k_r(T \to 0) = \frac{2J^2}{\hbar^2} e^{-S} \, \text{Re} \int_0^\infty dt \, e^{i\omega_{12}t} \sum_{k=0}^\infty \frac{S^k}{k!} e^{-ik\omega t}$$

$$k_r(T \to 0) = \frac{2\pi J^2}{\hbar^2} e^{-S} \sum_{k=0}^\infty \frac{S^k}{k!} \delta(\omega_{12} - k\omega) \tag{10.54}$$

This rate is the product of two factors times a delta function. The first factor, $2\pi J^2/\hbar^2$, is an electronic tunneling term and describes the mixing of the initial- and final-state parabolas of Fig. 10.2. The second term $e^{-S}S^k/k!$ is the Franck–Condon term, just the squared overlap of the initial harmonic ground state in the left well with the kth vibrational excited level of the product. Since in this model for H_{rel}, there are only two degrees of freedom (the electronic state and x), conservation of energy following the δ function in (4.12), requires that S quanta of harmonic vibration appear in the final state, to balance the energy lost by the electron.

The general form of (10.53) is always found for a linearly coupled electron–vibration system with harmonic vibrations; thus it also occurs in nonradiative decay theory, in Jahn–Teller dynamics, and in spin interconversion phenomena; with the frequency factor $e^{\pm i\omega_{ap}t}$ included under the integral, as would follow from (10.29), (10.53) is the correct form for optical absorption or emission at frequency ω_{ap}. All of these situations differ in the identification of J, which (for instance) is the transition dipole for optical spectra, the spin-orbit operator for intersystem crossing, the nonadiabatic operator $\partial/\partial Q$ for Jahn–Teller dynamics, and so on. In all of these cases the rate is a product of an electronic term proportional to J^2 and a vibrational overlap, or Franck–Condon, term [which is most easily evaluated in the $T \to 0$ limit, as in (10.54)].

For finite temperature, evaluation of (10.53) is more difficult. A useful and simple approximate evaluation may be obtained using the saddle-point method, first discussed in Section 4.3.2.

10.5

Steepest-Descents Evaluation of Franck–Condon Behavior: Energy Sharing

The rate expression (10.53) has the general form of an electronic matrix element times a general vibrational overlap. Such expressions occur in all vibronic coupling phenomena. Their evaluation is, in general, complicated. The exponentials can be expanded as was done in (10.54), but when several vibrational modes, rather than just one, are considered, these sums become unwieldy. A physically reasonable and computationally useful approximate result can often be obtained using steepest descents, as long as S remains substantially greater than zero.

For the single-mode molecular crystal model of (10.35), the rate constant expression is (10.53). We write

$$2 \, \mathrm{Re} \int_0^\infty e^{f(t)} \, dt \tag{10.55a}$$

$$= \mathrm{Re} \left\{ \int_0^\infty e^{f(t)} \, dt - \int_0^{-\infty} e^{f(-t)} \, dt \right\} \tag{10.55b}$$

$$= \mathrm{Re} \left\{ \int_0^\infty e^{f(t)} \, dt + \int_{-\infty}^0 e^{f*(t)} \, dt \right\} \tag{10.55c}$$

$$= \mathrm{Re} \int_{-\infty}^\infty e^{f(t)} \, dt \tag{10.55d}$$

The steps follow by substitution of $-t$ for t to find (10.55b), by noting from (10.53) that $f(-t) = f*(t)$ to find (10.55c) and finally obtaining (10.55d) since $\mathrm{Re}(Z) = \mathrm{Re}(Z*)$. To evaluate (10.55) using steepest descent, we approximate as

$$\int_{-\infty}^\infty dt \, e^{f(t)} \cong \int_{-\infty}^\infty dt \, \exp \left\{ f(t = t_s) + \frac{\partial f}{\partial t} \bigg|_{t_s} (t - t_s) + \frac{1}{2} \frac{\partial^2 f}{\partial t^2} \bigg|_{t_s} (t - t_s)^2 \right\}$$

If the saddle point t_s is chosen so that

$$\left.\frac{\partial f}{\partial t}\right|_{t_s} = 0 \qquad (10.56)$$

then we find, evaluating the gaussian integral,

$$\text{Re} \int_{-\infty}^{\infty} dt \; e^{f(t)} = e^{f(t_s)} \frac{(2\pi)^{1/2}}{\sqrt{|\partial^2 f/\partial t^2|_{t_s}}} \qquad (10.57)$$

Equations (10.56) and (10.57) define the steepest-descents approximation to the overlap factor in the rate.

For the particular vibronic form (10.53) we have

$$f = i\omega_{12}t - S(2n+1) + Sne^{i\omega t} + S(n+1)e^{-i\omega t} \qquad (10.58a)$$

$$f' = 0 = i\omega_{12} + i\omega Sne^{i\omega t_s} - i\omega S(n+1)e^{-i\omega t_s} \qquad (10.58b)$$

This last form is a transcendental equation, to be solved for the time t_s, which can then be substituted into (10.57) to obtain the rate. Note that if $S \to 0$, (10.58b) becomes meaningless, and the steepest-descent analysis fails.

Rewriting (10.58b), we find that

$$\omega_{12} = -nS\omega e^{i\omega t_s} + (n+1)S\omega e^{-i\omega t_s} \qquad (10.58c)$$

For $n \to 0$ (or $k_B T \ll \hbar\omega$), this just becomes

$$\omega_{12} = S\omega e^{-i\omega t_s} \qquad (10.58d)$$

so that

$$it_s = \frac{1}{\omega} \ln \frac{S\omega}{\omega_{12}} \qquad (10.58e)$$

$$k_r = \frac{J^2}{\hbar^2} e^{-S} e^{\omega_{12}/\omega} \left(\frac{\omega_{12}}{S\omega}\right)^{-\omega_{12}/\omega} \sqrt{\frac{2\pi}{\omega\omega_{12}}} \qquad (10.58f)$$

Note that t_s is pure imaginary. By differentiation, it is easy to see that k_r is maximized when

$$S = \frac{\omega_{12}}{\omega}$$

Then we find that

$$k_r^{(max)}(T \to 0) = \frac{J^2}{\hbar^2 \omega} \left(\frac{2\pi}{S}\right)^{1/2} \qquad (10.58g)$$

For this value of S, the function $f(t_s)$ in (10.57) and the activation energy E_A of (10.21) vanish as the exoergicity $|\Delta|$ is precisely the same as the reorganization energy λ.

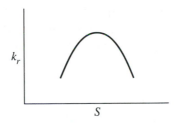

Figure 10.3. k_r versus S.

This result can be interpreted physically in three ways. Most simply, $S = \omega_{12}/\omega$ is equivalent to $|\Delta| = \lambda$, so that the crossing coordinate x_c of Fig. 10.2 is the same as x_L, so that there is no activation needed, and the rate maximizes. Alternatively, it is for the particular choice $S = \omega_{12}/\omega$ that the overlap integral between the ground vibrational level in the initial electronic state and the final vibrational state with quantum number $n_{final} = S$, corresponding to a perfect energy balance, is maximal. For S greater or smaller, the energy fit is not optimal and the rate drops. This is illustrated in Fig. 10.3. A third physical description of this result is that the energy of the final state $(b^+)^S a_2^+ |vac\rangle$ that is degenerate with the initial state $a_1^+ |vac\rangle$ is

$$S\hbar\omega + \varepsilon_2 - \varepsilon_1 = \frac{\omega_{12}}{\omega}(\hbar\omega) + \varepsilon_2 - \varepsilon_1 = 0.$$

Therefore, no motion of the nuclei is needed to achieve this degenerate situation. The initial and final state energies are coincident without any vibrational motion if $S = \omega_{12}/\omega$. For any other value of S, the initial state energy (zero) will not be the same as the final-state energy $\frac{1}{2}k(A/2k)^2 + \varepsilon_2 - \varepsilon_1$, so that vibrational displacement is required to make these energies coincident. When the nuclei have vibrated to a position x_c (for crossing, or coincident), such that

$$V_1(x_c) = V_2(x_c) \tag{10.59}$$

with V_1 and V_2 the harmonic curves of Eq. (10.36), the electron transfer can occur (with relative probability proportional to J^2). To attain the "coincidence event" geometry x_c requires [from (10.19) or Fig. 10.2] activation energy E_A. This interpretation of Franck–Condon restrictions, which originates in polaron theory, is useful to help interpret the activation process in vibronic coupling.

The energy-sharing interpretation of the low-temperature limit (10.58d) holds equally for the general one-mode harmonic situation of

(10.58c): the saddle-point time t_s is determined such that energy be shared optimally between the vibrational and electronic degrees of freedom.

It is now straightforward to generalize our treatment to the situation in which each site in the molecular crystal model has not one but many vibrational degrees of freedom. Then the hamiltonian is [see Eq. (10.35)]

$$H_{rel} = -J(a_1^+ a_2 + a_2^+ a_1) + \sum_{i=1}^{2} a_i^+ a_i \left\{ \varepsilon_i + \sum_q \left(b_q^+ b_q + \frac{1}{2} \right) \hbar \omega_q \right.$$

$$\left. + \sum_q \frac{A_q^{(i)}}{2} (b_q^+ + b_q) \sqrt{\frac{\hbar}{2 m_q \omega_q}} \right\} \quad (10.60)$$

with vibrational normal mode q of the dimer having frequency ω_q, reduced mass m_q, and linear coupling distance $A_q^{(i)}/2 m_q \omega_q^2$ in electronic state i. Then all of the canonical transformation follows through precisely as before; the rate constant is

$$k_r = \frac{J^2}{\hbar^2} \text{Re} \int_{-\infty}^{\infty} dt e^{i\omega_{12} t} \exp \left\{ - \sum_q S_q (2 n_q + 1) \right\}$$

$$\exp \left\{ \sum_q [n_q S_q e^{i\omega_q t} + (n_q + 1) S_q e^{-i\omega_q t}] \right\} \quad (10.61)$$

with

$$2 S_q^{1/2} = (A_q^{(1)} - A_q^{(2)}) \sqrt{\frac{1}{2 m_q \omega_q^3 \hbar}} \quad (10.62)$$

Evaluating (10.61) for many vibrational modes q is easiest using the saddle point scheme. Then the energy-sharing criterion becomes

$$\omega_{12} = - \sum_q \{ n_q S_q \omega_q e^{i\omega_q t_s} - (n_q + 1) S_q \omega_q e^{-i\omega_q t_s} \} \quad (10.63a)$$

with the saddle-point time t_s chosen so that this equation is satisfied. The weighting factor $e^{-i\omega_q t_s}$ determines how much each vibrational mode contributes to the energy balance; this is most clearly seen by taking the $k_B T \ll \hbar \omega_q$ limit of (10.63a),

$$\omega_{12} = \sum_q S_q \omega_q e^{-i\omega_q t_s} \quad (10.63b)$$

showing that the exoergicity $\varepsilon_1 - \varepsilon_2$ is spread among the vibrational modes, each accepting an energy equal to $\hbar \omega_q S_q e^{-i\omega_q t_s}$.

In the many-mode case the saddle-point expression for the rate constant is

$$k_r = \frac{J^2}{\hbar^2} \sqrt{\frac{2\pi}{|f''(t_s)|}} e^{f(t_s)} \quad (10.64)$$

with

$$f(t) = i\omega_{12}t - \sum_q S_q[(2n_q + 1) - (n_q + 1)e^{-i\omega_q t} - (n_q)e^{+i\omega_q t}]$$

Thus given the Hamiltonian parameters J, ω_q, $A_q^{(i)}$, m_q, and $\varepsilon_2 - \varepsilon_1$, along with the temperature T, the steepest-descents approximation to the rate constant is straightforwardly evaluated.

10.6

Solvent Contributions to Electron Transfer Rates

The description of ET processes discussed to this point has worked from the Holstein molecular crystal model, in which the nuclear motions include only intramolecular vibrations and frequency is assumed constant. The reorganization energy λ, which we will now rename λ_i, then refers to the reorganization of the molecular vibrations and is called the inner-sphere reorganization energy. Most ET processes are studied in solution, and the solvation of the reactant and product species will generally differ. This results in a second reorganization energy, referred to as the solvent, or outer sphere, and usually denoted λ_0. The actual writing of λ_0 in terms of solvent static and optical dielectric constants is a central part of ET theory. For our purposes, the most important observation is that, in the usual picture, the solvent is treated as a uniform continuous dielectric medium. In such a material, any local geometry change can be expressed as a sum over all the vibrations (phonons) of the medium. These vibrational modes will be damped in the liquid; more important, the characteristic frequencies are usually small compared to thermal energies.

We will assume, then, that the solvent is characterized by one frequency ω_{sol}, with a coupling strength S_{sol} defined, in analogy to (10.62) as proportional to the change of the solvation energy with displacement, in the solvent coordinates, between initial and final states. Then the solvent contribution to the vibrational overlap term of (10.50) can be written

$$\langle B_1^+(t)B_2(t)B_2^+ B_1\rangle_{\text{solv}} = \exp\{S_{\text{sol}}[(n_{\text{sol}} + 1)(e^{-i\omega_{\text{sol}}t} - 1) + n_{\text{sol}}(e^{+i\omega_{\text{sol}}t} - 1)]\} \quad (10.65)$$

To evaluate this, we consider the high-temperatuure limit $k_B T \gg \hbar\omega_{\text{sol}}$, which will generally be valid in solution ET. Then the Bose population becomes

$$n_{\text{sol}} = \frac{1}{e^{\hbar\omega_{\text{sol}}/k_B T} - 1} \cong \left(1 + \frac{\hbar\omega_{\text{sol}}}{k_B T} + \cdots - 1\right)^{-1} = \frac{k_B T}{\hbar\omega_{\text{sol}}} \quad (10.66)$$

and the frequency terms can be expanded as

$$e^{\pm i\omega_{\text{sol}}t} - 1 \cong 1 \pm i\omega_{\text{sol}}t - \frac{\omega_{\text{sol}}^2 t^2}{2} \cdots - 1$$

Then the vibrational overlap of (10.65) is

$$\exp\left\{ S_{\text{sol}} \left[1 + \frac{k_B T}{\hbar\omega_{\text{sol}}} \right] \left(-i\omega_{\text{sol}}t - \frac{\omega_{\text{sol}}^2 t^2}{2} \right) \right.$$

$$\left. + S_{\text{sol}} \frac{k_B T}{\hbar\omega_{\text{sol}}} \left[i\omega_{\text{sol}}t - \frac{\omega_{\text{sol}}^2 t^2}{2} \right] \right\}$$

$$= \exp\left\{ S_{\text{sol}} \left[-i\omega_{\text{sol}}t - \frac{\omega_{\text{sol}}t^2 k_B T}{\hbar} - \frac{\omega_{\text{sol}}^2 t^2}{2} \right] \right\}$$

which, since $k_B T \gg \hbar\omega_{\text{sol}}$, becomes

$$= \exp\left\{ S_{\text{sol}} \left[-i\omega_{\text{sol}}t - \frac{\omega_{\text{sol}}k_B T t^2}{\hbar} \right] \right\} \tag{10.67}$$

Defining now the outer-sphere reorganization energy by analogy with (10.51) as

$$\lambda_0 = \hbar\omega_{\text{sol}} S_{\text{sol}} \tag{10.68}$$

we can write the solvent vibrational overlap of (10.65) as

$$= \exp\left\{ \frac{-i\lambda_0 t}{\hbar} - \frac{\lambda_0 k_B T t^2}{\hbar^2} \right\} \tag{10.69}$$

This gaussian behavior in t^2 is mirrored in the overall behavior of the rate.

If the ET process involves no significant inner-sphere reorganization, insertion of (10.69) into (10.47) gives

$$k_r = \frac{J^2}{\hbar^2} \operatorname{Re} \int_{-\infty}^{\infty} dt \, e^{i\omega_{12}t} e^{-i\lambda_0 t/\hbar} e^{-\lambda_0 k_B T t^2/\hbar^2}$$

Using the steepest-descents, we have

$$f(t) = i\omega_{12}t - \frac{i\lambda_0 t}{\hbar} - \frac{\lambda_0 k_B T t^2}{\hbar^2}$$

$$\frac{\partial f(t_s)}{\partial t} = i\omega_{12} - \frac{i\lambda_0}{\hbar} - \frac{2\lambda_0 k_B T t_s}{\hbar^2}$$

or

$$t_s = \frac{i\hbar^2}{2\lambda_0 k_B T} \left(\omega_{12} - \frac{\lambda_0}{\hbar} \right) \tag{10.70a}$$

$$f(t_s) = \left(i\omega_{12} - \frac{i\lambda_0}{\hbar} \right)\left(\omega_{12} - \frac{\lambda_0}{\hbar} \right)\frac{i\hbar^2}{2\lambda_0 k_B T} - \frac{\lambda_0 k_B T}{\hbar^2} \left(\frac{i\hbar^2}{2\lambda_0 k_B T} \right)^2 \left(\omega_{12} - \frac{\lambda_0}{\hbar} \right)^2$$

Then

$$f(t_s) = \left(\omega_{12} - \frac{\lambda_0}{\hbar}\right)^2 \left(\frac{-\hbar^2}{4\lambda_0 k_B T}\right)$$

$$f''(t_s) = \frac{-2\lambda_0 k_B T}{\hbar^2}$$

(10.70b)

So

$$k_r = \frac{J^2}{\hbar} \sqrt{\frac{\pi}{\lambda_0 k_B T}} \exp\left\{\frac{-(\hbar\omega_{12} - \lambda_0)^2}{4\lambda_0 k_B T}\right\}$$

(10.70c)

This gaussian dependence on the difference between the outer-sphere reoganization energy and the exoergicity is a very important result of the original Marcus theory of ET rates—we saw in (10.21) that it arises simply from the height of the energy barrier in the small-J limit of the molecular crystal model, or in the configuration-coordinate diagram of Fig. 10.2. The simple form is valid only if $kT \gg \hbar\omega_{re}$, where ω_{re} is the relevant frequency for vibrations coupled to the electron motion. If this high-temperature limit does not hold, the rate should be calculated using (10.61), with the outer-sphere (solvent) coupling included in the sum. In this sense, (10.70c) and (10.21) are semiclassical results, valid only for the high-temperature limit of nonadiabatic ET.

10.7

Electron Transfer Reactions: Qualitative Remarks

Theoretical and experimental study of electron transfer has been one of the most active research topics in chemistry since the early 1960s. An enormous amount of work has gone into this area, and the simple discussion we have given in terms of vibronic coupling models really constitutes only the beginning of a correct theoretical description. For our purposes here, however, it illustrates nicely the ability of quantum-mechanical methodology to discuss both the quantum-mechanical tunneling (or nuclear tunneling) and the classical, activated regimes. The formulation, in terms of the small polaron picture within the Holstein molecular crystal model, is not quite standard in the field. Different formulations, using different nomenclature, notation, and mathematical techniques, have been presented by many groups. We would simply like to remark on some of the generalities that can be observed from our formulas.

At extremely low temperatures, where the outer-sphere reorganization energy is frozen because the solid reorientational motions are very slow, and the characteristic thermal energy is less than vibrational zero-point energy, reaction occurs largely by nuclear tunneling: In this case there is no temperature dependence to the rate, which is simply proportional to the Franck–Condon overlap between the initial state in the upper parabola of Fig. 10.2 and the vibrationally excited final-state degenerate with that initial state [Eq. (10.58f)].

In the very high temperature limit, as is clear from Eq. (10.70c), the rate is proportional to the activation energy involved in surmounting the barrier between the two parabolas in Fig. 10.2. The form derived in (10.70c) from quantum mechanics is precisely that obtained from simple geometry in Eq. (10.21). Thus the high-temperature quantum-mechanical limit gives precisely the transition-state theory result.

In the intermediate regime the overall rate can be computed using the steepest-descents formulas of Eqs. (10.64). The actual description of the energy transfer process, including approximately how much energy is accepted by each of the vibrations, follows from the steepest-descents formula or the steepest-descents condition (10.56). Explicit formulas in terms of reduced parameters are available in the literature, but the steepest-descents technique is particularly powerful for dealing with situations where many vibrational modes are active, and in which simple summation would yield computationally intractable forms. The entire treatment discussed in this chapter involves intramolecular electron transfer, in the sense that no considerations of the preequilibrium between donor and acceptor are included. In bimolecular electron transfer theory, the work involved in assembling the precursor complex can often dominate the overall kinetics. Our results really deal with electron transfer dynamics either within the precursor complex or, in the case of intramolecular electron transfer, from the initial state to the electron transfer product.

All our considerations apply in the so-called nonadiabatic electron transfer limit: This is the limit in which the electron tunneling matrix element J is small compared to characteristic vibrational frequencies and to the reorganization energy λ. Under these conditions, one can pictorially consider that (in the high-temperature limit) the results in Eq. (10.53) can be interpreted as the product of two factors: The first factor is simply the square of a small mixing element, the electron tunneling matrix element J. The second term is a Franck–Condon weighted vibrational density of states, which contains both the density of final product states and the vibrational activation terms that yield the Marcus–Hush activation energy (10.21) at high temperatures. In the high-temperature limit, then, the rate expression is given by (10.70c), with the second term simply being the activation energy involved in reaching the crossing point between the two parabolas. Ordinarily, this formula is in fact used with $\lambda = \lambda_0 + \lambda_i$, the total reorganization energy, replacing the outer-sphere reorganization energy. This is appropriate for very high temperatures in systems with relatively low vibrational frequencies. For high vibrational frequencies, one must use either the steepest-descents result or one of the many intermediate results available in the literature.

The term *tunneling* is appropriate for two separate processes in nonadiabatic electron transfer. The matrix elements J are electron tunneling matrix elements, describing electronic motion between the potential surfaces and arising simply from matrix elements of the electronic hamiltonian. At low temperatures, there is additionally a nuclear tunneling process through the barrier between the two parabolas of Fig. 10.2; this is automatically included in our analysis.

The other simple limit of electron transfer is that in which the splitting between the two dashed curves in Fig. 10.2 is large. When J is in fact large, electronic events are no longer rate controlling; rather, the overall rate is given by a characteristic frequency of passing the barrier top times (in the high-temperature limit) an exponential of exactly the form of Eq. (10.70c). Such transfers are called adiabatic and are essentially independent of electronic matrix elements in the hamiltonian.

Finally, the role of relaxation processes has been ignored entirely in our treatment. The matrix elements in the density-of-states-weighted Franck–Condon factor of Eq. (10.61) or (10.47) are evaluated in the equilibrium harmonic ensemble. When relaxation effects are important, one often finds that the key factor for electron transfer rates is related not to electronic state mixing but to the relaxation time that characterizes the trapping of the system trajectory in the final state, and is generally determined by the relaxation time spectrum of the solvent. Under these conditions, the calculations should really be carried out using a nonequilibrium, relaxing set of vibrational states.

The general spin boson problem, consisting of a two-level fermion system coupled linearly to vibrational displacements [Eq. (10.16)] is used to discuss a large number of other rate phenomena of importance in chemical and solid-state problems, including electrical conductivity in crystals, exciton motion, bimolecular reaction kinetics, dynamical Jahn–Teller motions, and photoconductivity. For each of these, analysis similar to that used in this chapter for electron transfer can be used to solve for the rates. Such concepts as density-of-states-weighted Franck–Condon factors, energy sharing among many coupled vibrations, nuclear tunneling, and activation processes are then critical for understanding the system behavior.

BIBLIOGRAPHY FOR CHAPTER 10

The treatment described here is mostly from the primary research literature. Papers include:

Fischer, S. F., and R. P. Van Duyne, *Chem. Phys.* 26, 9 (1977).

Holstein, T., *Ann. Phys.* 8, 325, 343 (1959).

Jortner, J., *J. Chem. Phys.* 64, 4860 (1976).

Marcus, R. A., and N. Sutin, *Biochem. Biophys. Acta* 811, 265 (1985).

Mikkelsen, K. V., and M. A. Ratner, *Chem. Rev.* 87, 113 (1987).

Newton, M. D., and N. Sutin, *Annu. Rev. Phys. Chem.* 35, 437 (1964).

Onuchic, J. N., and P. G. Wolynes, *J. Phys. Chem.* 92, 6495 (1988).

Scher, H., and T. Holstein, *Phil. Mag.* 44B, 343 (1981).

Siders, P., and R. A. Marcus, *J. Am. Chem. Soc.* 103, 741 (1981).

The original theory was given in:

Marcus, R. A., *J. Phys. Chem.* 67, 853, 2889 (1963); *J. Chem. Phys.* 24, 966, 979 (1956); N. S. Hush, *Trans. Faraday Soc.* 57, 557 (1968).

Good overviews of electron transfer theory include:

De Vault, D., *Quantum Mechanical Tunneling in Biological Systems* (Cambridge University Press, New York, 1984).

Ulstrup, J., *Charge Transfer Processes in Condensed Media* (Springer-Verlag, Berlin, 1979).

Articles that describe and apply the techniques used in this chapter are found in:

Jortner, J., and B. Pullman, eds., *Perspectives in Photosynthesis* (Kluwer, Dordrecht, The Netherlands, 1990).

Wagner, M., *Unitary Transformations in Solid State Physics* (North-Holland, Amsterdam, 1986).

Saddle-point methods are discussed in:

Mathews, J., and R. L. Walker, *Mathematical Methods of Physics* (W. A. Benjamin, New York, 1980).

PROBLEMS FOR CHAPTER 10

1. The treatment that we gave for even the two-site molecular crystal model ($H_2^+ - H_2$) is oversimplified. Defining q_1 and q_2 as the displacements of the diatomics on the left and the right, we can (assuming a Born–Oppenheimer separation and working within the diabatic representation) describe the electronic poten-

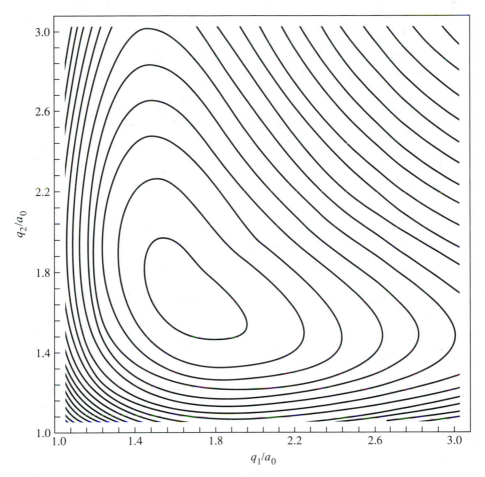

Figure 10.4a

tial surfaces as $V_i(q_1, q_2)$. Figure 10.4 shows results of a diatomics-in-molecules model potential surface for the ground electronic state; in Fig. 10.4a a single minimum is observed for a separation of 4 Å between the axes, while in Fig. 10.4b a double minimum is observed for a longer separation of 40 Å.

(a) Consider the double-minimum situation. Defining ϕ_1 as the diabatic electronic function with its minimum at (q_1^0, q_2^s) and ϕ_2 the function with its minimum at (q_1^s, q_2^0), expand the total potential $V(q_1, q_2)$ using electronic second quantization and nuclear coordinates. Make the Condon approximation for the transfer term, and stop after quadratic terms in the displacement. Expand V_1 and V_2 each about their own minimum.

(b) Now define sum and difference coordinates by

$$\sqrt{2}\, q_+ \equiv q_1 + q_2$$
$$\sqrt{2}\, q_- \equiv q_1 - q_2$$

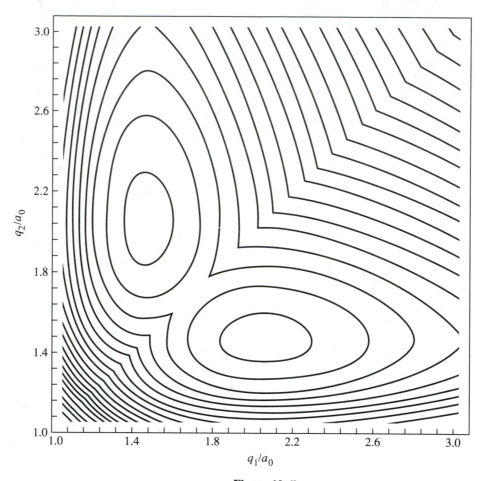

Figure 10.4b

Rewrite the potential in terms of q_+ and q_-. Note that the term linear in q_- is the linear electron–vibron coupling that we discussed in the text.

(c) Physically interpret the term linear in q_+. If the minima in Fig. 10.4b were round instead of egg-shaped, how would this term simplify? Why, physically, are the minima egg-shaped?

(*Note:* In actual calculations of electron transfer rates, the egg-shaped minima generally are assumed round.)

2. In the so-called inverted regime in which $|\Delta/\lambda| > 1$, the experimental tempera-ture dependence is often very weak, although the simple argument of acti-vated-complex theory, based on the intersection point of the two parabolas, would predict the usual activated form. To understand this weak temperature dependence, consider a simple one-mode picture. Then the rate is given by (10.53) and (10.57), with f and t_s given by (10.58).

(a) Show that in the low-temperature limit of the normal regime, where $|\Delta/\lambda| <$ 1, the saddle-point time it_s is positive.

(b) As the quotient $|\Delta/\lambda|$ increases, how does it_s change? What is the value of it_s for $|\Delta| = \lambda$?

(c) Assuming that the value of it_s found from the low-temperature limit remains valid, calculate an analytic form for $k(T)$ in terms of S, ω_{12}, ω, and n. Then by expanding, show that this form can be expressed, in the inverted regime, as

$$k(T) = k(T = 0) + k'n$$

(d) (Computer problem) Find it_s by numerical solution to the saddle-point condition (10.58b), plot it as a function of $|\Delta/\lambda|$. Then put it into (10.57) and compute (and graph) the rate as a function of temperature from $T = 0$ to $T = 400$ K. Use as parameters $\lambda = 0.7$ eV, $\omega = 2000$ cm^{-1}, $J = 2$ cm^{-1}. Do calculations in the inverted regime ($-\Delta = 1.0$ eV), in the normal regime ($-\Delta = 0.5$ eV), and at the crossing ($-\Delta = 0.7$ eV).

3. Hush, in his study of optical transitions in mixed valency systems, pointed out that in the limit where $|\lambda/J| \gg 1$, there should be a transition from a local minimum on the lower potential surface to the upper surface, and that this should correspond to intervalence transfer (i.e., the electronic wavefunction should switch from being localized on one site on the lower surface to the other site on the upper surface).

(a) What is the intervalence transfer frequency, ω_{IVT}, as a function of J, λ, and Δ for our one-mode model, in the limit $\lambda/J \gg 1$?

(b) How does the intervalence transfer frequency relate to the thermal activation energy for the special case $\Delta = 0$? Is this sensible, based on the potential curves?

(c) We can derive a useful approximate relation between the oscillator strength for the intervalence transfer band and the extent of delocalization. From Eq. (5.55), the oscillator strength is just

$$f = \frac{2m\omega}{\hbar e^2} \frac{\mu^2}{3}$$

For a broadband excitation, this is related to the decadic molar extinction coefficient ε by

$$f = 4.33 \times 10^{-9} \int \varepsilon \, d\bar{\nu}, \tag{P1}$$

where the integral runs over the band shape and $\bar{\nu}$ is the frequency in cm^{-1}. Using the relation $\omega = 2\pi c\bar{\nu}$, show that

$$f = 1.085 \times 10^{-5}\bar{\nu}D^2, \tag{P2}$$

where D is the so-called dipole strength ($e^2D^2 = \mu^2$) measured in Å. Then using the bare electronic hamiltonian of (10.7), write the ground-state electronic function in the left minimum of Fig. 10.2 as

$$\psi_s = (c_1a_1^+ + c_2a_2^+)|\text{vac}\rangle,$$

and show, by orthogonality, that the upper-curve state is

$$\psi_{ex} = (-c_2 a_1^+ + c_1 a_2^+)|vac\rangle.$$

Now evaluate $\langle \psi_{ex}|\mu|\psi_g \rangle = \mu_{ex,g}$ using the reasonable approximation (used extensively, following Pariser, in π-electron theory).

$$\langle u_i|\mu|u_j \rangle \cong \delta_{ij} e r_i,$$

with r_i the cartesian coordinate of site i. Next substitute this expression into that for f, obtaining (for a gaussian band shape)

$$f = 4.6 \times 10^{-9} \varepsilon_{max} \Delta_{1/2}, \tag{P3}$$

with ε_{max} the maximum extinction coefficient and $\Delta_{1/2}$ the full width at half height, in cm^{-1}.

Next combine (P3) and (P2) with your approximate expression for $\mu_{ex,g}$. This can be rewritten as an expression for $c_1 c_2$ in terms of $\Delta_{1/2}$, ε_{max}, and $\bar{\nu}$. Finally, since $c_1^2 + c_2^2 = 1$ and $c_1 \gg c_2$, we can write $c_1 c_2 \cong c_2$. Then solving the Hückel model of (10.7) for c_2 permits us to estimate J/ε.

(d) For the mixed-valency species Ti(III)–Cl–Ti(IV), Hush gives $f = 0.0026$, $\bar{\nu} = 20.41 \times 10^3 \, cm^{-1}$, $r = 5$ Å. From this, compute D, c_2, and J/ε. What percentage of the hole is localized on the (formally) Ti(III)?

4. Intramolecular electron transfer over long distances is often considered to arise from so-called superexchange coupling. To derive the simplest superexchange term, suppose that the electron transfer system is a three-site system comprising a donor/bridge/acceptor triple D–B–A. Neglecting nuclear motion initially, we represent the electronic hamiltonian (with $\varepsilon_A \equiv 0$) as

$$H_{el} = a_B^+ a_B \varepsilon_B + a_D^+ a_D \varepsilon_D + t_D(a_D^+ a_B + a_B^+ a_D) + t_A(a_A^+ a_B + a_B^+ a_A).$$

Using an expansion like that in Sections 4.3 and 4.5.1, and assuming both that the electron is initially localized on the donor and that $|t_D|, |t_A| \ll \varepsilon_B - \varepsilon_D$, derive an equation for the probability to find the electron localized on the acceptor at time t.

Use the usual assumptions of the golden rule type to rewrite one of the terms of your expressions in ordinary golden rule form: The effective matrix elements that enter this expression are of so-called superexchange type; interpret them physically with the aid of the perturbation-theoretic form [Eq. (1.23)] applied to the states $|D\rangle$ and $|A\rangle$. How might the bridge species be chosen to provide strong superexchange coupling? Assume orthogonality of the localized fragment orbitals ϕ_D, ϕ_A, and ϕ_B.

5. Consider the series of mixed-valency compounds $(NH_3)_5 RuBRu(NH_3)_5^{+5}$, with B a bridging dibasic ligand such as

NCCN,   N⟨○⟩N,   and   N⟨○⟩—⟨○⟩N

(We examined these in Problem 5 of Chapter 6.) Assume that this can be treated as a two-site electron transfer problem. Assume further that on each metal site only one mode is coupled, the symmetric sum of M—N stretches;

treat the M—B bond like a M—N bond, with similar frequency and mass ratio. The experimental Ru—N distances for $Ru(NH_3)_6^{+M}$ are 2.41 Å for M = 2, 2.37 Å for M = 3.

Calculate λ_i, the inner-sphere reorganization energy, for the intramolecular $M^{II} \rightarrow M^{III}$ transfer. At $T = 300$ K if $-J = -100$ cm^{-1}, calculate the rate of electron transfer assuming

(a) The Marcus form for the vibronic term.

(b) The steepest-descents form for the quantum-mechanical rate.

In each case, repeat the calculation twice, once using only the λ_i, and the other time using λ_0 calculated from the Marcus expression

$$\lambda_0 = (\Delta e)^2 \left(\frac{1}{2r_1} + \frac{1}{2r_2} - \frac{1}{r_{12}}\right)\left(\frac{1}{\varepsilon_\infty} - \frac{1}{\varepsilon_0}\right),$$

with Δe, r_1, r_2, r_{12}, ε_∞, and ε_0, respectively, the number of electrons transferred, the radius of site 1 at contact, that of site 2 at contact, the sum of r_1 and r_2, and the optical and static dielectric constants. Take $2r_1 = 2r_2 = r_{12}$, $\Delta e = e$, and (for H$_2$O) $\varepsilon_0 = 80$, $\varepsilon_\infty = 2$.

Comment qualitatively both on what changes the inclusion of λ_0 makes in the calculated rate and on what typical frequency ω_s corresponds [using (10.68)] to λ_0. Assume the local site functions to be orthogonal.

11

Density Matrices

11.1

Introduction

In classical mechanics, the dynamical variables such as $\mathbf{x}$ and $\mathbf{p}$ are determined precisely by the equations of motion. In quantum mechanics, by contrast, the idea of probability enters at the definition of a wavefunction $\psi(\mathbf{x}, t)$, whose physical interpretation is given by the statement that $|\psi(\mathbf{x}, t)|^2 \, d\mathbf{x}$ is the probability that the system is to be found in the interval $(\mathbf{x}, \mathbf{x} + d\mathbf{x})$. Even the exact solution of the Schrödinger equation yields wavefunctions that define only probabilities for observation. It is thus reasonable in the quantum context to extend these probability ideas to consider situations in which the wavefunction is not exactly known but some information is known about the system. von Neumann introduced the density operator as the fundamental statistical operator for quantum systems. In this chapter we define the density matrix, its relationship to wavefunctions, and its time evolution. We demonstrate its use to solve quantum problems for which complete wavefunction information is unavailable, to derive the behavior of subsystems of interest interacting with thermal baths, and to describe higher-order response to systematic driving fields.

Since this chapter is rather lengthy, references to the Bibliography are marked with the sections to which they are particularly relevant.

11.2

Density Operators and Density Matrices: Definitions and Averages

The density operator $\hat{\rho}$ is the quantum analog of the classical density ρ. Like all quantum operators with classical analogs, it can be expressed in Schrödinger, Heisenberg, or interaction representations, it is hermitian, and it has a matrix representation in any convenient chosen basis. The matrix representations of the density operator are called density matrices.

The density operator is the fundamental statistical operator for any given physical system. Statistics enters for quantal systems both because the wavefunction itself is a statistical quantity and because, in some systems, we may have limited knowledge about the wavefunction.

11.2.1 Wavefunctions and the Pure State Density Operator

We first consider the situation of a so-called pure state: This means simply that the physical system is characterized by a wavefunction. The projection operator onto this wavefunction, using Dirac notation, is written $|\psi\rangle\langle\psi|$. It is idempotent—that is,

$$(|\psi\rangle\langle\psi|)^2 = |\psi\rangle\langle\psi|\psi\rangle\langle\psi| = |\psi\rangle\langle\psi| \tag{11.1a}$$

If ψ itself is normalized

$$\langle\psi|\psi\rangle = 1 \tag{11.1b}$$

For a pure state the density operator is simply this projector.

$$\hat{\rho}_{\text{pure}} \equiv |\psi\rangle\langle\psi| \tag{11.2a}$$

It then follows that for a pure state the density operator is idempotent:

$$\hat{\rho}_{\text{pure}}^2 = \hat{\rho}_{\text{pure}} \tag{11.2b}$$

For any operator, we can define the trace as the sum of the diagonal elements in a given representation (see Section 8.5)

$$\text{Tr}\,\hat{B} \equiv \sum_k \langle\phi_k|\hat{B}|\phi_k\rangle \tag{11.2c}$$

We then have

$$\text{Tr}\,\hat{\rho}_{\text{pure}} = \sum_k \langle\phi_k|\hat{\rho}_{\text{pure}}|\phi_k\rangle = \sum_k \langle\phi_k|\psi\rangle\langle\psi|\phi_k\rangle = \sum_k \langle\psi|\phi_k\rangle\langle\phi_k|\psi\rangle$$

$$= \langle\psi|\psi\rangle = 1 \tag{11.2d}$$

where we have used the result $\Sigma_k |\phi_k\rangle\langle\phi_k| = 1$, and (11.1b). Note also from the definition that

$$\hat{\rho}_{\text{pure}}^+ = (|\psi\rangle\langle\psi|)^+ = ((\langle\psi|^+)(|\psi\rangle))^+ = |\psi\rangle\langle\psi| = \hat{\rho}_{\text{pure}} \qquad (11.2e)$$

For an operator $\hat{A}$ corresponding to an observable, we have

$$\langle\psi|\hat{A}|\psi\rangle = \sum_k \langle\psi|\phi_k\rangle\langle\phi_k|\hat{A}|\psi\rangle = \sum_k \langle\phi_k|\hat{A}|\psi\rangle\langle\psi|\phi_k\rangle$$

$$= \sum_k \langle\phi_k|\hat{A}\hat{\rho}_{\text{pure}}|\phi_k\rangle = \text{Tr}\{\hat{A}\hat{\rho}_{\text{pure}}\} \qquad (11.2f)$$

For the pure state situation, therefore, we have used simple algebra to prove the most relevant and important physical properties. The density operator is hermitian, has a trace of unity, and yields the expectation values for physical observables via a trace operation (11.2f). All of these properties, as we shall see, remain true for the more general density operator. The idempotency condition (11.2b), on the other hand, is true only for pure states.

For the pure state, from the viewpoint of experimental prediction, knowledge of $\hat{\rho}$ is equivalent to knowledge of ψ, since both permit calculation of observables, from the first or last results of (11.2f), respectively.

Note that for any operator product for which the trace exists,

$$\text{Tr}\{\hat{A}\hat{B}\} = \sum_{kl} \langle\phi_k|A|\chi_l\rangle\langle\chi_l|B|\phi_k\rangle$$

$$= \sum_{kl} \langle\chi_l|B|\phi_k\rangle\langle\phi_k|A|\chi_l\rangle = \text{Tr}\{\hat{B}\hat{A}\} \qquad (11.2g)$$

So that the trace is invariant to cyclic permutation of the operators, and we can write

$$\langle\psi|\hat{A}|\psi\rangle = \text{Tr}\{\hat{\rho}_{\text{pure}}\hat{A}\} = \text{Tr}\{\hat{A}\hat{\rho}_{\text{pure}}\} \qquad (11.2h)$$

11.2.2 Density Operators for Mixed States

We now generalize the notions of the preceding section to consider cases in which the system cannot be characterized by a unique wavefunction. Suppose that we know only that the relative probability of observing the wavefunction ψ is p_ψ, so that

$$\sum_\psi p_\psi = 1 \qquad (11.3a)$$

$$1 \geq p_\psi \geq 0 \qquad (11.3b)$$

and the expectation value becomes

$$\langle\hat{A}\rangle = \sum_\psi p_\psi\langle\psi|\hat{A}|\psi\rangle \qquad (11.4)$$

The sum rule over the set of projectors still holds so

$$\sum_{\psi} |\psi\rangle\langle\psi| = 1 \tag{11.5}$$

We now define the general density operator by

$$\hat{\rho} \equiv \sum_{\psi} p_\psi |\psi\rangle\langle\psi| \tag{11.6}$$

It is straightforward to show, generalizing the results of the preceding section, that $\hat{\rho}$ is hermitian and has unit trace. Similarly, we can see that

$$\langle \hat{A} \rangle = \sum_{\psi} p_\psi \langle\psi|\hat{A}|\psi\rangle = \sum_{\psi} \sum_{k} p_\psi \langle\psi|\hat{A}|\phi_k\rangle\langle\phi_k|\psi\rangle$$

$$= \sum_{k} \sum_{\psi} p_\psi \langle\phi_k|\psi\rangle\langle\psi|\hat{A}|\phi_k\rangle = \mathrm{Tr}\{\hat{\rho}\hat{A}\} \tag{11.7}$$

which is the generalization of (11.2f).

The situation that we are now considering, in which the wavefunction itself is known only with a certain probability, is usually called a mixed state (as distinguished from the pure state discussed in Section 11.2.1). For such mixed states, the expectation value in (11.7) really involves two separate averages: the matrix element $\langle\psi|\hat{A}|\psi\rangle$ averages over a given wavefunction, and the sum averages over the distribution of wavefunction probabilities. In some texts, the left side of (11.7) is written $\langle\overline{\hat{A}}\rangle$, with the overbar indicating the average over the wavefunction distribution. An important distinction between the mixed and pure state arises when we try to generalize (11.2b). We find that

$$\hat{\rho}^2 = \sum_{\psi,\psi'} p_\psi p_{\psi'} |\psi\rangle\langle\psi|\psi'\rangle\langle\psi'| = \sum_{\psi,\psi'} \delta_{\psi\psi'} p_\psi p_{\psi'} |\psi\rangle\langle\psi'| = \sum_{\psi} p_\psi^2 |\psi\rangle\langle\psi|$$

$$\tag{11.8}$$

And from the properties of the probability, (11.3), we have

$$p_\psi^2 \le p_\psi \tag{11.9a}$$

$$\mathrm{Tr}\,\hat{\rho}^2 = \sum_{k} \sum_{\psi} \langle\phi_k|\psi\rangle\langle\psi|\phi_k\rangle p_\psi^2 = \sum_{\psi} p_\psi^2 \le 1 \tag{11.9b}$$

with the inequality holding except for the pure state. Thus a mixed state is distinguished from a pure state because its density operator is not idempotent, and $\mathrm{Tr}\,\hat{\rho}^2 < 1$.

Two other properties of the trace will prove useful. We simply observe that

$$\mathrm{Tr}\{\hat{\rho}f(\hat{A})\} = \langle f(\hat{A})\rangle \tag{11.10}$$

for any reasonable function f of the operator $\hat{A}$. (This can be proved by expansion of the function f in a Taylor series.) Also,

$$\text{Tr}\{|\psi\rangle\langle\chi|\} = \sum_k \langle\phi_k|\psi\rangle\langle\chi|\phi_k\rangle = \sum_k \langle\chi|\phi_k\rangle\langle\phi_k|\psi\rangle = \langle\chi|\psi\rangle \quad (11.11)$$

so that there is a simple relationship between trace and scalar product.

We have defined the density operator and noted some of its properties—in particular, from (11.7), the density operator permits computation of observable quantities for both pure and mixed cases. We now consider how the density operator evolves in time, to obtain equations of motion whose solution yields $\hat{\rho}$.

Since the notation in the remainder of this chapter is fairly complicated, we will henceforth drop the "hat" notation and denote the density operator simply by ρ.

11.3

Representations and Equations of Motion

We begin in the Schrödinger representation. Then the evolution operator $U(t, t_0)$ is defined by

$$|\psi(t)\rangle = U(t, t_0)|\psi(t_0)\rangle \quad (11.12a)$$

with

$$U(t_0, t_0) = 1 \quad (11.12b)$$

Then the solution to the time-dependent Schrödinger equation yields the operator equation

$$i\hbar \frac{\partial}{\partial t} U(t, t_0) = HU(t, t_0) \quad (11.12c)$$

Then defining the density operator via (11.6), we find that

$$\rho(t) = \sum_\psi p_\psi U(t, t_0)|\psi(t_0)\rangle\langle\psi(t_0)|U^+(t, t_0) \quad (11.12d)$$

$$= U(t, t_0)\rho(t_0)U^+(t, t_0) \quad (11.12e)$$

Then the chain rule gives, with (11.12c),

$$i\hbar \frac{\partial}{\partial t} \rho(t) = [H, \rho(t)] \quad (11.13)$$

This is sometimes called the von Neumann equation, or the quantum Liouville equation; it is the quantum analog of the Liouville equation that gives the time evolution of classical phase space density in terms of a Poisson bracket of the hamiltonian and the density. This equation is fundamental for computations using density methods, just as the time-dependent Schrödinger equation and the Heisenberg equation of motion were key to solutions in the contexts of wavefunction or operator descriptions.

The density operator of (11.12) is given in the Schrödinger representation. In the Heisenberg picture, we have

$$\rho_H(t) = \Sigma\, p_\psi |\psi_H(t)\rangle\langle\psi_H(t)| = \rho_H(t_0) \tag{11.14}$$

Then for the (observable) expectation values we have

$$\langle A \rangle = \text{Tr}\{\rho_H(t)A_H(t)\} = \text{Tr}\{\rho_H(t_0)A_H(t)\}$$
$$= \text{Tr}\{\rho_H(t_0)U^+A_sU\} = \text{Tr}\{U\rho_H(t_0)U^+A_s\} = \text{Tr}\{\rho_s(t)A_s\} \tag{11.15}$$

where we have used the cyclic invariance of the trace and (11.12e), as well as the operator definition,

$$A_H(t) = U^+(t,\,t_0)A_s(t_0)U(t,\,t_0)$$

from Chapter 4. So we see, as expected, that observables can be calculated using density operators in either Schrödinger or Heisenberg pictures.

One particularly simple situation occurs when the hamiltonian is time independent. Then

$$U(t,\,t_0) = \exp\left\{-i(t-t_0)\frac{H}{\hbar}\right\} \tag{11.16a}$$

and in the representation of energy eigenstates

$$H|s\rangle = E_s|s\rangle \tag{11.16b}$$

we have

$$\langle s|\rho(t)|s'\rangle = \exp\{i(E_s - E_{s'})(t - t_0)\}\langle s|\rho(t_0)|s'\rangle \tag{11.16c}$$

So for time-independent hamiltonians the density matrix evolution in time involves simply an oscillatory phase factor.

11.4

Example: Spin-$\frac{1}{2}$ Particles

11.4.1 Representation of Density Matrix Using Pauli Spin Matrices

Density methods are particularly appropriate, and popular, for the description of quantal systems that have only a few available states. We shall examine such systems several times in this chapter, using various approximation schemes. Here we examine an ensemble of noninteracting spin-$\frac{1}{2}$ systems. We will show the relationship between measured quantities and the density operator.

It is convenient to work in the representation of Pauli spin operators, which were defined in Chapter 6. Recall that a spin-$\frac{1}{2}$ or two-level system may be described equivalently by creation and destruction operators for the spin states or by the Pauli matrices σ_x, σ_y, σ_z:

$$I = a_\uparrow^+ a_\uparrow + a_\downarrow^+ a_\downarrow \tag{11.17a}$$

$$\sigma_x = a_\uparrow^+ a_\downarrow + a_\downarrow^+ a_\uparrow \tag{11.17b}$$

$$\sigma_y = -i(a_\uparrow^+ a_\downarrow - a_\downarrow^+ a_\uparrow) \tag{11.17c}$$

$$\sigma_z = a_\uparrow^+ a_\uparrow - a_\downarrow^+ a_\downarrow \tag{11.17d}$$

$$I = \begin{pmatrix} 1 & 0 \\ 0 & 1 \end{pmatrix} \tag{11.18a}$$

$$\sigma_x = \begin{pmatrix} 0 & 1 \\ 1 & 0 \end{pmatrix} \tag{11.18b}$$

$$\sigma_y = \begin{pmatrix} 0 & -i \\ i & 0 \end{pmatrix} \tag{11.18c}$$

$$\sigma_z = \begin{pmatrix} 1 & 0 \\ 0 & -1 \end{pmatrix} \tag{11.18d}$$

These obey the commutation rules:

$$[\sigma_x, \sigma_y] = 2i\sigma_z \tag{11.19a}$$

and also

$$\sigma_x^2 = \sigma_y^2 = \sigma_z^2 = I \tag{11.19b}$$

$$\text{Tr}\{\sigma_i \sigma_j\} = 2\delta_{ij} \tag{11.19c}$$

The $\{\sigma_x, \sigma_y, \sigma_z, I\}$ form a linearly independent set of 2×2 matrices, so that any 2×2 can be expanded in this set. We can therefore write

$$\rho = C_0 I + C_x \sigma_x + C_y \sigma_y + C_z \sigma_z \tag{11.20a}$$

with the scalar C's to be determined. Quite clearly, from (11.18) and (11.19c), we have

$$\text{Tr}\{\rho I\} = 2C_0 \tag{11.20b}$$

$$\text{Tr}\{\rho \sigma_i\} = 2C_i = s_i \qquad i = x, y, z \tag{11.20c}$$

This permits a general rewriting as

$$\rho = \frac{1}{2} \begin{pmatrix} 1 + s_z & s_x - is_y \\ s_x + is_y & 1 - s_z \end{pmatrix} \tag{11.20d}$$

$$= \frac{1}{2}(I + \boldsymbol{\sigma} \cdot \mathbf{s}) \tag{11.20e}$$

where σ's are the sets of 2×2 matrices, and, for example, s_x is, from (11.20c), the expectation value of the x spin component.

We now consider a pure state. Then the wavefunction is written

$$|\psi\rangle = d_\alpha |\alpha\rangle + d_\beta |\beta\rangle \tag{11.21a}$$

with $|\alpha\rangle$, $|\beta\rangle$ the usual spin-up, spin-down kets and normalized such that $|d_\alpha|^2 + |d_\beta|^2 = 1$. Then, from (11.20d),

$$\rho = \begin{pmatrix} |d_\alpha|^2 & d_\alpha d_\beta^* \\ d_\alpha^* d_\beta & |d_\beta|^2 \end{pmatrix} \tag{11.21b}$$

(notice that $\mathrm{Tr}\,\rho = 1$, as it must be for a pure state). The magnetization $\mathbf{M}$ is then given by the average value of $\boldsymbol{\mu}$, the magnetic dipole moment operator. Since $\boldsymbol{\mu} = -\frac{1}{2}\gamma\hbar\boldsymbol{\sigma}$, where γ is the gyromagnetic ratio, we have

$$\mathbf{M} = \langle \boldsymbol{\mu} \rangle = \mathrm{Tr}\{\rho(t)\boldsymbol{\mu}\} = \frac{-\gamma\hbar}{2}\,\mathrm{Tr}\{\rho(t)\boldsymbol{\sigma}\} \tag{11.22a}$$

$$-2(\gamma\hbar)^{-1}M_x = d_\alpha d_\beta^* + d_\beta d_\alpha^* \tag{11.22b}$$

$$-2(\gamma\hbar)^{-1}M_y = i(d_\alpha d_\beta^* - d_\beta d_\alpha^*) \tag{11.22c}$$

$$-2(\gamma\hbar)^{-1}M_z = |d_\alpha|^2 - |d_\beta|^2 \tag{11.22d}$$

$$4(\gamma\hbar)^{-2}\mathbf{M}^2 = \frac{4(M_x^2 + M_y^2 + M_z^2)}{(\gamma\hbar)^2} \tag{11.22e}$$

$$= 1 \tag{11.22f}$$

Thus the complete magnetic information about the spin-$\frac{1}{2}$ system is contained either in its wavefunction (11.21a) or in its density matrix (11.21b).

11.4.2 Crude Description of Dephasing in Two-Site Systems

The spin-$\frac{1}{2}$ model can be used to describe a two-site tunneling problem as well as any two-level problem. Suppose that we now consider $|\alpha\rangle$, $|\beta\rangle$ not as spin states but as states localized on sites α and β. Then the potential will be diagonal in $|\alpha\rangle$, $|\beta\rangle$, but they will be mixed by the kinetic energy. We then define eigenstates of the two-site problem by

$$H = T + V_0 \tag{11.23a}$$

$$|+\rangle = \frac{|\alpha\rangle + |\beta\rangle}{\sqrt{2}} \tag{11.23b}$$

$$|-\rangle = \frac{|\alpha\rangle - |\beta\rangle}{\sqrt{2}} \qquad (11.23c)$$

$$H = E_+|+\rangle\langle+| + E_-|-\rangle\langle-| \qquad (11.23c)$$

$$= \tfrac{1}{2}(E_+ + E_-)I + \tfrac{1}{2}(E_+ - E_-)(|\alpha\rangle\langle\beta| + |\beta\rangle\langle\alpha|)$$

$$= \delta\sigma_x \qquad (11.23e)$$

where we have defined the energy origin by $E_+ + E_- = 0$, and where $2\delta = E_+ - E_-$; the last equation follows by consideration of the effects of the Pauli matrices on the state vector $\begin{pmatrix} |\alpha\rangle \\ |\beta\rangle \end{pmatrix}$.

Now suppose that at time t_0 the system is localized in one state, say $|\alpha\rangle$. Then we can find the solution at a later time from the solution to the Liouville equation, and from (11.20e),

$$i\hbar \frac{\partial\rho}{\partial t} = i\hbar\tfrac{1}{2}\boldsymbol{\sigma}\cdot\dot{\mathbf{s}} = [H, \rho] \qquad (11.24a)$$

$$= \delta[\sigma_x, \tfrac{1}{2}\boldsymbol{\sigma}\cdot\mathbf{s}] \qquad (11.24b)$$

$$= \delta\{is_y\sigma_z - i\sigma_y s_z\} \qquad (11.24c)$$

where we have used (11.20e) for the general form of ρ, and the commutation relations (11.19a). Taking the trace respectively with $\sigma_x, \sigma_y, \sigma_z$ as in (11.20c), we have

$$\dot{s}_x = 0 \qquad (11.25a)$$

$$\dot{s}_y = -\frac{2\delta}{\hbar} s_z \qquad (11.25b)$$

$$\dot{s}_z = \frac{2\delta}{\hbar} s_y \qquad (11.25c)$$

The initial condition of localization in $|\alpha\rangle$ means

$$\rho(t_0) = \begin{pmatrix} 1 & 0 \\ 0 & 0 \end{pmatrix} \qquad (11.26a)$$

or

$$s_x(t_0) = 0 \qquad (11.26b)$$

$$s_y(t_0) = 0 \qquad (11.26c)$$

$$s_z(t_0) = 1 \qquad (11.26d)$$

The solution of (11.25) with the initial conditions (11.26) is simply

$$s_x(t) = 0 \qquad (11.27a)$$

$$s_y(t) = -\sin\frac{2\delta(t - t_0)}{\hbar} \qquad (11.27b)$$

$$s_z(t) = \cos \frac{2\delta(t - t_0)}{\hbar} \qquad (11.27c)$$

These results are in accord with (11.16c). They simply state that the density matrix elements will evolve sinusoidally in time, with a phase factor that is simply the difference between two eigenfrequencies. Physically, if the state is originally localized in $|\alpha\rangle$, it will oscillate between $|\alpha\rangle$ and $|\beta\rangle$ with frequency $2\delta/\hbar$.

Now suppose that the Hamiltonian contains an extra term

$$H_1 = \sigma_z v_z(Q_B) \qquad (11.28)$$

that describes a relative change in the energies of $|\alpha\rangle$ and $|\beta\rangle$; here the potential v_z depends on some varible Q_B that is not part of the two-site space—for example, it might be a displacement coordinate in the host environment or bath. Note that if the ratio v_z/δ becomes very small, the $|+\rangle, |-\rangle$ states of (11.23) are the (delocalized) system eigenstates, whereas if δ/v_z becomes small, the localized functions $|\alpha\rangle, |\beta\rangle$ are eigenstates. Thus increasing values of v_z/δ will tend to localize the eigenstates. Often the effect of the H_1 term is called pure dephasing, since the states $|+\rangle, |-\rangle$ are coherent combinations of $|\alpha\rangle, |\beta\rangle$.

Now suppose that the role of the changes in $v_z(Q_B)$, caused by the temporal variations of Q_B, *can be understood as very short collisions that last a time τ_c and are separated by a time τ.* Suppose also that the mixing of $|\alpha\rangle, |\beta\rangle$ by H is relatively weak, so that

$$\hbar\delta^{-1} \gg \tau \gg \tau_c \qquad (11.29)$$

The interpretation is then that the system starts to tunnel from $|\alpha\rangle$ to $|\beta\rangle$ and does so smoothly and uninterruptedly for time τ, building up phase continuously. Then the collision occurs; we assume it to be a perfect dephasing collision, so that the coherence is completely interrupted, after which tunneling begins again.

Using this dephasing model, and the results of (11.27), we see that after time τ,

$$s_x(\tau) = 0$$

$$s_y(\tau) = -\sin 2\delta \frac{\tau}{\hbar} \qquad (11.30)$$

$$s_z(\tau) = \cos 2\delta \frac{\tau}{\hbar}$$

so that the system has started to evolve from $|\alpha\rangle$ towards $|\beta\rangle$, and has developed phase coherence, as shown by the finite value of s_y. Now after the perfect collision at time τ, the phase is completely lost, so that $s_y =$

$\text{Tr}\{\rho\sigma_y\} = 0$, but $s_z = \cos 2\delta\tau/\hbar$. Then the system again starts to tunnel. After N such perfect dephasing, short collisions we find that

$$s_z = \left(\cos 2\delta \frac{\tau}{\hbar}\right)^N \cong \left(1 - \frac{2\delta^2\tau^2}{\hbar^2}\right)^N \cong \exp\left(-2\frac{\delta^2\tau}{\hbar^2} N\tau\right) \quad (11.31a)$$

or since $N\tau = t - t_0$, the total elapsed time,

$$s_z = \exp\left\{\frac{-(t - t_0)}{\tau_R}\right\} \quad (11.31b)$$

with the relaxation time τ_R fixed by

$$\tau_R^{-1} = 2\delta^2 \frac{\tau}{\hbar^2} \quad (11.31c)$$

Notice that, following the physical assumptions (11.29),

$$\tau_R \gg \hbar\delta^{-1} \gg \tau \gg \tau_c \quad (11.32)$$

so that the probability to find the system in $|\alpha\rangle$ is increased to a time far longer than the tunneling period of H, that is, in turn much longer than the time between collisions.

Physically, this behavior can be viewed as a simple loss of coherence: As the system starts to tunnel from $|\alpha\rangle$ to $|\beta\rangle$ following the dynamics of H, the changes in local energy levels caused by H_1 cause loss of phase coherence, thus localizing the system in $|\alpha\rangle$ for times much longer than would have been expected based on the unperturbed tunneling frequency $2\delta/\hbar$. For observation times t_{obs} satisfying

$$\tau_R > \tau_{\text{obs}} \gg \hbar\delta^{-1} \quad (11.33)$$

the behavior of the $|\alpha\rangle$ population, $\frac{1}{2}(1 + s_z)$, becomes a decaying exponential in the presence of dephasing, whereas it would be oscillatory, following (11.27c), in the absence of dephasing.

A more extensive treatment of dephasing is included in Section 11.6.

11.5

Reduced Density Matrices

The density operator of (11.6) is defined in a multivariable space: If $\psi = \psi(\mathbf{x}_1, \mathbf{x}_2, \ldots, \mathbf{x}_n)$, then the operator $|\psi\rangle\langle\psi|$ contains $2n$ space components. For the Be atom, then, the electronic density operator would have eight space components. When we consider a quantum system composed of subsystems that are approximately or completely separable, such as a molecule in a solvent host or the valence and core electrons, it is useful to deal only with that part of the density operator, or density matrix, that is relevant to the subsystem that we wish to study. This leads to the notion of a reduced density matrix or reduced density operator, in which the process of reduction is carried out by integrating over the variables that are not of primary interest.

The reduced density matrices used in electronic structure theory and those used in quantum statistical mechanics and dynamics are conceptually identical, but the notations, methods, and applications associated with these two sorts of reduced density operators are entirely different. We therefore will deal with these subjects separately. Since most of our attention will be centered on statistical and dynamical applications, we deal first with electronic structure aspects; the analysis of dynamical process using density matrices is resumed in Section 11.6.

11.5.1 Reduced Density Matrices and Electronic Structure

The n-particle density matrix in space variables can be written

$$\rho(\mathbf{x}_1, \mathbf{x}_2, \ldots, \mathbf{x}_n; \mathbf{x}_1', \mathbf{x}_2', \ldots, \mathbf{x}_n') = \sum_k p_k \psi_k(\mathbf{x}_1 \cdots \mathbf{x}_n) \psi_k^*(\mathbf{x}_1' \cdots \mathbf{x}_n')$$

(11.34)

If the particles are fermions, then ρ is antisymmetric with respect to interchange of any two primed or unprimed coordinates; for bosons, ρ is symmetric. The matrix is also hermitian, and has trace equal to unity. The expectation value of any observable $\hat{A}$ is then just

$$\langle \hat{A} \rangle = \int d\mathbf{x}_1 \cdots d\mathbf{x}_n \int d\mathbf{x}_1'$$

$$\cdots d\mathbf{x}_n' \prod_{i=1}^n \delta(\mathbf{x}_i - \mathbf{x}_i') \hat{A} \rho(\mathbf{x}_1 \cdots \mathbf{x}_n; \mathbf{x}_1' \cdots \mathbf{x}_n') \quad (11.35)$$

The meaning of this (slightly heavy) notation is that the operator $\hat{A}$ acts only on the unprimed variables in ρ; it is equivalent to $\mathrm{Tr}[A\rho]$ in a continuous basis. After the operation $\hat{A}\rho$ has been completed, the primed variables and unprimed variables are set equal (effect of inner integration and delta functions), and then the integration is completed over unprimed variables.

The important operators in electronic structure theory are, from Chapter 6, of two types: One-electron operators, such as kinetic energy, are written (using $\boldsymbol{\pi}_i$ to represent the momentum corresponding to coordinate $\mathbf{x}_i$)

$$\hat{\theta}(q, p) = \sum_i \hat{\theta}_i(\mathbf{x}_i, \boldsymbol{\pi}_i) \qquad (6.52a)$$

and two-electron operators, including Coulomb repulsion, are

$$\hat{T} = \sum_{i<j} \hat{T}_{ij} \tag{6.52b}$$

Then the expectation value of a one-electron operator is, from (11.35),

$$\langle \theta \rangle = \int d\mathbf{x}_1 \cdots d\mathbf{x}_n \int d\mathbf{x}_1' \cdots d\mathbf{x}_n' \prod_{i=1}^{n} \delta(\mathbf{x}_i - \mathbf{x}_i')\hat{\theta}\rho(\mathbf{x}_1 \cdots \mathbf{x}_n; \mathbf{x}_1' \cdots \mathbf{x}_n')$$
$$\tag{11.36a}$$

Now from its form (6.52a), the operator $\hat{\theta}$ is symmetric with respect to interchange of the labels of any two particles; $\rho(\mathbf{x}_1 \cdots \mathbf{x}_n; \mathbf{x}_1' \cdots \mathbf{x}_n')$ is also symmetric if the interchange occurs in both primed and unprimed variables. Therefore, we can relabel the variables of integration, and it follows that

$$\langle \hat{\theta} \rangle = n \int d\mathbf{x}_1 \cdots d\mathbf{x}_n \int d\mathbf{x}_1'$$

$$\cdots d\mathbf{x}_n' \prod_{i=1}^{n} \delta(\mathbf{x}_i - \mathbf{x}_i')\hat{\theta}(\mathbf{x}_1, \boldsymbol{\pi}_1)\rho(\mathbf{x}_1 \cdots \mathbf{x}_n; \mathbf{x}_1' \cdots \mathbf{x}_n') \quad (11.36b)$$

The integrations over $\mathbf{x}_2 \cdots \mathbf{x}_n$ and $\mathbf{x}_2' \cdots \mathbf{x}_n'$ can then be carried out directly and are independent of the operator $\hat{\theta}$. It is then clear that if we define the one-electron reduced density matrix $\gamma(\mathbf{x}_1; \mathbf{x}_1')$ by

$$\gamma(\mathbf{x}_1; \mathbf{x}_1') \equiv n \int \rho(\mathbf{x}_1, \mathbf{x}_2, \ldots, \mathbf{x}_n; \mathbf{x}_1', \mathbf{x}_2, \mathbf{x}_3, \ldots, \mathbf{x}_n) \, d\mathbf{x}_2 \, d\mathbf{x}_3 \cdots d\mathbf{x}_n$$
$$\tag{11.36c}$$

the expectation value can be reexpressed from (11.35b) and (11.36c) as

$$\langle \theta \rangle = \int d\mathbf{x}_1 \int d\mathbf{x}_1'\delta(\mathbf{x}_1 - \mathbf{x}_1')\hat{\theta}(\mathbf{x}_1, \boldsymbol{\pi}_1)\gamma(\mathbf{x}_1; \mathbf{x}_1') \tag{11.36d}$$

Similarly, the two-electron reduced density matrix is defined by

$$\Gamma(\mathbf{x}_1, \mathbf{x}_2; \mathbf{x}_1', \mathbf{x}_2') \equiv \frac{n(n-1)}{2} \int \rho(\mathbf{x}_1, \mathbf{x}_2, \mathbf{x}_3, \ldots, \mathbf{x}_n;$$

$$\mathbf{x}_1', \mathbf{x}_2', \mathbf{x}_3, \ldots, \mathbf{x}_n) \, d\mathbf{x}_3 \cdots d\mathbf{x}_n \quad (11.37a)$$

and we then have

$$\langle \hat{T} \rangle = \int d\mathbf{x}_1 \, d\mathbf{x}_2 \int d\mathbf{x}_1' \, d\mathbf{x}_2' \, \delta(\mathbf{x}_1 - \mathbf{x}_1')\delta(\mathbf{x}_2 - \mathbf{x}_2')$$
$$\hat{T}(\mathbf{x}_1, \mathbf{x}_2, \boldsymbol{\pi}_1, \boldsymbol{\pi}_2)\Gamma(\mathbf{x}_1, \mathbf{x}_2; \mathbf{x}_1', \mathbf{x}_2') \quad (11.37b)$$

Therefore, knowledge of the reduced density matrices $\gamma(\mathbf{x}_1; \mathbf{x}_1')$ and $\Gamma(\mathbf{x}_1, \mathbf{x}_2; \mathbf{x}_1', \mathbf{x}_2')$ allows us to evaluate the expectation values of one- and two-particle operators, including the Coulomb hamiltonian in electronic structure calculations.

The electronic Born–Oppenheimer total energy for the Coulomb electronic structure problem is just

$$E_{\text{tot}} = \int dx_1 \, dx_2 \int dx_1' \, dx_2' \, \delta(x_1 - x_1')\delta(x_2 - x_2')$$

$$\times \left[\frac{2}{n-1} h_1(x_1, \pi_1) + V(x_1, x_2) \right] \Gamma(x_1, x_2; x_1', x_2') \quad (11.38a)$$

where the one- and two-electron parts are

$$h_1(x_1, \pi_1) = \frac{\pi_1^2}{2m_e} - \sum_\alpha \frac{Z_\alpha e^2}{|x_1 - R_\alpha|} \quad (11.38b)$$

$$V(x_1, x_2) = \frac{e^2}{|x_1 - x_2|} \quad (11.38c)$$

It is clear that just as knowledge of $\rho(x_1 \cdots x_n; x_1' \cdots x_n')$ yields both Γ and γ by integration, so we can obtain γ if Γ is known:

$$\gamma(x_1; x_1') = \frac{2}{n-1} \int dx_2 \int dx_2' \delta(x_2 - x_2')\Gamma(x_1, x_2; x_1', x_2') \quad (11.39)$$

It is also clear that, if desired, we can define reduced density matrices of three-particle, four-particle, . . . , $(n-1)$-particle type by appropriate integration of the full n-particle density matrix $\rho(x_1 \cdots x_n; x_1' \cdots x_n')$.

11.5.2 Hartree–Fock Density Matrices; Natural Orbitals

Since the one- and two-particle reduced density matrices give the expectation values of the one- and two-particle parts of the electronic hamiltonian, they can be used to describe the dominant features of the electronic structure of atoms, molecules, or solids.

The reduced density matrix can also be expressed directly using the occupation number formulation (second quantization) developed in Chapter 6, particularly Section 6.6.2. Since we have, from (11.36d),

$$\langle \hat{\theta} \rangle = \text{Tr}\{\gamma \hat{\theta}\} = \sum_{st} \gamma_{st} \theta_{ts} \quad (11.40a)$$

but also, from (6.55),

$$\langle \hat{\theta} \rangle = \left\langle \sum_{ts} \theta_{ts} a_t^+ a_s \right\rangle \quad (11.40b)$$

It follows by comparison of the last two results that

$$\gamma_{st} = \langle a_t^+ a_s \rangle \quad (11.40c)$$

The order of subscripts in (11.40b) is unexpected, but is consistent both with the definition of the density operator ρ and with the definition of expectation values in terms of traces, given in (11.7).

Up to this point, we have used the subscripts r, s, . . . to denote arbitrary basis functions. We can associate both space and spin variables with each subscript and thus define

$$\gamma_{st}^{\mu} = \langle a_{t\mu}^{+} a_{s\mu} \rangle \tag{11.41}$$

Then the electron number in the sth basis function is

$$n_s = n_s^{\alpha} + n_s^{\beta} = \langle a_{s\alpha}^{+} a_{s\alpha} + a_{s\beta}^{+} a_{s\beta} \rangle = \gamma_{ss}^{\alpha} + \gamma_{ss}^{\beta} \tag{11.42}$$

If, in particular, we interpret the $a_{\lambda\mu}^{+}$ as creating an electron in a molecular orbital ψ_{λ}, and take the orthonormal molecular orbitals in LCAO form

$$\psi_{\lambda} = \sum_{t} C_{\lambda t} u_{t} \tag{11.43a}$$

with u_t an atomic orbital basis function, then

$$\sum_{\lambda} \psi_{\lambda} C_{\lambda s}^{*} = \sum_{t\lambda} C_{\lambda s}^{*} C_{\lambda t} u_{t} = \sum_{t} \delta_{st} u_{t} = u_{s} \tag{11.43b}$$

Therefore, the density matrix in AO representation becomes

$$\gamma_{st}^{\mu} = \langle a_{t\mu}^{+} a_{s\mu} \rangle = \sum_{\lambda\sigma} C_{\lambda t}^{*} C_{\sigma s} \langle a_{\lambda\mu}^{+} a_{\sigma\mu} \rangle$$

$$= \sum_{\lambda\sigma} n_{\sigma}^{\mu} \delta_{\lambda\sigma} C_{\lambda t}^{*} C_{\sigma s} = \sum_{\sigma} n_{\sigma}^{\mu} C_{\sigma t}^{*} C_{\sigma s} = p_{ts}^{\mu} \tag{11.44}$$

with p_{ts}^{μ} the Coulson bond order between the atomic orbitals u_t and u_s, defined in (6.112). The Hartree–Fock equations can then be interpreted as iterative equations to determine γ_{st}^{μ}, as described in Section 6.6.2.

In the representation of molecular orbitals, the Hartree–Fock ground-state ψ_{HF} can be represented as (occ denoting occupied)

$$\psi_{HF} = \prod_{\lambda}^{occ} a_{\lambda\mu}^{+} | vac \rangle \tag{11.45}$$

and therefore, in the MO representation, with the HF ground state, γ^{μ} is diagonal:

$$\gamma_{\lambda\sigma}^{\mu} = \langle \psi_{HF} | a_{\sigma\mu}^{+} a_{\lambda\mu} | \psi_{HF} \rangle = \delta_{\lambda\sigma} n_{\lambda}^{\mu} \tag{11.46}$$

When the many-electron wavefunction is not simply the HF state (e.g., if it is a CI function), then γ^{μ} might not be diagonal. Since, however, $\gamma_{\lambda\sigma}^{\mu}$ is a hermitian matrix, it can always be diagonalized by a unitary transformation. The orbitals that diagonalize the one-particle reduced density matrix for an arbitrary state ψ are called the natural orbitals of the system; they represent the most obvious extension of the molecular orbital concept for a correlated state (they also give the most rapidly convergent CI expansion). If we denote the MOs by ψ_{λ} and the natural orbitals by η_m, we have

$$\psi_\lambda = \sum_k Z^*_{\lambda k} \eta_k \qquad (11.47)$$

and thus

$$\gamma(\mathbf{x}_1, \mathbf{x}'_1) = \sum_{\sigma\mu} \psi_\sigma(\mathbf{x}_1)\gamma_{\sigma\mu}\psi^*_\mu(\mathbf{x}'_1) = \sum_{\sigma\mu kl} Z^*_{\sigma k}\gamma_{\sigma\mu}\eta^*_l \eta_k Z_{\mu l}$$

$$= \sum_{kl} \eta^*_l \eta_k \sum_{\sigma\mu} Z^+_{k\sigma}\gamma_{\sigma\mu}Z_{\mu l} = \sum_{kl} \eta^*_l \eta_k (Z^+\gamma Z)_{kl}$$

$$= \sum_{kl} \eta_k(\mathbf{x}_1)L_{kl}\eta^*_l(\mathbf{x}'_1) \qquad (11.48)$$

Here we have defined

$$L = Z^+\gamma Z \qquad (11.49a)$$

Now we can choose Z to diagonalize L, so that

$$L_{ij} = \delta_{ij}L_i$$

and then

$$\gamma(\mathbf{x}_1, \mathbf{x}'_1) = \sum_k L_k\eta_k(\mathbf{x}_1)\eta^*_k(\mathbf{x}'_1) \qquad (11.49b)$$

The value of $0 \le L_k \le 1$ is the occupation number of the natural orbital. In the HF limit, L_k is zero or unity, and the natural orbitals η_k become the MO's ψ_λ.

11.6

Reduced Density Matrices for Dynamical Statistical Systems

The reduction schemes defined in (11.36c) and (11.37a) apply to a situation in which reduction of many-particle properties to fewer-particle properties (i.e., the reduction of the *n*-particle to the one-particle or two-particle density matrix) is complicated by the antisymmetry requirements of the Pauli principle. The factors n and $n(n-1)/2$ arise from indistinguishability of particles. Another situation of substantial importance and generality occurs when a physical system consists of a subsystem of primary interest interacting with another system, with which it can exchange energy. For example, one might be interested in studying a molecule in a solvent, or a nuclear spin in a solid, or a chromophore group in a molecule. In such situations the overall density matrix describing the interacting subsystems can yield a reduced density matrix for the subsystem of interest, and the behavior of that reduced density matrix can be solved to deduce the desired information or experimental predictions.

In this section we define subsystem reduced density matrices and write their equations of motion using a generalized interaction representation. We then develop perturbation-theoretic methods for obtaining them, and finally, examine particular cases of spin dynamics using these methods.

11.6.1 Reduced Density Matrices for Subsystems

Suppose that the hamiltonian for the problem of interest can be written

$$H = H_s + H_b + V \tag{11.50}$$

with the three terms on the right, respectively, the hamiltonians of the subsystem of interest, the bath (or other subsystem) with which the primary subsystem interacts, and the coupling between subsystems. The overall density operator $\rho(t)$ evolves according to the Liouville equation

$$i\hbar \frac{\partial \rho(t)}{\partial t} = [H, \rho(t)] \tag{11.51}$$

Just as we did for the time dependence of operators and wavefunctions in Chapter 4, so here we can separate the time evolution due to $H_s + H_b$ from that due to V. We thus define the interaction representation density operator, $\rho^I(t)$, by

$$\rho^I(t) \equiv e^{iH_0 t/\hbar} \rho(t) e^{-iH_0 t/\hbar} \tag{11.52}$$

with $H_0 = H_s + H_b$. Then the Liouville equation becomes

$$i\hbar \frac{\partial \rho^I}{\partial t} = i\hbar \left\{ \frac{i}{\hbar} [H_0, \rho^I] + e^{iH_0 t/\hbar} \frac{\partial \rho}{\partial t} e^{-iH_0 t/\hbar} \right\}$$

$$= i\hbar \left\{ \frac{i}{\hbar} [H_0, \rho^I] - \frac{i}{\hbar} e^{iH_0 t/\hbar} [H_0 + V, \rho] e^{-iH_0 t/\hbar} \right\}$$

$$i\hbar \frac{\partial \rho^I(t)}{\partial t} = [V^I, \rho^I] \tag{11.53}$$

Here we have defined

$$V^I = e^{iH_0 t/\hbar} V e^{-iH_0 t/\hbar} \tag{11.54}$$

Physically, Eq. (11.53) states that the density operator in the interaction representation evolves in time due only to the interaction V^I.

Notationally, we can define so-called liouvillian operators $\mathcal{L}$ and $\mathcal{L}_I$ by

$$\mathscr{L}S \equiv \frac{1}{\hbar}[H, S] \qquad (11.55a)$$

$$\mathscr{L}_I(t)S \equiv \frac{1}{\hbar}[V^I(t), S] \qquad (11.55b)$$

These might be, and occasionally are, called superoperators because they operate not on wavefunctions but rather on other operators. Using this notation, we rewrite the equations of motion as

$$\frac{\partial \rho(t)}{\partial t} = -i\mathscr{L}\rho(t) \qquad (11.55c)$$

$$\frac{\partial \rho^I(t)}{\partial t} = -i\mathscr{L}_I\rho^I(t) \qquad (11.55d)$$

A formal solution of the last equation can be obtained by simple integration as

$$\rho^I(t) = \rho^I(0) - i\int_0^t dt_1 \mathscr{L}_I(t_1)\rho^I(t_1) \qquad (11.56)$$

(This can be checked by differentiation.)

This can then be iterated, to yield a series reminiscent of Eq. (4.29):

$$\rho^I(t) = \rho^I(0) - i\int_0^t dt_1 \, \mathscr{L}_I(t_1)\rho^I(0)$$

$$+ (-i)^2\int_0^t dt_1 \int_0^{t_1} dt_2 \, \mathscr{L}_I(t_1)\mathscr{L}_I(t_2)\rho^I(0) + \cdots \qquad (11.57)$$

Solution of this equation gives a perturbative development for $\rho^I(t)$.

Now we define a reduced density operator for the subsystem as

$$\tilde{\rho}^I(t) \equiv \mathrm{Tr}_b \, \rho^I(t) \qquad (11.58)$$

where the Tr_b means that a partial trace is performed over the states of the bath. Then assuming that initially the bath was unperturbed (that is, that at $t = 0$ there is no correlation between system and bath), we can write, at $t = 0$,

$$\rho^I(t = 0) = \rho_b^{(0)}\tilde{\rho}^I(0) \qquad (11.59a)$$

Here $\rho_b^{(0)}$ denotes the equilibrium bath density operator, which is simply

$$\rho_b^{(0)} = \frac{e^{-\beta H_b}}{\mathrm{Tr}_b \, e^{-\beta H_b}} \qquad (11.59b)$$

for a thermal bath (with $\beta = 1/k_B T$). Performing the analogous trace operation to (11.58) on the perturbation series (11.57), we find that

$$\tilde{\rho}^I(t) = \tilde{\rho}^I(0) - i \int_0^t dt_1 \, \text{Tr}_b(\mathcal{L}_I(t_1)\rho_b^{(0)})\tilde{\rho}^I(0)$$

$$+ (-i)^2 \int_0^t dt_1 \int_0^{t_1} dt_2 \, \text{Tr}_b(\mathcal{L}_I(t_1)\mathcal{L}_I(t_2)\rho_b^{(0)})\tilde{\rho}^I(0) + \cdots \quad (11.60)$$

This, in turn, can be rewritten as an evolution equation

$$\tilde{\rho}^I(t) = U(t)\tilde{\rho}^I(0) \quad (11.61)$$

with the generalized evolution operator

$$U(t) = 1 - i \int_0^t dt_1 \, \text{Tr}_b(\mathcal{L}_I(t_1)\rho_b^{(0)})$$

$$+ (-i)^2 \int_0^t dt_1 \int_0^{t_1} dt_2 \, \text{Tr}_b(\mathcal{L}_I(t_1)\mathcal{L}_I(t_2)\rho_b^{(0)}) + \cdots \quad (11.62a)$$

$$= \sum_{n=0}^{\infty} \frac{M^{(n)}}{n!} \quad (11.62b)$$

with the generalized moments

$$M^{(n)} = (-i)^{(n)}n! \int_0^t dt_1 \int_0^{t_1} dt_2 \cdots \int_0^{t_{n-1}} dt_n \, \text{Tr}_b(\mathcal{L}_I(t_1)\mathcal{L}_I(t_2) \cdots \mathcal{L}_I(t_n)\rho_b^{(0)}) \quad (11.63)$$

Due to the coupling with the system, the evolution operator U involves the behavior due to the bath term $\rho_b^{(0)}$, arising from the coupling with the system. Taking derivatives in (11.61), we find that

$$\dot{\tilde{\rho}}^I(t) = \dot{U}(t)\tilde{\rho}^I(0)$$

$$= \dot{U}(t)U^{-1}\tilde{\rho}^I(t) \quad (11.64a)$$

$$\dot{\tilde{\rho}}^I = \hat{R}(t)\tilde{\rho}^I(t)$$

where we have defined

$$\hat{R}(t) \equiv \dot{U}(t)U^{-1}(t) \quad (11.64b)$$

The system density matrix ρ^I will generally be described by two indices, so that $\hat{R}$ is a fourth-rank tensor, and in component form (11.64a) is written

$$\dot{\tilde{\rho}}^I_{\alpha\alpha'} = \sum_{\beta\beta'} \hat{R}_{\alpha\alpha';\beta\beta'}(t)\tilde{\rho}^I_{\beta\beta'}(t) \quad (11.64c)$$

Note that this result is exact if U is evaluated to all orders in (11.62a). Other formal methods, including projection operator techniques, can be used to obtain the same result.

11.6.2 Zero-Order Density Matrix: Equilibrium

If we consider the zero-order term in (11.57), we find that $\rho^{(0)}(t) = \rho^{(0)}(t = 0)$, so that $\rho^{(0)}(t) = $ constant. Then by comparison of the result (11.7) with that from simple Boltzmann statistics, we find that

$$\rho^{(0)I}(t) = \rho^{(0)I} = \rho^{(0)} = \frac{e^{-\beta H_0}}{\text{Tr } e^{-\beta H_0}} \tag{11.65}$$

This result simply states that the zero-order density operator gives the Boltzmann weighting at equilibrium.

11.6.3 First-Order Expressions for the Density Matrix: Linear Response Theory

Most uses of density matrix methods are indeed for systems characterized by weak interactions between subsystems. In such situations the mixing term V in (11.50) is in some sense small, and it is therefore appropriate to treat (11.57) as a perturbation series. We will therefore develop explicitly formulas for two special cases of real importance. In this subsection we consider the situation in which the "bath" is an applied field. Then use of the density formalism permits writing the system behavior solely in terms of the field-free dynamics of the system itself; this is the so-called linear response scheme. In the next subsection we extend this to consider behavior to second order in the interactions: This permits discussion of relaxation phenomena and generalizations of simple rate equations.

One of the most important situations in experimental study of molecular and solid-state systems involves the response of the system to a time-dependent spatial potential. In Chapters 5 and 9 we studied this problem using the golden rule and time-dependent perturbation theory of the wavefunction. Here we consider a slightly more general approach based on the perturbed density.

To be specific, we assume the system to be perturbed by a time-dependent electric field $\mathbf{E}(\omega)$, and wish to calculate the conductivity tensor $g_{\mu\nu}(\omega)$. It is related to the current density $\mathbf{j}$ by the linear relationship

$$\langle \mathbf{j}(\omega) \rangle = \mathbf{g}(\omega)\mathbf{E}(\omega) \tag{11.66a}$$

or in components

$$\langle j_\mu(\omega) \rangle = \sum_\nu g_{\mu\nu}(\omega)E_\nu(\omega) \tag{11.66b}$$

The current density expectation value is in turn given by

$$\langle \mathbf{j}(\omega) \rangle = \text{Tr}\{\rho(\omega)\mathbf{j}\} \tag{11.66c}$$

To find an approximate solution, we assume that a sinusoidal perturbation is turned on adiabatically starting at $t = -\infty$:

$$H = H_0 + V = H_0 + \lim_{\alpha \to 0} e\mathbf{E} \cdot \mathbf{r} \, \exp(-i\omega t + \alpha t) \qquad (11.67a)$$

and expanding ρ to first order:

$$\rho^I = \rho^{(0)I} + \rho^{(1)I} \qquad (11.67b)$$

Now we examine the Liouville equation in the interaction representation, (11.53), to first order:

$$i\hbar\dot{\rho}^{(1)I} = [V^I, \rho^{(0)I}] \qquad (11.68a)$$

Integrating from $-\infty$ to $t = 0$,

$$i\hbar\rho^{(1)I}(0) = \int_{-\infty}^{0} d\tau \, e^{iH_0\tau/\hbar}[e\mathbf{E} \cdot \mathbf{r}, \rho^{(0)}]e^{-iH_0\tau/\hbar}e^{-i\omega\tau}e^{\alpha\tau} \qquad (11.68b)$$

where $\rho^{(1)I}(-\infty) = 0$ (because $V = 0$ at $\tau = -\infty$). But by the definition (11.52), $\rho^I(0) = \rho(0)$, so that we have

$$\rho^{(1)}(0) = \frac{1}{i\hbar}\int_{-\infty}^{0} d\tau \, e^{iH_0\tau/\hbar}e\mathbf{E} \cdot [\mathbf{r}, \rho^{(0)}]e^{-iH_0\tau/\hbar}e^{-i\omega\tau}e^{\alpha\tau} \qquad (11.68c)$$

Then, from (11.66c),

$$\langle \mathbf{j}(\omega) \rangle = \text{Tr}\{\mathbf{j}\rho^{(1)}(0)\}$$

$$= \lim_{\alpha \to 0} \frac{1}{i\hbar}\int_{-\infty}^{0} d\tau \, e^{+\alpha\tau}e^{-i\omega\tau}\text{Tr}\{\mathbf{j}e^{iH_0\tau/\hbar}e\mathbf{E} \cdot [\mathbf{r}, \rho^{(0)}]e^{-iH_0\tau/\hbar}\} \qquad (11.69)$$

Then, by comparison with (11.66a), we have

$$g_{\mu\nu}(\omega) = \lim_{\alpha \to 0}\int_{-\infty}^{0} d\tau \, e^{-i\omega\tau}e^{\alpha\tau}\hat{\kappa}_{\mu\nu}(\tau) \qquad (11.70)$$

$$\hat{\kappa}_{\mu\nu}(\tau) \equiv \frac{1}{i\hbar}\text{Tr}\{j_{\mu}e^{iH_0\tau/\hbar}[er_{\nu}, \rho^{(0)}]e^{-iH_0\tau/\hbar}\} \qquad (11.71a)$$

It is more convenient to define $t = -\tau$, so that we have

$$g_{\mu\nu}(\omega) = \lim_{\alpha \to 0}\int_{0}^{\infty} dt \, e^{i\omega t}e^{-\alpha t}\kappa_{\mu\nu}(t) \qquad (11.71b)$$

where

$$\kappa_{\mu\nu}(t) = \hat{\kappa}_{\mu\nu}(-t) = \frac{1}{i\hbar}\text{Tr}\{e^{iH_0t/\hbar}j_{\mu}e^{-iH_0t/\hbar}[er_{\nu}, \rho^{(0)}]\} \qquad (11.71c)$$

An alternative form arises from $-e\dot{r}_{\nu} = j_{\nu} \cdot$ volume and from the identity (Problem 4)

$$\frac{[er_\nu, \rho^{(0)}]}{\text{volume}} = i\hbar\rho^{(0)} \int_0^{1/k_BT} d\lambda \, \exp(\lambda H_0)j_\nu \exp(-\lambda H_0) \qquad (11.72)$$

as

$$\kappa_{\mu\nu}(t) = \int_0^{1/k_BT} d\lambda \langle j_\nu(-t - i\hbar\lambda)j_\mu\rangle \text{volume} \qquad (11.73)$$

The results (11.73), or (11.71c) combined with (11.71b), are often referred to as the linear response result, or the Kubo formula, for $g_{\mu\nu}(\omega)$. Rewriting this once more, the Kubo formula is

$$\frac{g_{\mu\nu}(\omega)}{\text{volume}} = \lim_{\alpha\to 0} \int_0^\infty dt \, e^{i\omega t}e^{-\alpha t} \int_0^{1/k_BT} d\lambda \, \langle j_\nu(-t - i\hbar\lambda)j_\mu(0)\rangle \qquad (11.74)$$

Several important comments might be made about this Kubo formula. First, as is stressed in Problem 3, it is the correct generalization of the correlation function schemes that were discussed in Chapters 9 and 10. Second, it incorporates both quantum mechanics and statistics (from the density operator). Third, if we have the more general form for the perturbation hamiltonian

$$H_1 = -\sum_j A_j E_j(t) \qquad (11.75)$$

with A_j a system property and $E_j(t)$ a time-dependent field component, then the arbitrary current $\dot{B}_j = J_j$ is related to the transport coefficient $C_{jl}(\omega)$ by

$$\langle J_j(\omega)\rangle = C_{jl}(\omega)E_l(\omega) \qquad (11.76)$$

and then the transport coefficients are given by

$$C_{jl}(\omega) = \lim_{\alpha\to 0} \int_0^\infty dt \, e^{-\alpha t} \, e^{i\omega t} \int_0^{1/k_BT} d\lambda\langle e^{\lambda H_0}\dot{A}_l e^{-\lambda H_0}\dot{B}_j(t)\rangle \qquad (11.77)$$

For example, one can thus calculate electrical conductivity, magnetic or electric susceptibility, viscosity, neutron scattering cross section, diffusion coefficient, absorption coefficient, and (as we saw in Chapter 8), reaction rate constants. Even the special form for the frequency-dependent conductivity, (11.74), is useful in chemistry. Of course, it applies directly if one wishes to find the conductivity of solid-state materials, but it is also directly related to the optical properties via the relationships

$$\varepsilon(\omega) = 1 + \frac{4\pi i}{\omega} \sigma(\omega) \qquad (11.78)$$

$$\varepsilon = \varepsilon_1 + i\varepsilon_2 = (n + ik)^2$$

$$k = \frac{\varepsilon_2}{2n}$$

These arise from Maxwell's equations. The tensor quantities $\varepsilon(\omega)$ and $\sigma(\omega)$ are, respectively, the complex dielectric function and conductivity, while n and k are, respectively, the refractive index and extinction coefficient. The latter relates the incident intensity $\mathscr{E}_z^0$ of an electromagnetic field along the z axis to its value at another point:

$$\mathscr{E}_z(z) = \mathscr{E}_z^0 e^{i\omega(t-nz/c)} e^{-\omega kz/c} \tag{11.79}$$

with c the speed of light. Thus knowledge of $\sigma(\omega)$, from (11.74), permits computation of the extinction coefficient.

11.6.4 Second-Order Response and the Density Matrix: Redfield Equations and Relaxation Processes

In such situations as nuclear spin dynamics in molecules, or vibrational dynamics of a particular chromophore, or polarization relaxation in a dielectric, the coupling V of (11.50) is generally taken (to lowest order at least) as a linear function of the coordinates of the "bath" subsystem. In such cases, the trace operation over the bath, which occurs in each term of (11.62a), means that all odd-order terms (those with an odd number of $\mathscr{L}$ operators) will vanish by symmetry. (Recall that $\langle x^n \rangle = 0$ for odd n in a harmonic oscillator system.) In such cases the lowest nonvanishing terms in the perturbation expansion of (11.62a) are zeroth- and second-order terms.

We will then consider a situation in which the bath, described by H_b of (11.50), is simply a set of harmonic oscillators. These could, for instance, be the normal modes of a molecule or the phonons of a crystal host. Then we can write

$$H_s = \sum_{x,y} |x\rangle\langle y| h_{xy}^{(s)} \tag{11.80}$$

$$H_b = \sum_q \hbar\omega_q (b_q^+ b_q + \tfrac{1}{2}) \tag{11.81a}$$

$$\Lambda_q = \omega_q (b_q^+ + b_q) \tag{11.81b}$$

$$V = \sum_q \sum_{x,y} |x\rangle\langle y| K_{xy}^q \Lambda_q \hbar \tag{11.82}$$

with ω_q the frequency of the qth oscillator, in which a quantum is created by the b_q^+ operator. The system states $|x\rangle, |y\rangle$ are mixed by the V operator, and K_{xy}^q is a dimensionless coupling constant, similar to those of Chapter 10.

Rewriting (11.64), we have for the reduced density operator of the system

$$\dot{\rho}^I(t) = \hat{R}(t)\tilde{\rho}^I(t) \qquad (11.83a)$$

$$\hat{R}(t) = \dot{U}(t)U^{-1}(t) \qquad (11.83b)$$

Since we have the explicit development of (11.62a), the operators $\dot{U}$ and U^{-1} can each be written directly as a power series in V^I. To lowest order, we can neglect the possible noncommutation of $\dot{U}$ and U^{-1}, and therefore (11.64b) becomes

$$\hat{R}(t) = \frac{\partial}{\partial t} \ln U(t) \qquad (11.84)$$

From (11.62a), we have (to second order)

$$U(t) \cong 1 + \tfrac{1}{2} M^{(2)} \qquad (11.85a)$$

$$\hat{R}(t) = \frac{\partial}{\partial t} \ln\left(1 + \frac{1}{2} M^{(2)}\right)$$

$$= \frac{\partial}{\partial t}\left(\frac{1}{2} M^{(2)}\right) = \frac{1}{2} \dot{M}^{(2)}(t)$$

$$= \frac{1}{2}\frac{\partial}{\partial t}\left\{(-i)^2 \cdot 2 \cdot \int_0^t dt_1 \int_0^{t_1} dt_2\, Tr_b\left(\mathcal{L}_I(t_1)\mathcal{L}_I(t_2)\rho_b^{(0)}\right)\right\} \qquad (11.85b)$$

$$= -\int_0^t dt_2\, Tr_b(\mathcal{L}_I(t)\mathcal{L}_I(t_2)\rho_b^{(0)}) \qquad (11.85c)$$

Combining the formal expressions for $\dot{\rho}^I$ and $\hat{R}(r)$, Eqs. (11.83a) and (11.85c), we can rewrite the equation of motion of the system density to second order, in component form, as

$$\hbar^2\, \dot{\tilde{\rho}}^I_{\alpha\alpha'} = -\sum_{\beta\beta'}\int_0^t dt_2\, Tr_b\left\{\sum_\gamma V^I_{\alpha\gamma}(t_2)V^I_{\gamma\beta}(t)\rho_b\tilde{\rho}^I_{\beta\alpha'}\right.$$

$$+ \sum_\gamma \tilde{\rho}^I_{\alpha\beta'}\rho_b V^I_{\beta'\gamma}(t)V^I_{\gamma\alpha'}(t_2)$$

$$\left. - V^I_{\alpha\beta}(t)\rho_b V^I_{\beta'\alpha'}(t_2)\tilde{\rho}^I_{\beta\beta'} - V^I_{\alpha\beta}(t_2)\rho_b V^I_{\beta'\alpha'}(t)\tilde{\rho}^I_{\beta\beta'}\right\} \qquad (11.86)$$

With particular assumptions about the heat bath, these equations can be written in a far more succinct fashion. Defining

$$C_{qq'}(t,t_2) \equiv \hbar^2\langle\Lambda^I_q(t_2)\Lambda^I_{q'}(t)\rangle = \hbar^2\, Tr_b\{\rho_b\Lambda^I_q(t_2)\Lambda^I_{q'}(t)\} \qquad (11.87a)$$

We assume that this correlation function is time invariant and symmetric, so that

$$C_{qq'}(t,t_2) = C_{qq'}(t_2 - t) = C_{qq'}(t - t_2) \qquad (11.87b)$$

We also define weighted correlation functions by

$$G_{ijkl}(t') = \sum_{qq'} K^q_{lk}\, K^{q'}_{ij}\, C_{qq'}(t') \qquad (11.88)$$

with $t' = t - t_2$. Then the equation for $\dot{\rho}^I$ can be written, after some algebra, as (with ε_α the eigenvalue of H_s)

$$\hbar^2 \dot{\rho}^I_{\alpha\alpha'}(t) = \sum_{\beta\beta'} \exp\left\{\frac{it}{\hbar}[\varepsilon_\alpha - \varepsilon_\beta - \varepsilon_{\alpha'} + \varepsilon_{\beta'}]\right\}$$

$$\int_0^t dt' \left\{-\sum_\gamma G_{\gamma\beta\gamma\alpha}(t')\delta_{\alpha'\beta'}\exp[i\omega_{\gamma\alpha'}t']\right.$$

$$- \sum_\gamma G_{\gamma\alpha'\alpha\beta'}(t')\delta_{\alpha\beta}\exp[i\omega_{\alpha'\gamma}t']$$

$$+ G_{\alpha\beta\alpha'\beta'}(t')\exp[i\omega_{\alpha'\beta'}t']$$

$$\left.+ G_{\alpha\beta\alpha'\beta'}(t')\exp[i\omega_{\beta\alpha}t']\right\}\dot{\rho}^I_{\beta\beta'}(t) \qquad (11.89)$$

Our treatment to this point has been formal, but for further progress it is necessary to make a crucial physical argument. The bath motions that enter into the correlation function $C_{qq'}(t')$ become independent of one another, on average, when the time t' exceeds a typical correlation time τ_c of the bath. Such typical behavior of the correlations is sketched in Fig. 11.1. If this condition holds, then the magnitudes of all the integrals in (11.89) will be very small for times $t' \gg \tau_c$. This means that the upper limit in (11.89) can be taken to infinity. Defining spectral densities by

$$J_{\alpha\alpha'\beta\beta'}(\omega) \equiv \int_0^\infty dt' e^{i\omega t'} G_{\alpha\alpha'\beta\beta'}(t') \qquad (11.90)$$

We can rewrite the second-order result for the system reduced density operator as

$$\dot{\rho}^I_{\alpha\alpha'}(t) = \sum_{\beta\beta'} \exp\left\{\frac{it}{\hbar}[\varepsilon_\alpha - \varepsilon_\beta - \varepsilon_{\alpha'} + \varepsilon_{\beta'}]\right\} R_{\alpha\alpha';\beta\beta'}\dot{\rho}^I_{\beta\beta'}(t) \quad (11.91)$$

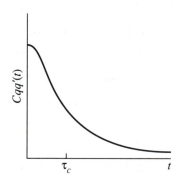

Figure 11.1 Typical correlation function.

with the definition of the four-index (tetradic) relaxation matrix

$$\hbar^2 R_{\alpha\alpha';\beta\beta'} \equiv J_{\alpha\beta\alpha'\beta'}(\omega_{\alpha'\beta'}) + J_{\alpha\beta\alpha'\beta'}(\omega_{\alpha\beta})$$

$$- \sum_\gamma J_{\gamma\beta\gamma\alpha}(\omega_{\alpha\gamma})\delta_{\alpha'\beta'} - \sum_\gamma J_{\gamma\alpha'\gamma\beta'}(\omega_{\alpha'\gamma})\delta_{\alpha\beta} \quad (11.92)$$

Equation (11.91) is generally called the Redfield equation; it was first derived in the context of magnetic resonance, but has been subsequently used extensively in a number of other areas of condensed-phase chemical physics. We will give a simple application in the next subsection, but first we should make a few general remarks. First, notice that the Redfield equation is not a generally valid result for problems with weak coupling between subsystems: It depends crucially on the timescale separation that permitted the extension of the upper limit on the integrals in Eq. (11.89) to infinity. Thus we generally expect the Redfield equations to be valid if V is in some sense small, and if the following condition is satisfied:

$$\frac{1}{R_{\alpha\alpha';\beta\beta'}} \gg \tau \gg \tau_c \quad (11.93)$$

where τ is the upper limit in (11.89). The $R_{\alpha\alpha';\beta\beta'}$ is usually called the Redfield relaxation matrix; the equations are valid (or, at least, our derivation is appropriate) only if the density matrix elements do not change significantly during times of order τ. The coupling V is almost always that between system and bath; thus the Redfield formulation is ideal for examining the effects of heat baths (solvents) on chemical systems.

A second point concerns the relative sizes of differing terms in the relaxation matrix $R_{\alpha\alpha';\beta\beta'}$. There are then two classes of terms. If the energy match condition

$$\varepsilon_\alpha - \varepsilon_{\alpha'} + \varepsilon_{\beta'} - \varepsilon_\beta = 0 \quad (11.94a)$$

holds, then the $R_{\alpha\alpha';\beta\beta'}$ connects elements of the density matrix that have the same unperturbed frequency, while if it does not hold, the elements $\tilde{\rho}^I_{\alpha\alpha'}$ and $\tilde{\rho}^I_{\beta\beta'}$ have different frequencies. One can then anticipate by analogy with the arguments raised in Chapter 4 in connection with the golden rule derivation that the terms for which (11.94a) holds should be much larger than those for which it does not hold. To be more precise, for time intervals $\Delta t \gtrsim \hbar/[\varepsilon_\alpha - \varepsilon_{\alpha'} + \varepsilon_{\beta'} - \varepsilon_\beta]$, terms for which the condition (11.94a) does not hold will undergo a number of oscillations in time, and therefore their average effect on the evolution of $\tilde{\rho}^I_{\alpha\alpha'}$, will tend to average out. This is not true for terms for which (11.94a) holds. This argument really is valid only if

$$\hbar R_{\alpha\alpha'\beta\beta'} \ll \varepsilon_\alpha - \varepsilon_{\alpha'} + \varepsilon_{\beta'} - \varepsilon_\beta \quad (11.94b)$$

This is the actual condition under which particular $R_{\alpha\alpha'\beta\beta'}$ terms can be neglected. Frequently, this is equivalent to the neglect of all terms for which the degeneracy condition (11.94a) holds, but not always. Terms for which the condition (11.94b) holds are called nonsecular terms and can be ignored in calculating the evolution of $\tilde{\rho}^I_{\alpha\alpha'}$. Terms for which (11.94b) does not hold, including degeneracy situations for which (11.94a) is valid, are called secular perturbations and should be retained. Often the neglect of terms for which (11.94a) does not hold is referred to as the rotating wave approximation, because these terms have an extra rotating phase component due to the exponentials in (11.91). Although this is usually valid, the more general condition based on (11.94b) is really preferable (or one can retain the small nonsecular terms).

A third important point is that while the golden rule is used to study $|C_\alpha(t)|^2$, the probability that at time t the system will be in one of a complete set of states indexed by α, the Redfield equations contain significantly more information. They describe not only the time evolution of the populations (relaxation) that are the diagonal elements of $\tilde{\rho}^I$, but also of the transition rates that are the off-diagonal elements of $\tilde{\rho}^I$. This is important, for example, in the decay of coherence in quantal systems and in the special example of dephasing. This will be shown explicitly in the next section.

A fourth point is the nature of these relaxation equations: time-inversion symmetry is present in the Schrödinger, Heisenberg, and Newton equations, but is gone from the Redfield results: time moves forward as relaxation occurs. This irreversibility enters because we average over the bath, which is assumed to have multiple degrees of freedom in which phase memory is lost. In this sense, both relaxation and dephasing enter our description only because we seek information about a subsystem of interest—the actual total system dynamics fixed by H of Eq. (11.50) is multiply periodic and (on a *very* long time scale) does not show relaxation or dephasing. On any experimental timescale of interest, though, dephasing and relaxation will occur.

A fifth and very important point is that while the Redfield equations (11.91) hold under a wide set of conditions (even to powers of V^I beyond second), the specific form (11.92) holds only if the symmetrization assumption (11.87b) is valid. For many situations, such as that of the quantum boson bath, the assumption (11.87b) does not hold, so that, while (11.91) is correct, (11.92) is not, and the relaxation matrix $R_{\alpha\alpha';\beta\beta'}$ must be evaluated directly from (11.85). Then the time-independent relaxation matrix is defined from $\hat{R}(t)$ [from (11.91) and (11.83a)]

$$R_{\alpha\alpha';\beta\beta'} \equiv \hat{R}_{\alpha\alpha';\beta\beta'} \exp\left[\frac{-i(\varepsilon_\alpha + \varepsilon_{\beta'} - \varepsilon_\beta - \varepsilon_{\alpha'})t}{\hbar}\right] \qquad (11.95)$$

We now examine just such a situation.

11.6.5 Example: Relaxation and Dephasing in a Simple Spin System

To illustrate the power of density matrix methods, we will examine the relaxation dynamics of a simple spin, or two-level, system interacting with a boson bath of oscillators. The hamiltonian is then that of Eqs. (11.80) and (11.81), and we anticipate that the bath correlation times are sufficiently short that the Redfield equations (11.91) hold. We then wish to examine how an initial ensemble decays.

If the energy of one state, called $|0\rangle$, in our two-level system is defined to be zero, we have

$$H_s = E_1 |1\rangle\langle 1| \qquad (11.96a)$$

$$V = \hbar(\Lambda|0\rangle\langle 1| + \Lambda^+|1\rangle\langle 0|) \qquad (11.96b)$$

$$\Lambda = \sum_k h_k(b_k^+ + b_k) \qquad (11.96c)$$

with E_1 and h_k, respectively the energy of state $|1\rangle$ and a coupling frequency. The hamiltonian specified by (11.96) is very closely related to the two-site system for electron transfer that we studied in Chapter 10. Indeed, defining the H_s eigenstates in terms of local states $|l\rangle$, $|r\rangle$ as

$$|0\rangle = \frac{1}{\sqrt{2}} (|l\rangle + |r\rangle)$$
$$\qquad (11.97)$$
$$|1\rangle = \frac{1}{\sqrt{2}} (|l\rangle - |r\rangle)$$

we can rewrite the hamiltonian as

$$H_s = -\tfrac{1}{2}E_1 \{|l\rangle\langle r| + |r\rangle\langle l|\} + \tfrac{1}{2}E_1 \qquad (11.98a)$$

$$V = \frac{\hbar}{2} (\Lambda + \Lambda^+)(|l\rangle\langle l| - |r\rangle\langle r|) + \frac{\hbar}{2} (\Lambda - \Lambda^+)(|l\rangle\langle r| - |r\rangle\langle l|) \qquad (11.98b)$$

Thus if rewritten in terms of local site operators $|r\rangle$, $|l\rangle$ rather than state operators $|1\rangle$ and $|0\rangle$, the hamiltonian consists of a tunneling term (11.98a) and a term coupling linear solvent displacements, Λ, to diagonal and nondiagonal local operators. This is just what we had for electron transfer in Chapter 10 if Λ is hermitian.

In the interaction representation, we find

$$V^I(t) = \hbar\{\Lambda^I(t) \, |1\rangle\langle 0| + \Lambda^{I+}(t) \, |0\rangle\langle 1|\} \tag{11.98c}$$

with

$$\Lambda^I(t) \equiv e^{iE_1 t/\hbar}\Lambda(t) \tag{11.98d}$$

We have chosen to write $\Lambda(t)$ in real form. Then the Redfield equation for the diagonal elements of the reduced density operator is

$$\dot{\tilde\rho}_{11}(t) = R_{1100}\tilde\rho_{00}(t) + R_{1111}\tilde\rho_{11}(t) \tag{11.99a}$$

$$\dot{\tilde\rho}_{00}(t) = R_{0000}\tilde\rho_{00}(t) + R_{0011}\tilde\rho_{11}(t) \tag{11.99b}$$

while the off-diagonal terms become (defining $\hbar\omega_{ij} = E_i - E_j$)

$$\dot{\tilde\rho}_{01} = (i\omega_{10} + R_{0101})\tilde\rho_{01} + R_{0110}\tilde\rho_{10} \tag{11.99c}$$

$$\dot{\tilde\rho}_{10} = (-i\omega_{10} + R_{1010})\tilde\rho_{10} + R_{1001}\tilde\rho_{01} \tag{11.99d}$$

We deal first with the evolution of populations. Adding (11.99a,b), and remembering that $\tilde\rho_{00} + \tilde\rho_{11} = 1$, because ρ_{11} is the population in the $|1\rangle$ state, we find that

$$R_{0000} = -R_{1100} \equiv -k_{10} \tag{11.100a}$$

$$R_{1111} = -R_{0011} \equiv -k_{01} \tag{11.100b}$$

and therefore the populations follow a simple master equation form

$$\dot{\tilde\rho}_{11} = k_{10}\tilde\rho_{00} - k_{01}\tilde\rho_{11} \tag{11.100c}$$

$$\dot{\tilde\rho}_{00} = -k_{10}\tilde\rho_{00} + k_{01}\tilde\rho_{11} \tag{11.100d}$$

To evaluate the rate constants k, we start with the definition of R from (11.85), (11.86), and (11.83a). Then, to second order in the interaction V,

$$k_{10} = \int_0^t dt_2 \, \mathrm{Tr}_b[\Lambda^{I+}(t)\Lambda^I(t_2)\rho_b + \rho_b\Lambda^{I+}(t_2)\Lambda^I(t)]$$
$$\lim t\to\infty$$

Then defining correlation functions by

$$C_1(t - t') = \mathrm{Tr}_b[\rho_b\Lambda^+(t)\Lambda(t')] \tag{11.101a}$$

we have

$$k_{10} = 2 \, \mathrm{Re} \int_0^\infty dt_2 \, e^{-iE_1(t - t_2)/\hbar}C_1(t - t_2)$$
$$= 2 \, \mathrm{Re} \int_0^\infty dt' \, e^{-iE_1 t'/\hbar}C_1(t') \tag{11.101b}$$

Now a few definitions are useful. The Fourier, or spectral, representation is given by

$$\hat{C}_1(\omega) = \int_{-\infty}^{\infty} e^{i\omega t} C_1(t)\, dt \qquad (11.102a)$$

and, using its inverse,

$$C_1(t) = \frac{1}{2\pi} \int_{-\infty}^{\infty} e^{-i\omega t} \hat{C}_1(\omega)\, d\omega \qquad (11.102b)$$

we find, from (11.101b),

$$k_{10} = \hat{C}_1\left(-\frac{E_1}{\hbar}\right) \qquad (11.103)$$

Now the explicit form for $C_1(t - t')$ follows from the Λ expression of (11.96c) and the definition (11.101a), using the same vibrational average discussed in connection with Eq. (10.50) as

$$C_1(t - t') = \sum_k |h_k|^2 \{[n(\omega_k) + 1]e^{-i\omega_k(t - t')} + n(\omega_k)e^{i\omega_k(t - t')}\}$$

$$(11.104a)$$

with $n(\omega)$ the boson population of (10.52). Thus the correlation function is

$$\hat{C}_1(\omega) = 2\pi \sum_k |h_k|^2 \{[n(\omega_k) + 1]\delta(\omega - \omega_k) + n(\omega_k)\delta(\omega + \omega_k)\}$$

$$(11.104b)$$

where we have used the definition for the δ function (see Appendix A):

$$2\pi\delta(\omega) = \int_{-\infty}^{\infty} e^{i\omega t}\, dt \qquad (11.105)$$

The form

$$\Gamma(\omega) = \pi \sum_k |h_k|^2 \delta(\omega - \omega_k) \qquad (11.106)$$

can be interpreted as a density of states weighted by the coupling constant $|h_k|^2$. Then the Fourier-transformed correlation function is

$$\hat{C}_1(\omega) = 2\{\Gamma(\omega)[n(\omega) + 1] + \Gamma(-\omega)n(-\omega)\}$$

and the rate constant, from (11.103), is just $[\Gamma(-E_1/\hbar)$ vanishes from (11.106), since all real frequencies ω_k are positive]

$$k_{10} = 2\Gamma\left(\frac{E_1}{\hbar}\right) n\left(\frac{E_1}{\hbar}\right) \qquad (11.107)$$

This is essentially just what would be obtained from the golden rule and is easily interpreted as the rate due to absorption of one quantum, with frequency $E_1/\hbar$. If $T \to 0$, so that $n(E_1/\hbar) \to 0$, there are no such quanta present, and the transition rate vanishes.

The rate constant k_{01}, describing decay of $\bar{\rho}_{11}$, is easy to obtain, by repeating the arguments of (11.101)–(11.107), with state $|0\rangle$ replaced by $|1\rangle$ in all cases. But this simply means, in the expressions, that $E_1 \to -E_1$, $\Lambda \to \Lambda^+$. The last change has no effect, since $\Lambda^+(t)$ and $\Lambda(t)$ differ only by $h_k \to h_k^*$, and only $|h|^2$ enters the final expression. Thus we have

$$k_{01} = \hat{C}\left(\frac{E_1}{\hbar}\right) = 2\Gamma\left(\frac{E_1}{\hbar}\right)\left[n\left(\frac{E_1}{\hbar}\right) + 1\right] \tag{11.108}$$

The $n + 1$ factor includes both stimulated and spontaneous emission. The spontaneous emission is absent if the assmption (11.87b) is made—that is, if the proper dynamical behavior of the bath is ignored.

The analysis of $\dot{\bar{\rho}}_{10}(t)$ from (11.99d) is similar to that for the population evolution. Because the density matrix is hermitian, $\bar{\rho}_{01}(t) = \bar{\rho}_{10}^*(t)$, and this means, from (11.99c,d), that

$$
\begin{aligned}
R_{0101} &= R_{1010}^* \\
R_{0110} &= R_{1001}^*
\end{aligned}
\tag{11.109}
$$

The terms involving R_{0110}, R_{1001} in (11.99c,d) are nonsecular terms and would be neglected in the rotating wave approximation. (In fact, they can be included without much trouble.) We assume that such neglect is justified, and write

$$-\text{Re } R_{0101} = -\text{Re } R_{1010} \equiv \frac{1}{T_2} \tag{11.110}$$

where $1/T_2$ is, for now, just a convenient notation.

To second order, from (11.86), (11.95), and (11.101a),

$$\frac{1}{T_2} = \text{Re}\left[\int_0^t dt_2 \ \text{Tr}_b\{\rho_b \Lambda^I(t)\Lambda^{I+}(t_2) + \rho_b \Lambda^{I+}(t_2)\Lambda^I(t)\}\right] \tag{11.111a}$$

$$= \text{Re}\left[\int_0^\infty dt' \ e^{iE_1 t'/\hbar}[C_1(t') + C_1(-t')]\right] \tag{11.111b}$$

Then in the Fourier representation, we have

$$\frac{1}{T_2} = \frac{1}{2}\left(\hat{C}_1\left(\frac{E_1}{\hbar}\right) + \hat{C}_1\left(\frac{-E_1}{\hbar}\right)\right) \tag{11.111c}$$

To interpret these results for k_{01} and $1/T_2$, we go back to the population evolution, or master equation (11.100c,d). Defining the expectation value of the Pauli matrices, following (111.17), (11.20c) by

$$I = \tilde{\rho}_{11} + \tilde{\rho}_{00}$$

$$S_z = \tilde{\rho}_{11} - \tilde{\rho}_{00}$$

$$S_x = \tilde{\rho}_{01} + \tilde{\rho}_{10} \qquad (11.112)$$

$$S_y = -i(\tilde{\rho}_{01} - \tilde{\rho}_{10})$$

we write the master equations as

$$\frac{d}{dt}(I + S_z) = (I - S_z)k_{10} - (I + S_z)k_{01}$$

$$\frac{d}{dt}(I - S_z) = -(I - S_z)k_{10} + (I + S_z)k_{01}$$

or

$$\dot{S}_z = (k_{10} - k_{01}) - (k_{10} + k_{01})S_z \qquad (11.113\text{a})$$

Then defining

$$\frac{1}{T_1} \equiv k_{01} + k_{10} \qquad (11.113\text{b})$$

$$n_0 \equiv \frac{k_{10} - k_{01}}{k_{10} + k_{01}} \qquad (11.113\text{c})$$

we have

$$\dot{S}_z = \frac{n_0}{T_1} - \frac{S_z}{T_1} \qquad (11.113\text{d})$$

$$S_z(t) = n_0 + (S_z(t = 0) - n_0)e^{-t/T_1} \qquad (11.113\text{e})$$

The population difference S_z thus relaxes from its initial value $S_z(t = 0)$ toward its long-time equilibrium value n_0 with the characteristic relaxation time T_1. The value n_0, from (11.113c), is just what would be expected at steady state from the solution to (11.100c,d), for $n_0 = \tilde{\rho}_{11} - \tilde{\rho}_{00}$.

The time T_1 is called the spin-lattice relaxation time in magnetic resonance; more generally, it is known as the longitudinal relaxation time. It characterizes the decay of the population difference, or spin polarization, according to (11.113d). From (11.113b), the longitudinal relaxation rate, which is the inverse of T_1, is the sum of the rate constants for transitions between the populations $\tilde{\rho}_{11}$ and $\tilde{\rho}_{00}$.

The time T_2 is usually called the spin-spin relaxation time in magnetic resonance, or more generally, the transverse relaxation time. From our results [Eqs. (11.108), (11.103), (11.111c), and (11.113b)] we find that

$$\frac{1}{T_2} = \frac{1}{2}\{k_{01} + k_{10}\} = \frac{1}{2T_1} \qquad (11.114)$$

so that, to second order in the system field interaction V, the population difference relaxation rate constant T_1^{-1} is twice as large as the phase relaxation rate T_2^{-1}. Thus in second order, energy relaxation and dephasing both arise from the same physical term, the coupling V, which is off-diagonal in the $|1\rangle$, $|0\rangle$ delocalized basis, but corresponds to a fluctuation in the local energy levels $|l\rangle$, $|r\rangle$ of (11.98). This fluctuation leads to dephasing, as modeled simply in Section 11.3.2, using an interrupted collision model. The rates themselves are, from (11.108), (11.107), and (10.52),

$$\frac{1}{T_1} = 2\Gamma_1\left(\frac{E_1}{\hbar}\right) \coth\left(\frac{E_1}{k_B T}\right) = 2\pi \coth\left(\frac{E_1}{k_B T}\right) \sum_q |h_q|^2 \delta\left(\frac{E_1}{\hbar} - \omega_q\right)$$

(11.115)

That is, they depend on the strength h_q coupling the qth vibration to the spin system, and on the density of vibrational states at the frequency $E_1/\hbar$ defining the energy difference for spin relaxation.

Physically, the T_1 process describes transfer of energy between system and bath, while T_2 is simply a loss of system phase. The evolution of the off-diagonal terms can, from (11.99c,d) and (11.110), together with the definition

$$\Delta\omega = -\mathrm{Im}\, R_{1010} = \mathrm{Im}\, R_{0101}$$

(11.116)

be written as

$$\dot{\tilde{\rho}}_{01} = (i\omega_{10} + i\Delta\omega)\tilde{\rho}_{01} - \frac{\tilde{\rho}_{01}}{T_2}$$

(11.117a)

$$\dot{\tilde{\rho}}_{10} = -(i\omega_{10} + i\Delta\omega)\tilde{\rho}_{10} - \frac{\tilde{\rho}_{10}}{T_2}$$

(11.117b)

where the nonsecular terms from R_{0110}, R_{1001} have been omitted. The solutions, clearly, are

$$\tilde{\rho}_{01}(t) = \tilde{\rho}_{01}(t = 0)e^{i(\omega_{10} + \Delta\omega)t}e^{-t/T_2}$$

(11.118)

Therefore the off-diagonal terms oscillate, and decay toward an equilibrium value of zero.

Equations (11.100c), (100d), and (117) are referred to as Bloch equations and were first introduced by Bloch into the study of NMR. We have derived them from the hamiltonian of a single spin interacting with a bath of harmonic oscillators and have deduced values for the characteristic relaxation times T_1 and T_2, expressions that are valid to second order in the mixing perturbation V. To this order the relationship (11.114) holds, relating dephasing time, or transverse relaxation time, to T_1, the energy, or longitudinal relaxation time. To higher order in V, the relationship (11.114) fails, and often one writes

$$\frac{1}{T_2} = \frac{1}{2T_1} + \frac{1}{T_2^*} \tag{11.119}$$

where T_2^* is the so-called "pure dephasing" time. We can interpret the real parts of the Redfield tensors directly: R_{iijj} are population relaxation terms, and R_{ijij} are dephasing rates for the $\bar{\rho}_{ij}$ coherences ($i \neq j$). The other terms R_{ijkl} describe coherence transfer and couplings between populations and coherences.

The Bloch equations are special cases of the Redfield relaxation equation (11.91). In our case they have been shown to be valid in second order if nonsecular terms are neglected. In the general case it is best to use the full Redfield description.

11.6.6 Second-Order Corrections to the Density: Application to Molecular Nonlinear Optics

Molecular response to applied electromagnetic fields has already been discussed (Chapter 5) in the context of perturbation theory, and again (Chapter 9) in connection with the correlation function formalism. When considering higher-order response, however, especially when relaxation terms are important, it is convenient to use a density matrix formalism. We will derive the term for the second-order susceptibility (first hyperpolarizability) following the approach of Bloembergen and Shen.

The treatment is semiclassical, in that the field is not quantized. The aim is to describe contributions to the polarization of a chromophore induced by interactions with the field. Since the resultant polarization can itself interact with applied fields to give absorption, scattering, or emission, knowledge of the induced polarization describes the nonlinear spectral response of the molecule. We will then define the hamiltonian as

$$H = H_0 + V + H_{\text{random}} \tag{11.120}$$

with H_0 the system hamiltonian, having eigenfunctions determined by

$$H_0|n\rangle = E_n|n\rangle \tag{11.121}$$

and V the electric dipole interaction between molecular electrons at point $\mathbf{r}$ and the field $\mathbf{E}$,

$$V = e\mathbf{r} \cdot \mathbf{E} \tag{11.122}$$

The H_{random} term leads to relaxation processes, as is clear from solution to the Liouville equation

$$\frac{\partial \rho}{\partial t} = \frac{1}{i\hbar} [H_0 + V, \rho] + \left(\frac{\partial \rho}{\partial t}\right)_{\text{relax}} \tag{11.123}$$

The relaxation part $(\partial \rho / \partial t)_{\text{relax}}$ arises, formally, from the commutator of ρ with the random term in H. It has longitudinal and transverse components (T_1 and T_2, or energy relaxation and dephasing) as discussed in the preceding section. Generally, we expect from the previous discussion of relaxation,

$$\frac{\partial}{\partial t} (\rho_{nn} - \rho_{nn}^{(\text{eq})})_{\text{relax}} = \sum_{n'} [W_{n' \rightarrow n}(\rho_{n'n'} - \rho_{n'n'}^{(\text{eq})})$$

$$- W_{n \rightarrow n'}(\rho_{nn} - \rho_{nn}^{(\text{eq})})] \tag{11.124}$$

$$\left(\frac{\partial \rho_{nn'}}{\partial t}\right)_{\text{relax}} = -\left(\frac{1}{T_2}\right)_{nn'} \rho_{nn'} \equiv -\Gamma_{nn'} \rho_{nn'} \qquad n \neq n' \tag{11.125}$$

with T_2 the transverse relaxation time, $W_{n \rightarrow n'}$ transition rates and $\Gamma_{nn'}$ relaxation rates. The superscript (eq) denotes the equilibrium value.

Now we expand the density operator in terms of the perturbation V, and write for the polarization response

$$\langle \mathbf{P} \rangle = \langle \mathbf{P}^{(0)} \rangle + \langle \mathbf{P}^{(1)} \rangle + \langle \mathbf{P}^{(2)} \rangle + \cdots \tag{11.126}$$

$$\rho = \rho^{(0)} + \rho^{(1)} + \rho^{(2)} + \cdots \tag{11.127}$$

$$\langle \mathbf{P}^{(n)} \rangle \equiv \text{Tr}\{\rho^{(n)} \mathbf{P}\} \tag{11.128}$$

with $\rho^{(0)}$ the density operator at thermal equilibrium. Assuming that $\mathbf{P} = -N e \mathbf{r}$, $\langle \mathbf{P}^{(0)} \rangle = \text{Tr}\{\rho^{(0)} \mathbf{P}\} = 0$, we write the equation of motion for the different orders of ρ as

$$\frac{\partial \rho^{(1)}}{\partial t} = \frac{1}{i\hbar} \{[H_0, \rho^{(1)}] + [V, \rho^{(0)}]\} + \left(\frac{\partial \rho^{(1)}}{\partial t}\right)_{\text{relax}} \tag{11.129}$$

$$\frac{\partial \rho^{(2)}}{\partial t} = \frac{1}{i\hbar} \{[H_0, \rho^{(2)}] + [V, \rho^{(1)}]\} + \left(\frac{\partial \rho^{(2)}}{\partial t}\right)_{\text{relax}} \tag{11.130}$$

and so on.

Optical phenomena, experimentally, are nearly always described in the frequency regime, and we find it convenient to construct the theory in the same way. We therefore expand in Fourier components

$$\mathbf{E} = \sum_i \boldsymbol{\epsilon}_i \exp\{i\mathbf{k}_i \cdot \mathbf{r} - i\omega_i t\} \tag{11.131}$$

$$V = \sum_i V_i(\omega_i) = e\mathbf{r} \cdot \sum_i \boldsymbol{\epsilon}_i \exp\{i\mathbf{k}_i \cdot \mathbf{r} - i\omega_i t\} \tag{11.132}$$

$$\rho^{(n)} = \sum_j \rho^{(n)}(\omega_j)$$

$$\frac{\partial \rho^{(n)}(\omega_j)}{\partial t} = -i\omega_j \rho^{(n)}(\omega_j) \tag{11.133}$$

Then the solutions to first and second order are

$$\rho_{nn'}^{(1)}(\omega_j) = \frac{[V(\omega_j)]_{nn'}}{\hbar[\omega_j - \omega_{nn'} + i\Gamma_{nn'}]}(\rho_{n'n'}^{(0)} - \rho_{nn}^{(0)}) \tag{11.134}$$

$$\rho_{nn'}^{(2)}(\omega_j + \omega_k) = \frac{[V(\omega_j), \rho^{(1)}(\omega_k)]_{nn'} + [V(\omega_k), \rho^{(1)}(\omega_j)]_{nn'}}{\hbar[\omega_j + \omega_k - \omega_{nn'} + i\Gamma_{nn'}]} \tag{11.135}$$

$$= \left(\frac{1}{\hbar[\omega_j + \omega_k - \omega_{nn'} + i\Gamma_{nn'}]}\right) \cdot \sum_{n''} \{V_{nn''}(\omega_j)\rho_{n''n}^{(1)}(\omega_k)$$

$$- \rho_{nn''}^{(1)}(\omega_k)V_{n''n'}(\omega_j) + V_{nn''}(\omega_k)\rho_{n''n'}^{(1)}(\omega_j)$$

$$- \rho_{nn''}(\omega_j)V_{n''n'}(\omega_k)\} \tag{11.136}$$

Explicit formulas for $\rho^{(2)}$ are found by inserting $\rho^{(1)}$ from (11.134) into the last expression.

The nonlinear susceptibilities $\chi^{(n)}$ are defined just as are the usual linear susceptibilities.

$$\chi_{ij}^{(1)}(\omega) \equiv \frac{P_i^{(1)}(\omega)}{E_j(\omega)} \tag{11.137}$$

$$\chi_{ijk}^{(2)}(\omega = \omega_1 + \omega_2) = \frac{P_i^{(2)}(\omega)}{E_j(\omega_1)E_k(\omega_2)} \tag{11.138}$$

with i, j, k labeling cartesian directions. Then with $\mathbf{P} = -N e \mathbf{r}$ and $V = e \mathbf{r} \cdot \mathbf{E}$, and using the polarizations from (11.126) and the density matrix elements from (11.136), we can write detailed expressions for the susceptibilities as

$$\chi_{ij}^{(1)}(\omega) = \frac{Ne^2}{\hbar} \sum_{g,n} \rho_{gg}^{(0)} \left\{ \frac{(r_i)_{ng}(r_j)_{gn}}{\omega + \omega_{ng} + i\Gamma_{ng}} - \frac{(r_j)_{ng}(r_i)_{gn}}{\omega - \omega_{ng} + i\Gamma_{ng}} \right\} \tag{11.139}$$

$$\chi_{ijk}^{(2)}(\omega = \omega_1 + \omega_2)$$

$$= \frac{P_i^{(2)}(\omega)}{E_j(\omega_1)E_k(\omega_2)}$$

$$= -N \frac{e^3}{\hbar^2} \sum_{g,n,n'} \left[\frac{(r_i)_{gn}(r_j)_{nn'}(r_k)_{n'g}}{(\omega - \omega_{ng} + i\Gamma_{ng})(\omega_2 - \omega_{n'g} + i\Gamma_{n'g})} \right.$$

$$+ \frac{(r_i)_{gn}(r_k)_{nn'}(r_j)_{n'g}}{(\omega - \omega_{ng} + i\Gamma_{ng})(\omega_1 - \omega_{n'g} + i\Gamma_{n'g})}$$

$$+ \frac{(r_k)_{gn'}(r_j)_{n'n}(r_i)_{ng}}{(\omega + \omega_{ng} + i\Gamma_{ng})(\omega_2 + \omega_{n'g} + i\Gamma_{n'g})}$$

$$+ \frac{(r_j)_{gn'}(r_k)_{n'n}(r_i)_{ng}}{(\omega + \omega_{ng} + i\Gamma_{ng})(\omega_1 + \omega_{n'g} + i\Gamma_{n'g})}$$

$$- \frac{(r_j)_{ng}(r_i)_{n'n}(r_k)_{gn'}}{(\omega - \omega_{nn'} + i\Gamma_{nn'})}\left(\frac{1}{\omega_2 + \omega_{n'g} + i\Gamma_{n'g}} + \frac{1}{\omega_1 - \omega_{ng} + i\Gamma_{ng}}\right)$$

$$\left.- \frac{(r_k)_{ng}(r_i)_{n'n}(r_j)_{gn'}}{(\omega - \omega_{nn'} + i\Gamma_{nn'})}\left(\frac{1}{\omega_2 - \omega_{ng} + i\Gamma_{ng}} + \frac{1}{\omega_1 + \omega_{n'g} + i\Gamma_{n'g}}\right)\right]\rho_{gg}^{(0)}$$

$$(11.40)$$

The relaxation terms $\Gamma_{n'g}$ are important only near resonance. They lead to a broadening of the $\chi^{(2)}$ response, reminiscent of the Lorentzian lineshapes in Chapter 9.

Clearly, these terms become rather messy to write down explicitly: for $\chi^{(3)}$, there are 48 terms, as opposed to the two for $\chi^{(1)}$ and the eight for $\chi^{(2)}$. As with other forms of perturbation theory, diagram techniques have been developed to keep track of higher-order terms. The important point is that density matrix methods provide a rigorous, useful, straightforward scheme for computing all orders of response of the system to applied fields. The formulas for nonlinear response so derived provide a useful way actually to calculate the nonlinear optical response of molecules and polymers.

11.7

Higher-Order Corrections to the Density Matrix:
Pulsed Spectroscopy

Density matrix methods are especially useful when we wish to study dynamics of a subsystem induced by controlled, time-dependent perturbations. Particularly elegant and useful applications appear in pulsed NMR, where the dynamics in finite-dimensional spin manifolds are controlled by shaped pulses of power. More recent applications of the same concept have been used in multiple-pulse pump/probe optical spectroscopy, but for both instrumental and physical reasons, pulsed methods in NMR remain the prime domain for "spin alchemy," the manipulation of nuclear spins by time-dependent external fields to yield information about these spins, their environment, and their interactions. Several recent books, as well as the *Advances in Magnetic Resonance* volumes, trace the conceptual and methodological sophistication of these techniques, as well as their immense power for learning about structure and dynamics in condensed phases.

Density matrix methods are the standard theoretical approach to the field of pulsed spectroscopy in the time domain. We will give simple examples of the sorts of manipulation involved.

11.7.1 Spins: Rotations and Angular Momentum

Formally, the basis of pulsed NMR is simply the fact that spins are angular momenta and that external fields can redirect these angular momenta. To be more explicit, the operator $\exp(-i\alpha I_x)$ can be used to generate rotations by angle α about the x axis; here we define I_x, I_y, I_z as (spin) angular momentum operators. Explicitly, we expand in a power series

$$\exp\{-i\alpha I_x\} = \sum_{m=0} \frac{(-i\alpha)^m}{m!} I_x^m = \cos\frac{\alpha}{2} - 2i\sin\frac{\alpha}{2} I_x \quad (11.141)$$

(This is most easily seen using σ_x, the Pauli matrix, for $2I_x$, and employing (11.19).) We then have

$$\exp\{-i\alpha I_x\}I_y \exp\{i\alpha I_x\} = \left[\cos\frac{\alpha}{2} - 2i\sin\frac{\alpha}{2} I_x\right]I_y\left[\cos\frac{\alpha}{2} + 2i\sin\frac{\alpha}{2} I_x\right]$$

$$= I_y\cos\alpha + I_z\sin\alpha \quad (11.142)$$

where, again, we have used (11.19).

The physical meaning of (11.142) is that rotation of α about the x axis transforms I_y into $I_y\cos\alpha + I_z\sin\alpha$; in particular, for $\alpha = 0$, $I_y \rightarrow I_y$; for $\alpha = \pi/2$, $I_y \rightarrow I_z$; for $\alpha = \pi$, $I_y \rightarrow -I_y$.

It is then useful to collect all of these rotations as

$$e^{-i\alpha I_x}\begin{pmatrix} I_x \\ I_y \\ I_z \end{pmatrix}e^{i\alpha I_x} = \begin{cases} I_x \\ I_y\cos\alpha + I_z\sin\alpha \\ I_z\cos\alpha - I_y\sin\alpha \end{cases} \quad (11.143a)$$

$$e^{-i\alpha I_y}\begin{pmatrix} I_x \\ I_y \\ I_z \end{pmatrix}e^{i\alpha I_y} = \begin{cases} I_x\cos\alpha - I_z\sin\alpha \\ I_y \\ I_z\cos\alpha + I_x\sin\alpha \end{cases} \quad (11.143b)$$

$$e^{-i\alpha I_z}\begin{pmatrix} I_x \\ I_y \\ I_z \end{pmatrix}e^{i\alpha I_z} = \begin{cases} I_x\cos\alpha + I_y\sin\alpha \\ I_y\cos\alpha - I_x\sin\alpha \\ I_z \end{cases} \quad (11.143c)$$

Note that the product of consecutive rotations can be cast as simple rotations. For example,

$$e^{-i\alpha I_x} = e^{-i\frac{\pi}{2} I_y} e^{-i\alpha I_z} e^{i\frac{\pi}{2} I_y} \quad (11.144)$$

11.7.2 Rotating Frame Transformation

Suppose that a nuclear spin of $\frac{1}{2}$ (e.g., a proton) is subject to an external static magnetic field B_{0z} along the z axis (in this section we use B for magnetic field and $\mathscr{H}$ for hamiltonian, in agreement with the dominant usage in the field). Then the Zeeman hamiltonian describing the interaction is just

$$\mathscr{H}_{Zee} = -\gamma\hbar B_{0z}I_z \qquad (11.145a)$$

$$= \hbar\omega_0 I_z \qquad (11.145b)$$

with γ the gyromagnetic ratio and $\hbar I_z$ the spin angular momentum along z; here we have introduced the Larmor frequency

$$\omega_0 \equiv -\gamma B_{0z} \qquad (11.146)$$

which is simply the classical frequency of the spin in the applied field B_{0z}. In an NMR spectrometer, a second, radio-frequency field $\mathbf{B}_{rf}$ is applied in the (x, y) plane. Then the spin interaction with the applied field can be written

$$\mathscr{H}_{rf}(t) = -B_1\hbar\gamma\{I_x \cos(\omega_{rf}t + \phi) + I_y \sin(\omega_{rf}t + \phi)\} \quad (11.147)$$

with B_1, ω_{rf}, and ϕ, respectively, the strength, frequency, and phase of the radio-frequency field.

The total hamiltonian for the spin is then

$$\mathscr{H}(t) = \mathscr{H}_{Zee} + \mathscr{H}_{rf}(t), \qquad (11.148)$$

and it is convenient to define a new reference frame that makes the hamiltonian time independent. This can be done by choosing a frame that rotates about the z axis with frequency ω_{rf}. Formally, then, we have the so-called rotating-frame hamiltonian as

$$\mathscr{H}^r = \exp(+i\omega_{rf}tI_z)\mathscr{H}(t)\exp(-i\omega_{rf}tI_z) \qquad (11.149)$$

$$= -B_1\gamma\hbar\{I_x \cos\phi + I_y \sin\phi\} + \hbar\omega_0 I_z \qquad (11.150)$$

The rotating frame accounts for the motion induced by the radio frequency ω_{rf}, and renders the resulting hamiltonian $\mathscr{H}^r$ time independent. The equation of motion (Liouville equation) in the rotating frame becomes

$$i\hbar\dot{\bar{\rho}}^r = [\mathscr{H}^r - \hbar\omega_{rf}I_z, \bar{\rho}^r] \qquad (11.151)$$

Thus the effective hamiltonian has a shift of $-\omega_{rf}$ in all Larmor frequencies, and has a new frequency origin at $\omega = \omega_{rf}$. In the rotating frame, the Larmor frequency becomes

$$\Omega = \omega_0 - \omega_{rf} \qquad (11.152)$$

11.7.3 Simple Pulse Experiment: Carr–Purcell Spin Echo for Uncoupled Spins

Here we illustrate the principles of pulse manipulation using one of the earliest spin-echo experiments. The equilibrium density matrix for the spins subject to $\mathcal{H}$ of (11.150) (we drop the r notation for convenience) is

$$\rho = \frac{e^{-\mathcal{H}/kT}}{\text{Tr}\{e^{-\mathcal{H}/kT}\}} \tag{11.153}$$

The Zeeman term is much greater than the radio-frequency term, but both are quite small compared to ordinary temperatures. Then we can expand the exponentials, to find

$$\rho \cong \frac{(1 - \mathcal{H}_{\text{Zee}}/kT)}{\text{Tr}\{1\}} \tag{11.154}$$

Since we wish to calculate spin polarization as $\text{Tr}\{\rho I_\alpha\}$, the 1 in the numerator is unimportant, since $\text{Tr}\{1 I_\alpha\} = 0$. Therefore, the truncated density operator of the equilibrated system before perturbing fields are applied is simply

$$\rho_0 = I_z \frac{-\hbar\omega_0}{kT\,\text{Tr}\{1\}} \equiv \frac{I_z}{A} \tag{11.155}$$

with $A \equiv -kT\,\text{Tr}\{1\}/\hbar\omega_0$.

We now consider the application of the pulse train sketched in Fig. 11.2 to the density, in order to calculate the resultant spin as $\langle I_x \rangle =$

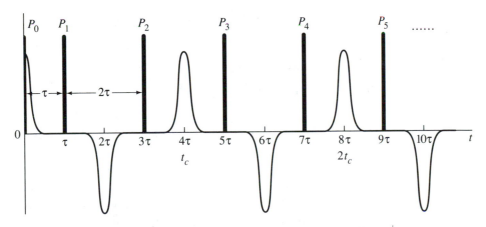

Figure 11.2. Schematic drawing of a Carr–Purcell sequence, generated by P_0 being a $\pi/2$ pulse and the π pulse P_i, $i \geq 1$. Notice that the phase of the spin echoes is alternating.

$\text{Tr}\{\rho I_x\}$. At time $t = 0$, a pulse of radio-frequency power is applied along the y axis. We assume that this pulse is essentially instantaneous, and that during its application it entirely determines the dynamics of the spin. Then we have

$$A\rho(t = 0+) = e^{-i\frac{\pi}{2}I_y}I_z e^{i\frac{\pi}{2}I_y} \tag{11.156}$$

as the density matrix at time just larger than zero, after application of a pulse of magnitude $\pi/2$ along the y axis. After this pulse is turned off, the system evolves under $\mathcal{H}$ of (11.150) for the period τ, so that

$$\rho(t = \tau) = e^{-i\mathcal{H}\tau/\hbar}\rho(t = 0+)e^{i\mathcal{H}\tau/\hbar} \tag{11.157}$$

Next a π-pulse is applied, so that

$$\rho(t = \tau+) = e^{-i\pi I_y}\rho(t = \tau)e^{i\pi I_y} \tag{11.158}$$

followed by free evolution to 2τ:

$$\rho(t = 2\tau) = e^{-i\mathcal{H}\tau/\hbar}\rho(t = \tau+)e^{i\mathcal{H}\tau/\hbar} \tag{11.159}$$

All of the evolution can then be combined to yield

$$A\rho(t = 2\tau) = e^{-i\mathcal{H}\tau/\hbar}e^{-i\pi I_y}e^{-i\mathcal{H}\tau/\hbar}e^{-i\pi/2 I_y}I_z e^{i\pi/2 I_y}e^{i\mathcal{H}\tau/\hbar}e^{i\pi I_y}e^{i\mathcal{H}\tau/\hbar} \tag{11.160}$$

Fortunately, spin algebra permits facile computation of the resultant density. Simply using the definitions of (11.143), we find (assuming for simplicity that $B_1 \lll B_{0z}$) that

$$A\rho(0+) = I_x$$
$$A\rho(\tau) = I_x \cos \omega_0\tau + I_y \sin \omega_0\tau$$
$$A\rho(\tau+) = -I_x \cos \omega_0\tau + I_y \sin \omega_0\tau$$
$$A\rho(2\tau) = -I_x$$

We can extend the analysis to 4τ as

$$A\rho(4\tau) = e^{-i\mathcal{H}\tau/\hbar}e^{-i\pi I_y}e^{-i\mathcal{H}\tau/\hbar}\rho(2\tau)e^{i\mathcal{H}\tau/\hbar}e^{i\pi I_y}e^{i\mathcal{H}\tau/\hbar} = I_x \tag{11.161}$$

Thus as illustrated in Fig. 11.2, the expectation value $\langle I_x \rangle = \text{Tr}\{\rho I_x\}$ assumes values of $-\langle I_x^2 \rangle \hbar\omega_0/kT \, \text{Tr}\{1\}$ for $t = 4\tau, 8\tau$, and so on, and values of opposite sign for $t = 2\tau, 6\tau$, and so on.

This spin manipulation is very much overspecialized. For instance, suppose now that the Zeeman term is generalized to include the chemical shift terms (i.e., the local fields differ for each of several spins in the sample). Then for nuclear spins I, $\text{Tr}\{1\}$ is $(2I + 1)^N$, and the hamiltonian becomes

$$\mathcal{H} \equiv \hbar\omega_0 I_z + \sum_k \delta_k i_{kz} = \hbar\omega_0 I_z + \mathcal{H}_1 \tag{11.162}$$

where

$$I_z = \sum_k i_{kz}$$

and i_{kz} is the zth component of the kth spin. The value of δ_k is the local fluctuation of the resonance frequency—the chemical shift. For this hamiltonian, the Carr–Purcell sequence can be written exactly as in (11.160), but with the full $\mathcal{H}$ of (11.150) used. We can then write

$$A\rho(2\tau) = D(\tau)P_2 D(\tau)P_1 I_z P_1^{-1}D^{-1}(\tau)P_2^{-1}D^{-1}(\tau) \qquad (11.163)$$

with

$$D(\tau) = e^{-i\mathcal{H}\tau/\hbar} \qquad (11.164)$$

$$P_1 = e^{-i\pi/2I_y} \qquad (11.165)$$

$$P_2 = P_1^2 \qquad (11.166)$$

Then defining

$$\tilde{D}(\tau) = P_2 D(\tau)P_2^{-1} \qquad (11.167)$$

we find

$$A\rho(2\tau) = D(\tau)\tilde{D}(\tau)P_2 P_1 I_z P_1^{-1}P_2^{-1}\tilde{D}^{-1}(\tau)D^{-1}(\tau) \qquad (11.168)$$

Now

$$\tilde{D}(\tau) = e^{-i\pi I_y}D(\tau)e^{i\pi I_y} = e^{-i\pi I_y}e^{-i\mathcal{H}\tau/\hbar}e^{i\pi I_y} \qquad (11.169)$$

But we have from (11.143) that

$$e^{-i\pi I_y}I_z e^{i\pi I_y} = -I_z$$

Since this also holds for i_{kz}, we have, for the $\mathcal{H}$ of (11.162),

$$P_2 \mathcal{H} P_2^{-1} = -\mathcal{H} \qquad (11.170)$$

That is, the pulse flips the z component of spin. Then from (11.163) we find that

$$\begin{aligned} A\rho(2\tau) &= e^{-i\mathcal{H}\tau/\hbar}e^{i\mathcal{H}\tau/\hbar}P_2 P_1 I_z P_1^{-1}P_2^{-1}e^{-i\mathcal{H}\tau/\hbar}e^{i\mathcal{H}\tau/\hbar} \\ &= P_2 P_1 I_z P_1^{-1}P_2^{-1} = -I_z \end{aligned} \qquad (11.171)$$

Thus the alternating sign echo of the Carr–Purcell sequence is not affected at all by differing chemical shifts of differing nuclei—it is a technique in which the observed signal is unaffected by broadening due to chemical shift. Experimentally, this is the great advantage of multiple-pulse spectroscopy: By manipulation of pulses, the evolution of $\rho(t)$ can be designed to feature, or to circumvent, particular parts of the spin hamiltonian.

The treatment here is a simple one and is valid for noninteracting spins. (There is no spin-spin coupling in $\mathcal{H}$). Physically, real molecules do have spin-spin and other sorts of couplings, and the development of NMR techniques to deal with them, using pulse sequences, underlies the great utility of NMR in structural chemistry.

11.7.4 Average Hamiltonian Theory for Multiple-Pulse NMR*

The explicit calculation just carried out for the time evolution of the density matrix can be generalized nicely for arbitrary pulse sequences. The essential idea of multiple-pulse NMR is to design a particular set of pulses such that certain terms in the spin hamiltonian can be made to vanish in the evolution of the spin system under study. To do this efficiently, it is necessary to have a convenient scheme for calculating what effect particular pulse sequences will have on the time evolution of the system. Just such a convenient formalism, called average Hamiltonian Theory, has been developed by Waugh and coworkers. We will finish our discussion of density matrix time evolution with this topic, which is representative of problems whose hamiltonians contain explicit time dependence. The average hamiltonian method that we will present incorporates a so-called Magnus expansion of the propagator, as do schemes for propagating both wavefunctions and density matrices in such situations as pump-probe spectroscopy and time-dependent Hartree calculations.

Our development in this section will actually be based on the use of the evolution operator W that solves the quantum Liouville equation (propagates the density matrix in time). The average hamiltonian that is derived generates this evolution operator, and the trick of choosing pulse sequences is simply to generate an average hamiltonian in which particular complicating parts of the true hamiltonian vanish. In the Carr–Purcell sequence just discussed, for example, the chemical shift broadening is not present in the echo signal. The average hamiltonian that we will derive then provides the effective evolution operator, that yields the density operator via (by analogy with (11.157)) Eqs. (11.173, 11.172b), to follow. The density operator $\bar{\rho}(t)$ is then used to compute the spectrum (from 11.15)

$$\langle I_j(t) \rangle = \mathrm{Tr}\{\bar{\rho}(t)I_j\}, \qquad j = x, y, z \qquad (11.171)$$

The system itself will be governed by a time-independent hamiltonian h, and the experiment subjects it to a second, time-dependent hamiltonian

---

* We follow closely the text by Ernst, Bodenhausen, and Wokaun, but have modified their notation.

$\mathcal{H}_0(t)$. We will assume that the perturbation $\mathcal{H}_0$ (which is at the control of the experimenter) is periodic, and is synchronized with the observations made on the system. The average hamiltonian $\overline{\mathcal{H}}(t_1, t_2)$ is that hamiltonian that describes the evolution between t_1 and t_2. If the system is periodic, we can define t_c, the period of the perturbation, and then we wish to find $\overline{\mathcal{H}}(t_c)$ as the average hamiltonian that characterizes the motion of the system in the interval $(0, t_c)$. If there were no perturbation, $\overline{\mathcal{H}} = h$.

For most pulse sequences of interest, such as the Carr–Purcell sequence discussed in the preceding section, the hamiltonian is piecewise constant throughout the period $(0, t_c)$. Suppose that during the period the hamiltonian assumes n different constant values, i.e., $\mathcal{H}_0(t) = \mathcal{H}_l$ for the interval from τ_l to τ_{l+1} for the n intervals $l = 1, 2, \ldots, n$. Then we can define an evolution operator $W(t_c)$ such that the Liouville equation

$$i\hbar\dot{\tilde{\rho}} = [\mathcal{H}(t), \tilde{\rho}] \tag{11.172a}$$

can be solved to yield

$$\tilde{\rho}(t_c) = W(t_c)\tilde{\rho}(0)W^+(t_c) \tag{11.172b}$$

By analogy with (11.161), a solution exists in the form

$$W(t_c) = \exp\left\{\frac{-i\mathcal{H}_n\tau_n}{\hbar}\right\} \exp\left\{\frac{-i\mathcal{H}_{n-1}\tau_{n-1}}{\hbar}\right\} \cdots \exp\left\{\frac{-i\mathcal{H}_1\tau_1}{\hbar}\right\} \tag{11.172c}$$

and with the total period fixed by

$$t_c = \sum_{l=1}^{n} \tau_l \tag{11.172d}$$

Now, with the total hamiltonian defined by,

$$\mathcal{H} = \mathcal{H}_0(t) + h \tag{11.172e}$$

the desired average hamiltonian is determined by the condition

$$W(t_c) = \exp\left\{\frac{-i\overline{\mathcal{H}}(t_c)t_c}{\hbar}\right\} \tag{11.173}$$

and there is an exact solution within any finite basis: One simply evaluates the matrix product on the right of Eq. (11.172c), diagonalizes the result, and takes logarithms to fix the eigenvalues in $\exp\{-i\overline{\mathcal{H}}t_c/\hbar\}$.

An easier, more intuitive approach is an approximate one that is often employed in quantum time evolution problems. For any two operators A and B, the series

$$e^B e^A = \exp\{A + B + \tfrac{1}{2}[B,A] + \tfrac{1}{12}[B,[B,A]] + \tfrac{1}{12}[[B,A],A] + \cdots\} \tag{11.174}$$

is an exact relationship, known as the Baker–Campbell–Hausdorff series [we used the first two terms in connection with harmonic oscillator hamiltonians in Eq. (10.48)]. Applying a generalization of (11.174) to $W(t_c)$ defined by (11.173) and (11.172c), one obtains the form

$$\overline{\mathcal{H}}(t_c) = \overline{\mathcal{H}}^{(0)} + \overline{\mathcal{H}}^{(1)} + \overline{\mathcal{H}}^{(2)} + \cdots \tag{11.175}$$

with

$$\overline{\mathcal{H}}^0(t_c) = \frac{1}{t_c}\{\mathcal{H}_1\tau_1 + \mathcal{H}_2\tau_2 + \cdots + \mathcal{H}_n\tau_n\} \tag{11.175a}$$

$$\overline{\mathcal{H}}^{(1)} = \frac{-i}{2t_c\hbar}\{[\mathcal{H}_2, \mathcal{H}_1]\tau_1\tau_2 + [\mathcal{H}_3, \mathcal{H}_1]\tau_1\tau_3 + [\mathcal{H}_3, \mathcal{H}_2]\tau_2\tau_3 + \cdots\} \tag{11.175b}$$

$$\overline{\mathcal{H}}^{(2)} = -\frac{1}{6\tau_c\hbar^2}\{[\mathcal{H}_3, [\mathcal{H}_2, \mathcal{H}_1]]\tau_1\tau_2\tau_3 + [[\mathcal{H}_3, \mathcal{H}_2], \mathcal{H}_1]\tau_1\tau_2\tau_3$$

$$+ [\mathcal{H}_2, [\mathcal{H}_2, \mathcal{H}_1]]\tau_1\tau_2^2 + [[\mathcal{H}_2, \mathcal{H}_1], \mathcal{H}_1]\tau_2\tau_1^2 + \cdots\} \tag{11.175c}$$

The forms (11.175) are valid for piecewise constant hamiltonians and can be generalized to continuously varying hamiltonians, yielding

$$\overline{\mathcal{H}}^{(0)} = \frac{1}{t_c}\int_0^{t_c} dt_1\, \mathcal{H}(t_1) \tag{11.176a}$$

$$\overline{\mathcal{H}}^{(1)} = \frac{-i}{2t_c\hbar}\int_0^{t_c} dt_2 \int_0^{t_2} dt_1\, [\mathcal{H}(t_2), \mathcal{H}(t_1)] \tag{11.176b}$$

$$\overline{\mathcal{H}}^{(2)} = \frac{(-i)^2}{3!\, t_c\hbar^2}\int_0^{t_c} dt_3 \int_0^{t_3} dt_2 \int_0^{t_2} dt_1\, \{[\mathcal{H}(t_3), [\mathcal{H}(t_2), \mathcal{H}(t_1)]]$$

$$+ [[\mathcal{H}(t_3), \mathcal{H}(t_2)], \mathcal{H}(t_1)]\} \tag{11.176c}$$

The expansion in (11.176) is usually called the Magnus expansion and is a convenient way to employ the average hamiltonian theory. Note that if the hamiltonian were constant in time, or if $\mathcal{H}(t_1)$ always commuted with $\mathcal{H}(t_2)$, we would have

$$\overline{\mathcal{H}}(t_c) = \overline{\mathcal{H}}^{(0)} = \frac{1}{t_c}\int_0^{t_c} dt_1\, \overline{\mathcal{H}}(t_1) \tag{11.177}$$

the usual definition of a simple average.

The average hamiltonian describes the evolution of the system density matrix over a time t_c. Now if we choose $\mathcal{H}_0(t)$ properly, a particular part of h, the system hamiltonian, can be made to vanish in $\overline{\mathcal{H}}$, at least up to a certain order. Perturbations such as pulse sequences and sample spinning are examples of $\mathcal{H}_0(t)$ properly so chosen.

The construction from (11.175) or (11.176) of $\overline{\mathcal{H}}(t_c)$, the average hamiltonian that describes propagation over the cycle $(0,t_c)$, is facilitated by using a special sort of interaction representation, called the toggling frame. The propagator $W(t)$ can generally be written (see Problem 6)

$$W(t) = P \exp\left\{-i \int_0^t \frac{dt_1(h + \mathcal{H}_0(t_1))}{\hbar}\right\} \tag{11.178}$$

with P the Dyson time-ordering operator of Eq. (4.30). We then define $\hat{h}(t)$ as the interaction frame (toggling frame) hamiltonian, by

$$\hat{h}(t) = W_0^+(t)hW_0(t) \tag{11.179}$$

with

$$W_0(t) = P \exp\left\{-i \int_0^t \frac{dt_1 \mathcal{H}_0(t_1)}{\hbar}\right\} \tag{11.180}$$

Then by evaluating the toggling frame propagator

$$W_1(t) = P \exp\left\{-i \int_0^t dt_1 \frac{\hat{h}(t_1)}{\hbar}\right\} \tag{11.181}$$

the propagator $W(t)$ can be expressed exactly as

$$W(t) = W_0(t)W_1(t) \tag{11.182}$$

These equations are simply a rewriting of the usual interaction representation of Eq. (11.54), with a time-dependent term $\mathcal{H}_0(t)$ replacing what was there called H_0, and the h term here replacing what was there called V.

It is convenient to choose the $\mathcal{H}_0(t)$ term such that it is cyclic, so that $W_0(t_c) = 1$. This means that the average direct effect of the perturbation vanishes. Then we have, from (11.182),

$$W(t_c) = W_1(t_c) = \exp\left\{\frac{-i\overline{h}t_c}{\hbar}\right\} \tag{11.183}$$

This expression holds for t_c, not for arbitrary time, but because observations are made only after $t_c, 2t_c, 3t_c, \ldots$, $W_1(t_c)$ itself is sufficient to describe the experiment. For the relevant times $(\tau_c, 2\tau_c, 3\tau_c, \ldots)$, $\mathcal{H}$ becomes $\overline{h}(t)$. For several (m) periods,

$$\overline{\mathcal{H}}(mt_c) = \overline{h}(mt_c) \tag{11.184a}$$

$$W(mt_c) = [W(t_c)]^m = \exp\{-im\overline{h}t_c/\hbar\} \tag{11.184b}$$

Here $\overline{h}$ is the relevant average hamiltonian, which must be determined. But from (11.175) and (11.176), we can write

$$\overline{h} = \overline{h}^{(0)} + \overline{h}^{(1)} + \overline{h}^{(2)} + \cdots \tag{11.185}$$

$$\bar{h}^{(0)} = \frac{1}{t_c} \int_0^{t_c} dt_1\, \hat{h}(t_1) \tag{11.186a}$$

$$\bar{h}^{(1)} = \frac{-i}{2t_c\hbar} \int_0^{t_c} dt_2 \int_0^{t_2} dt_1\, [\hat{h}(t_2), \hat{h}(t_1)] \tag{11.186b}$$

$$\bar{h}^{(2)} = \frac{(-i)^2}{6t_c\hbar^2} \int_0^{t_c} dt_3 \int_0^{t_3} dt_2 \int_0^{t_2} dt_1\, \{[\hat{h}(t_3), [\hat{h}(t_2), \hat{h}(t_1)]]\}$$

$$+ \,[[\hat{h}(t_3), \hat{h}(t_2)], \hat{h}(t_1)]\} \tag{11.186c}$$

The toggling frame hamiltonian $\hat{h}$ of (11.179) thus defines the effective average hamiltonian $\bar{h}$ to all orders.

The $\bar{h}^{(0)}$ term is simply the average of the toggling-frame hamiltonian. The entire aim of a properly designed pulse sequence $\mathcal{H}_0(t)$ is to make these time averages vanish for certain undesired terms in h. Even then, of course, such terms can still enter in higher-order parts $\bar{h}^{(1)}$, $\bar{h}^{(2)}$, ... of the average hamiltonian $\bar{h}$. If the overall cycle time t_c is reduced, clearly the magnitude of such terms drops; shorter pulse sequences therefore lead to better averaging.

Many pulse sequences consist, as did the Carr–Purcell sequence of Section 11.7.3, of a series of short radio-frequency pulses separated by periods of free precession. The toggling frame rotates during the pulses but remains static during the free precession. We can define $\bar{h}_{(k)}$ as the value of $\bar{h}(t)$ in the interval τ_k after the kth pulse. Then stepwise transformations, as in Eq. (11.171), can give $\bar{h}_{(k)}$ as

$$\bar{h}_{(0)} = h \tag{11.187a}$$

$$\bar{h}_{(1)} = U_1^+ h U_1 \tag{11.187b}$$

$$\bar{h}_{(2)} = U_1^+ U_2^+ h U_2 U_1 \tag{11.187c}*$$

and then the average hamiltonian, in lowest order, is

$$\bar{h}^{(0)} = \frac{1}{t_c} \sum_{k=0}^{n} \tau_k U_1^+ \cdots U_k^+ h U_k \cdots U_1 \tag{11.188}*$$

The transformation U_k describes the rotation caused by the kth pulse.

A particularly important pulse sequence, the WHH-4 sequence, is illustrated in Fig. 11.3. The sequence is both periodic and cyclic, so that $W_0(t_c) = 1$. The pulses occur over intervals $(\tau, \tau, 2\tau, \tau, \tau)$, so that $t_c = 6\tau$. All pulses are $\pi/2$. Then the cartoons in Fig. 11.3 correspond to the

---

* This unexpected ordering is correct. See R. R. Ernst, G. Bodenhausen, and A. Wokaun, *Principles of Nuclear Magnetic Resonance in One and Two Dimensions* (Clarendon Press, Oxford, 1990), p.78.

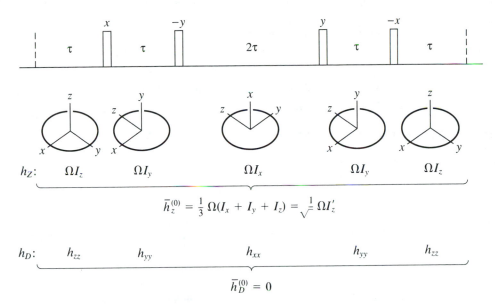

$$\bar{h}_z^{(0)} = \frac{1}{3}\,\Omega(I_x + I_y + I_z) = \frac{1}{\sqrt{3}}\,\Omega I_z'$$

$$\bar{h}_D^{(0)} = 0$$

Figure 11.3. WHH–4 multiple-pulse sequence for homonuclear dipolar decoupling. In each cycle of total length $\tau_c = 6\tau$, the four pulses spaced by τ and 2τ lead to rotated coordinate systems, known as a toggling frame. The average hamiltonian $\bar{h}$ is found by averaging the transformed hamiltonians in the toggling frame, as illustrated for Zeeman and dipolar terms.

transformations in Eq. (11.188); in the sequence of pulses $z \to y \to x \to y \to z$, $zz \to yy \to xx \to yy \to zz$. Then the average Zeeman term is (Ω = radio frequency)

$$\bar{h}_Z^{(0)} = \frac{1}{6\tau}\{\Omega I_z\tau + \Omega I_y\tau + 2\Omega I_x\tau + \Omega I_y\tau + \Omega I_z\tau\}$$

$$= \frac{1}{3}\,\Omega(I_x + I_y + I_z) = \left(\frac{1}{3}\right)^{1/2}\Omega I_z' \qquad (11.189)$$

That is, the average Zeeman interaction has a Larmor frequency $\Omega/\sqrt{3}$, with a new quantization axis $(1, 1, 1)$.

More interesting behavior occurs for the dipolar terms. The dipolar hamiltonian is generally (with k, l labeling spins and b_{kl} a coupling strength)

$$h_{xx} = \sum_{k<l} \tfrac{1}{2}b_{kl}(1 - 3\cos^2\theta_{kl})(3I_{kx}I_{lx} - \mathbf{I}_k \cdot \mathbf{I}_l) \qquad (11.190)$$

and so on, for the nine components. The average dipolar hamiltonian for the WHH-4 sequence is [easily proved using (11.190)]

$$\bar{h}_D^{(0)} = \frac{1}{6\tau} \{\tau h_{zz} + \tau h_{yy} + 2\tau h_{xx} + \tau h_{yy} + \tau h_{zz}]$$

$$= 0 \qquad (11.191)$$

So the dipolar term makes no contribution to $\bar{h}^{(0)}$, and therefore, to this order, the NMR signal observed is not sensitive to dipolar coupling.

BIBLIOGRAPHY FOR CHAPTER 11*

Abragam, A., *Principles of Nuclear Magnetism* (Clarendon Press, Oxford, 1961). [11.2, 3, 4, 6]

Ernst, R. R., G. Bodenhausen, and A. Wokaun, *Principles of Nuclear Magnetic Resonance in One and Two Dimensions* (Clarendon Press, Oxford, 1990). [11.2, 3, 6, 7]

Harriman, J. E., in J. Keller and J. L. Gasquez, eds., *Density Functional Theory* (Springer-Verlag, Berlin, 1983). [11.5]

Jones, W. H., and N. March, *Theoretical Solid-State Physics* (Dover, New York, 1985). [11.6.4]

Levine, R. D., *Quantum Mechanics of Molecular Rate Processes* (Clarendon University Press, Oxford, 1969). [11.2, 3, 4]

Louisell, W. H., *Radiation and Noise in Quantum Electronics* (McGraw-Hill, New York, 1974). [11.2, 3, 4]

Madelung, O., *Introduction to Solid-State Theory* (Springer-Verlag, Berlin, 1981). [11.6.3]

Mehring, M., *Principles of High-Resolution NMR in Solids* (Springer-Verlag, Berlin, 1983). [11.7]

Shen, Y. R., *The Principles of Nonlinear Optics* (Wiley, New York, 1984). [11.6.6]

Slichter, C. P., *Principles of Magnetic Resonance* (Springer-Verlag, Berlin, 1990). [11.6.4]

Tolman, R. C., *Principles of Statistical Mechanics* (Oxford University Press, Oxford, 1938). [11.2.3]

Some leading journal articles are:

Harris, R. A., and R. Silbey, *J. Phys. Chem.* 93, 7062 (1989). [11.4.2]

Jean, J., R. A. Friesner, and G. R. Fleming, *J. Chem. Phys.* 96, 5827 (1992). [11.6.4]

Laird, B. B., J. Budimir, and J. L. Skinner, *J. Chem. Phys.* 94, 4391 (1991). [11.6]

Redfield, A. G., *Adv. Magn. Reson.* 1, 1 (1966). [11.6.4]

Zwanzig, R., *Annu. Rev. Phys. Chem.* 16, 67 (1966). [11.6.3]

* The citations in brackets are text sections to which these references are particularly applicable.

PROBLEMS FOR CHAPTER 11

1. For the homogeneous two-site Hubbard problem discussed extensively in Chapter 6, the hamiltonian is

$$H = \sum_{i=1}^{2} U a_{i\alpha}^{+} a_{i\alpha} a_{i\beta}^{+} a_{i\beta} + \sum_{\mu} \beta(a_{1\mu}^{+} a_{2\mu} + a_{2\mu}^{+} a_{1\mu})$$

with 1,2 the site labels, and the zero of energy fixed as the one-site energy (Hückel α). We showed that the symmetry-determined molecular orbitals are

$$\phi_{+,\alpha} = \frac{1}{\sqrt{2}}(a_{1\alpha}^{+} + a_{2\alpha}^{+})|\text{vac}\rangle$$

$$\phi_{-,\alpha} = \frac{1}{\sqrt{2}}(a_{1\alpha}^{+} - a_{2\alpha}^{+})|\text{vac}\rangle$$

corresponding, respectively, to bonding and antibonding MOs.

Consider the special case $U = -3\beta$. Using the ground and doubly excited states as a basis (of two-electron functions), diagonalize the hamiltonian, and express the resulting CI ground-state function Ψ_0 in terms of ϕ_+ and ϕ_-. Now express the one-electron reduced density matrix γ using the molecular orbitals. Do this for the HF state $|\phi_{+\alpha}\phi_{+\beta}\rangle$ and for the CI state Ψ_0. Remark on the differences for the particular situation. How are the natural orbitals related to the MOs? Why? Would you expect this to be true in general? Why?

2. We have derived the Redfield equation in the interaction representation for $\bar{\rho}^I$, as defined by Eq. (11.52). Additional insight can be gained in the Schrödinger representation. Show that

$$\dot{\bar{\rho}}_{\alpha\alpha'} = \frac{1}{i\hbar}(\varepsilon_\alpha - \varepsilon_{\alpha'})\bar{\rho}_{\alpha\alpha'} + \sum_{\beta\beta'} R_{\alpha\alpha';\beta\beta'}\bar{\rho}_{\beta\beta'}(t)$$

with α,α' labeling eigenstates of H_s. This is then interpreted by saying that the time evolution of the reduced system density matrix $\bar{\rho}_{\alpha\alpha'}$ has two components. The first term on the right of $\dot{\bar{\rho}}_{\alpha\alpha'}$ is just the oscillatory part arising from evolution due to H_s. The second term is the relaxation part, that arises due to the coupling of the system of interest to the thermal bath.

3. The Kubo-type linear response formulas of Section 11.6.3 are reminiscent of the correlation-function formulas that we used in Chapters 9 and 10. There are differences in the form of the integrals and in the quantity being calculated (the golden rule calculates a transition probability per unit time, the linear response forms find a generalized susceptibility or a linear response coefficient like conductivity, susceptibility, diffusion coefficient, viscosity, etc.). They also start from different assumptions—in particular, the golden rule assumes that the final state lies in a dense manifold such that a transition probability per unit time can be defined.

Formally, show that in the limit of high temperature, the Kubo result can be made to look very similar to the correlation function forms that appeared in Chapters 9 and 10. In particular, show that (11.74) becomes

$$\frac{g_{\mu\nu}(\omega)}{\text{volume}} = \frac{1}{kTi\hbar} \int_0^\infty e^{-i\omega t}\langle j_\mu(t)j_\nu(0)\rangle \, dt$$

4. The form

$$[x,\rho^{(0)}] = -i\hbar\rho^{(0)} \int_0^\beta d\lambda \, e^{\lambda H_0}\dot{x}e^{-\lambda H_0}$$

is sometimes called the Kubo identity. To prove it, it suffices to show that it holds at any single value and that the derivative of both sides is the same.

(a) Show, using the results of Section 11.6.3, that the Kubo identity holds for $T \to \infty$.

(b) Take the derivative of both sides of the Kubo identity with respect to β. Then show, using the identity itself, that the equality indeed holds.

5. The Redfield equations, and the Bloch equations, are given in the representation of the eigenstates of the system hamiltonian H_s. Therefore, there are no transitions among levels (α, α') caused by H_s, since H_s is diagonal in (α, α'), its eigenstates.

To gain some understanding of how different the picture looks in terms of some zero-order set of states, consider yet again the spin boson problem of Eq. (11.96). Write the Redfield equation for $\dot{\rho}_{\alpha\alpha'}$, using the analysis of Section 11.6.5, in particular the Bloch equations (11.117) and (11.100c,d). Then rewrite these same equations in terms of the localized states $|r\rangle$, $|l\rangle$ of (11.97); in particular, write a second-order differential equation in $\Delta = \tilde{\rho}_{ll} - \tilde{\rho}_{rr}$. Now interpret the meanings of the terms T_2, k_{01}, k_{10}, $\Delta\omega$, and E_0 in terms of the system's evolution.

6. Verify through second order, by use of Dyson time-ordering operators and explicit power series expansion, the validity of the Magnus expansion (11.174)–(11.175) for piecewise constant hamiltonians. Start from the requirements (11.172c)–(11.173).

7. Often formal operator manipulations are described by so-called superoperators. These are defined as an operator that acts upon operators. Two particularly important linear superoperators are the so-called Liouville operator defined by

$$\hat{\hat{\mathscr{L}}}\tilde{\rho} \equiv \frac{[H, \tilde{\rho}]}{\hbar} \equiv \hbar^{-1}\hat{\hat{H}}\tilde{\rho} \qquad \text{(P1)}$$

and the unitary transformation superoperator $\hat{\hat{R}}$ defined by

$$\hat{\hat{R}}A = \exp\{-i\hat{\hat{G}}\}A = \exp(-iG)A\exp(iG) \qquad \text{(P2)}$$

The double-carat notation is one way to write a superoperator.

(a) Prove the second identity of (P2).

(b) For the particular choice $\hbar G = H\tau$, show that

$$\hat{\hat{R}}\tilde{\rho}(t_0) = \tilde{\rho}(\tau + t_0) \qquad \text{(P3)}$$

(c) In so-called two-dimensional NMR, the spin system of interest is manipulated by multiple pulses and observations are made in two separate time inter-

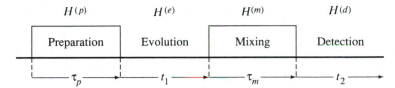

Figure 11.4. Basic scheme for 2D time-domain spectroscopy, with four distinct intervals leading to a time domain signal $S(t_1, t_2, \tau_P, \tau_M)$.

vals t_1 and t_2. The so-called preparation and mixing times τ_P and τ_M are shown in Fig. 11.4. The observed signal can then be described as

$$S(t_1, t_2, \tau_P, \tau_M) = \text{Tr}\{F^+ \bar{\rho}(t_1, t_2)\} \tag{P4}$$

where F is a spin operator and $\bar{\rho}(t_1, t_2)$ is a density operator depending on the two times t_1, t_2.

Show that, in the evolution times t_1, t_2 between pulses, the von Neumann equation

$$\hbar \dot{\bar{\rho}}(t) = \left\{ -i\hat{H}^{(k)} - \hat{\Gamma}^{(k)} \right\} \bar{\rho}(t) + \hat{\Gamma}^{(k)} \bar{\rho}_0^{(k)} \tag{P5}$$

(with $\hat{\Gamma}^{(k)}$ the relaxation operator in the kth interval, $k = 1, 2$) has the solution

$$\bar{\rho}(t_k) = \exp\left\{ \frac{-(i\hat{H}^{(k)} + \hat{\Gamma}^{(k)})t_k}{\hbar} \right\} (\bar{\rho}(t_k = 0) - \bar{\rho}_0^{(k)}) + \bar{\rho}_0^{(k)} \tag{P6}$$

Interpret this result physically.

(d) Combine the result (P5) with the form for evolution under a given pulse

$$\bar{\rho}(t_k = 0) = \hat{P}^{(k)} \bar{\rho}(t_{k-1}) = P_k \bar{\rho}(t_{k-1}) P_k^{-1} \tag{P7}$$

Then neglect all terms of the form $\bar{\rho}_0^{(k)}$, $k \neq 0$, to derive the formal result

$$\bar{\rho}(t_1, t_2) = \prod_{k=1}^{2} \exp\left\{ \frac{-(i\hat{H}^{(k)} + \hat{\Gamma}^{(k)})t_k}{\hbar} \right\} \hat{P}_k \bar{\rho}_0 \tag{P8}$$

or for the sequence in Fig. 11.4,

$$\bar{\rho}(t_1, t_2) = \exp\left\{ \frac{-(i\hat{H}^{(d)} + \hat{\Gamma}^{(d)})t_2}{\hbar} \right\} \hat{R} \exp\left\{ \frac{-(i\hat{H}^{(e)} + \hat{\Gamma}^{(e)})t_1}{\hbar} \right\} \hat{P} \bar{\rho}_0 \tag{P9}$$

with

$$\hat{R} = \exp\left\{ \frac{-i\hat{H}^{(m)}\tau_m}{\hbar} \right\}$$

$$\hat{P} = \exp\left\{ \frac{-i\hat{H}^{(p)}\tau_p}{\hbar} \right\}$$

Interpret the form (P9) physically.

The terms $\bar{\rho}_0^{(k)}$, $k \neq 0$, are the result of relaxation toward equilibrium in the course of free precession during time t_k; generally, two-dimensional NMR experiments are designed to eliminate their effects.

8. Useful approximate formulas for the linear and first nonlinear susceptibilities can be derived using a so-called two-level model, first used for frequency-doubling response by J. L. Oudar and D. S. Chemla [*J. Chem. Phys.* 66, 2664 (1977)]. To derive these formulas, start with (11.139) and (11.140). Ignore the imaginary terms, which are important only near resonances. Assume that there is only one ground state $|g\rangle$ and one excited level $|n\rangle$ (hence the name "two-level model"). For simplicity, assume that only the $\chi_{zz}^{(1)}$ and $\chi_{zzz}^{(2)}$ terms are important, and that for $\chi^{(2)}$, we are interested in the frequency-doubling response, so that $\omega_1 = \omega_2$.

Rewrite (11.139) and (11.140) making these simplifications. Collect the terms so that each χ is expressed as a single fraction. Then use the definition of oscillator strength [from Eq. (5.65)] to simplify the expression further. For the nonlinear susceptibility, the resulting formula suggests three characteristics that a molecule must have for efficient frequency doubling (large $\chi^{(2)}$). What are these three properties? If one wanted to double the NdYAG laser line at $\lambda = 1.06$ μm, would the benzene molecule, the *p*-nitroaniline molecule,

$$H_2N - \langle\bigcirc\rangle - NO_2,$$

or the *p,p'*-nitroaminostilbene molecule,

$$H_2N - \langle\bigcirc\rangle - \!\!\!=\!\!\! - \langle\bigcirc\rangle - NO_2,$$

be most efficient? Why?

APPENDIX A

Dirac Delta Function

The Dirac delta function $\delta(x)$ is defined by

$$\delta(x) = 0 \qquad x \neq 0 \tag{A.1a}$$

$$\int_{-a}^{+b} \delta(x) \, dx = 1 \qquad a, b > 0 \tag{A.1b}$$

These equations imply that

$$\int_a^b f(y) \, \delta(y - x) \, dy = f(x) \qquad a < x < b \tag{A.2}$$

for any function $f(x)$. Note from (A.1b) that $\delta(ax) = |a|^{-1} \delta(x)$.

There are many different ways to represent the delta function mathematically. For example, from the theory of Fourier transforms, if

$$f(x) = \frac{1}{\sqrt{2\pi}} \int_{-\infty}^{\infty} g(y) e^{ixy} \, dy \tag{A.3}$$

and

$$g(y) = \frac{1}{\sqrt{2\pi}} \int_{-\infty}^{\infty} f(x) e^{-ixy} \, dx \tag{A.4}$$

then

$$f(x) = \int_{-\infty}^{\infty} dx' \, f(x') \frac{1}{2\pi} \int_{-\infty}^{\infty} e^{i(x-x')y} \, dy \tag{A.5}$$

Comparing (A.5) and (A.2), we see that

$$\delta(x) = \frac{1}{2\pi} \int_{-\infty}^{\infty} e^{ixy} \, dy \tag{A.6}$$

The delta function may also be expressed as a sharply peaked function in the limit that the peak height becomes infinite while the peak width goes to zero such that the area is constant. Thus we can write [see Eq. (4.65)]:

$$\delta(x) = \lim_{t \to \infty} \frac{2 \sin^2 \frac{1}{2}xt}{\pi x^2 t} \tag{A.7}$$

and [Eq. (7.8)]

$$\delta(x) = \lim_{t \to \infty} \left(\frac{t}{\pi}\right)^{1/2} e^{-tx^2} \tag{A.8}$$

Derivatives of delta functions can also be useful. For example, an expression analogous to (A.2), which can be derived by integration by parts, is

$$\int_a^b f(y) \frac{d}{dy} \delta(y - x) \, dy = - \frac{df(x)}{dx} \tag{A.9}$$

APPENDIX B

Laplace Transforms

The Laplace transform technique is a very useful method for solution of single and coupled ordinary differential equations. Formally, for any two variables s and t, one defines the Laplace transform of a function $f(t)$ by

$$L(f(t)) = \int_0^\infty e^{-st} f(t) \, dt = \tilde{F}(s) \tag{B.1}$$

where the $\tilde{F}$ is a useful notation for the transform.

The function $f(t)$ can be recovered by inverting the transform, to obtain

$$f(t) = L^{-1}(\tilde{F}(s)) = \frac{1}{2\pi i} \int_{a-i\infty}^{a+i\infty} e^{st} \tilde{F}(s) \, ds \tag{B.2}$$

where a is chosen to the right of any singularity in $\tilde{F}(s)$.

One great advantage of Laplace transforms is the availability in closed form and in many sources of $\tilde{F}(s)$ for many functions $f(t)$, and of $f(t)$ for many forms of $\tilde{F}(s)$. The utility of Laplace transforms for solving differential equations comes from the results

$$L\left(\frac{df(t)}{dt}\right) = s\tilde{F}(s) - f(t = 0) \tag{B.3}$$

$$L\left(\frac{d^2f}{dt^2}\right) = s^2\tilde{F}(s) - sf(t = 0) - \left.\frac{df}{dt}\right|_{t=0} \tag{B.4}$$

$$L\left(\int_0^t f(\theta) \, d\theta\right) = s^{-1}\tilde{F}(s) \tag{B.5}$$

Results (B.3) and (B.4) can be verified with (B.1) using integration by parts, and (B.5) can be derived from (B.3) with $f = \int_0^t g(\theta) \, d\theta$.

For example, the damped harmonic oscillator equation

$$\ddot{x} = -\gamma\dot{x} - \omega^2 x \tag{B.6}$$

can be solved as follows:

317

1. Use (B.3) and (B.4) to find

$$s^2\tilde{x}(s) - sx(0) - \dot{x}(0) = -\gamma s\tilde{x}(s) + \gamma x(0) - \omega^2\tilde{x}(s)$$

so that

$$\tilde{x} = \frac{\dot{x}(0) + x(0)(\gamma + s)}{s^2 + \omega^2 + \gamma s} = \frac{\dot{x}(0) + x(0)(s + \gamma)}{(s - a)(s - b)} \tag{B.7}$$

with a and b defined by $(s - a)(s - b) = s^2 + \omega^2 + \gamma s$.

2. Then, using the inverse formulas

$$L^{-1}\left[\frac{1}{(s - a)(s - b)}\right] = \frac{1}{a - b}(e^{at} - e^{bt}) \tag{B.8}$$

$$L^{-1}\left[\frac{s}{(s - a)(s - b)}\right] = \frac{1}{a - b}(ae^{at} - be^{bt}) \tag{B.9}$$

we write

$$x(t) = L^{-1}[\tilde{x}(s)] = \frac{\dot{x}(0) + \gamma x(0)}{a - b}(e^{at} - e^{bt}) + \frac{x(0)}{a - b}(ae^{at} - be^{bt}) \tag{B.10}$$

with

$$a = \tfrac{1}{2}\left[-\gamma + \sqrt{\gamma^2 - 4\omega^2}\right]$$

$$b = \tfrac{1}{2}\left[-\gamma - \sqrt{\gamma^2 - 4\omega^2}\right] \tag{B.11}$$

Equations (B.10) and (B.11) represent the general solution to the damped oscillator.

Often we are interested in the Laplace transforms themselves: for instance, in Chapter 9, we used Laplace transforms to write the lineshape as

$$I(\omega) = 2\,\mathrm{Re}\int_0^\infty e^{-i\omega t}C(t)\,dt = 2\,\mathrm{Re}\,L[C(t)] = 2\,\mathrm{Re}\,\tilde{C}(s) \tag{B.12}$$

with $s = i\omega$.

Index